Hot Thin Plasmas in Astrophysics

NATO ASI Series

Advanced Science Institutes Series

A Series presenting the results of activities sponsored by the NATO Science Committee, which aims at the dissemination of advanced scientific and technological knowledge, with a view to strengthening links between scientific communities.

The Series is published by an international board of publishers in conjunction with the NATO Scientific Affairs Division

A	Life Sciences	Plenum Publishing Corporation
B	Physics	London and New York
C	Mathematical and Physical Sciences	Kluwer Academic Publishers Dordrecht, Boston and London
D	Behavioural and Social Sciences	
E	Applied Sciences	
F	Computer and Systems Sciences	Springer-Verlag
G	Ecological Sciences	Berlin, Heidelberg, New York, London,
H	Cell Biology	Paris and Tokyo

Series C: Mathematical and Physical Sciences - Vol. 249

Hot Thin Plasmas in Astrophysics

edited by

R. Pallavicini
Osservatorio Astrofisico di Arcetri,
Florence, Italy

Kluwer Academic Publishers

Dordrecht / Boston / London

Published in cooperation with NATO Scientific Affairs Division

Proceedings of the NATO Advanced Study Institute on
Hot Thin Plasmas in Astrophysics
Cargèse, Corsica, France
September 8–18, 1987

Library of Congress Cataloging in Publication Data

```
Hot thin plasmas in astrophysics : proceedings of a NATO Advanced
   Study Institute, held in Cargèse (Corsica, France) on September
   8-18, 1987 / edited by R. Pallavicini.
       p.    cm. -- (NATO advanced science institutes. Series C,
   Mathematical and physical sciences ; 249)
       "NATO Advanced Study Institute on Hot Thin Plasmas in
   Astrophysics"--Pref.
       ISBN 9027728127
       1. Plasma astrophysics--Congresses.  2. High temperature plasmas-
   -Congresses.  3. X-ray astronomy--Congresses.   I. Pallavicini,
   Roberto.  II. NATO Advanced Study Institute on Hot Thin Plasmas in
   Astrophysics (1987 : Cargèse, Corsica)  III. Series: NATO ASI
   series.  Series C, Mathematical and physical sciences ; no. 249.
   QB462.7.H67 1988
   523.01--dc19                                              88-13470
                                                                 CIP
```

ISBN 90–277–2812–7

Published by Kluwer Academic Publishers,
P.O. Box 17, 3300 AA Dordrecht, The Netherlands.

Kluwer Academic Publishers incorporates the publishing programmes of
D. Reidel, Martinus Nijhoff, Dr W. Junk, and MTP Press.

Sold and distributed in the U.S.A. and Canada
by Kluwer Academic Publishers,
101 Philip Drive, Norwell, MA 02061, U.S.A.

In all other countries, sold and distributed
by Kluwer Academic Publishers Group,
P.O. Box 322, 3300 AH Dordrecht, The Netherlands.

All Rights Reserved
© 1988 by Kluwer Academic PublishersNo part of the material protected by this copyright notice may be reproduced or utilized in any form or by any means, electronic or mechanical, including photocopying, recording or by any information storage and retrieval system, without written permission from the copyright owner.

Printed in The Netherlands

TABLE OF CONTENTS

Preface ix

List of Participants xi

1. Radiative Processes

RADIATION FROM HOT, THIN PLASMAS
    J.C. Raymond 3

DIAGNOSTIC TECHNIQUES FOR HOT THIN ASTROPHYSICAL PLASMAS
    F. Bely-Dubau 21

2. X-Ray Instrumentation

INSTRUMENTATION FOR THE STUDY OF COSMIC X-RAY PLASMAS
    H.W. Schnopper 35

SYSTEMATIC ERRORS IN THE DETERMINATION OF X-RAY SPECTRUM PARAMETERS AND THE CALIBRATION OF SPACE EXPERIMENTS
    M.V. Zombeck 65

3. Solar and Stellar Coronae

THE SOLAR CORONA
    A.H. Gabriel 79

ULTRAVIOLET STELLAR SPECTROSCOPY
    C. Jordan 97

X-RAY EMISSION FROM NORMAL STARS
    J.H.M.M. Schmitt 109

SOME EXOSAT RESULTS ON STELLAR CORONAE
    R. Pallavicini 121

THE MAGNETIC FIELDS ON COOL STARS AND THEIR CORRELATION WITH CHROMOSPHERIC AND CORONAL EMISSION
    S.H. Saar 139

RE-ANALYSIS OF THE CORONAL EMISSION FROM RS CVN TYPE
BINARIES
    O. Demircan   147

4. Related Optically-Thick Stellar Sources

HIGH RESOLUTION SOFT X-RAY SPECTROSCOPY OF HOT
WHITE DWARFS
    F. Paerels and J. Heise   157

PULSATING WHITE DWARFS
    M.A. Barstow   167

RADIATION FROM GAS ENVELOPES AROUND BE STARS
    K.M.V. Apparao and S.P. Tarafdar   177

5. Supernova Remnants and the Hot Interstellar Medium

X-RAY OBSERVATIONS OF HOT THIN PLASMA IN SUPERNOVA
REMNANTS
    B. Aschenbach   185

THE HOT INTERSTELLAR MEDIUM: OBSERVATIONS
    R. Rothenflug   197

THEORY OF SUPERNOVA REMNANTS AND THE HOT
INTERSTELLAR MEDIUM
    R.A. Chevalier   213

6. Galaxies and Galactic Halos

OBSERVATIONS OF GALAXIES AND GALACTIC HALOS
    G. Trinchieri   235

A THEORETICAL UNDERSTANDING OF HOT GAS AROUND
GALAXIES
    J.N. Bregman   247

RAM PRESSURE STRIPPING AND GALACTIC FOUNTAINS
    E.E. Salpeter   261

7. Clusters of Galaxies

X-RAY OBSERVATIONS OF CLUSTERS OF GALAXIES
    R. Mushotzky   273

THEORY OF INTRACLUSTER GAS
    A.C. Fabian   293

INTERGALACTIC PLASMA IN CLUSTERS: EVOLUTION
A. Cavaliere and S. Colafrancesco   315

EXOSAT OBSERVATIONS OF THE VIRGO CLUSTER
A.C. Edge, G.C. Stewart and A. Smith   335

8. Future X-Ray Missions

FIRST RESULTS FROM GINGA
Y. Tanaka   343

THE ROSAT MISSION
J. Truemper   355

THE WIDE FIELD CAMERA FOR ROSAT: OBSERVING STARS
M.A. Barstow and K.A. Pounds   359

THE SAX X-RAY ASTRONOMY MISSION
G.C. Perola   369

THE ADVANCED X-RAY ASTROPHYSICS FACILITY
H. Tananbaum   379

THE XMM MISSION
J.A.M. Bleeker and A. Peacock   391

THE USSR SPACE ASTRONOMY PROGRAMME
A. Smith   415

FUTURE SPACE ASTRONOMY PROGRAMME OF JAPAN
Y. Tanaka   429

PREFACE

This volume contains all but one of the lectures and seminars presented at the NATO Advanced Study Institute on Hot Thin Plasmas in Astrophysics held in Cargèse, Corsica, from September 8 to 18, 1987.

The meeting was planned in collaboration with the members of the Scientific Organizing Committee, to whom I am grateful for suggesting a comprehensive and well balanced program. The SOC was comprised of Prof. J. Bleeker (Space Research Institute, Utrecht, The Netherlands), Dr. C. Cesarsky (CEN Saclay, France), Dr. R. Mushotzky (GSFC, USA), Prof. K. Pounds (University of Leicester, UK), Prof. H. Schnopper (Danish Space Research Laboratory, Denmark), Dr. H. Tananbaum (Center for Astrophysics, USA), Dr. G. Trinchieri (Arcetri Observatory, Italy), and Prof. J. Truemper (MPE, Garching, Germany). The ASI, fully supported by the NATO Scientific Affairs Division, was organized with the intent of providing a critical and up-to-date overview of our present kowledge and understanding of the properties of hot thin plasmas in astrophysics as they are revealed by X-ray observations from space.

The X-ray and UV emission from optically thin thermal plasmas is a common feature of many astrophysical systems. This type of emission occurs in the solar corona and in the coronae of other stars, in supernova remnants and in the hot interstellar medium, in normal galaxies and galactic halos, and in the intergalactic gas in clusters. Over the past few years, observations obtained with satellites such as EINSTEIN, IUE, EXOSAT, TEMNA, GINGA, SKYLAB, SMM, etc., have increased enormously our knowledge of hot astrophysical plasmas and have provided significant insights into the physical processes at work. A new generation of X-ray instruments with increased sensitivity and higher spatial and spectral resolution is under study for use in future missions (AXAF, XMM). New X-ray satellites (ROSAT, SAX, ASTRO-D, SOHO, SOLAR-A) are under construction or approved for launch in the next few years.

The Institute reviewed the results obtained recently and discussed throrougly the capabilities of future X-ray missions. A global approach was adopted that went beyond the usual subdivisions of the field into galactic and extragalactic astronomy or solar and stellar physics. Astronomers from different disciplines were brought together to discuss the physical implications of recent observations and to address the fundamental questions to be answered by future space missions. Basic physical processes (radiation, plasma diagnostics, magnetohydrodynamics) were discussed in a series of introductory lectures, of genuinely tutorial character, that were devoted to young researchers just entering the field (we regret that the lectures on MHD processes could not be included in this volume). The principles of X-ray instrumentation were also discussed in detail and considerable attention was paid to the systematic uncertainties that affect measurements. The remaining lectures treated more specifically Solar and Stellar Coronae, Supernova Remnants, the Hot Phase of the Interstellar Medium, Galaxies, Galactic Halos, and the Intergalactic Gas in Clusters. More specialized topics (including a few which touched upon related, although optically thick, stellar sources) were presented in a series of seminars distributed throughout the conference.

Until 1975, the Sun was the only star known to possess an X-ray corona. Results of solar studies at X-ray and UV wavelengths were reviewed and it was shown that they are of great importance for many other areas of astronomy. The Sun is the only star for which surface details can be observed and which is bright enough for the application of sophisticated plasma diagnostic techniques. The recent results from the EINSTEIN and EXOSAT Observatories on the coronae of other stars were also discussed at length and the importance of magnetic fields for understanding the observed emission was clearly demonstrated. In supernova remnants, the X-ray emitting gas is heated by shock waves produced in supernova explosions. The dynamics of the explosion was discussed in detail as well as the relationship between supernova remnants and the hot phase of the interstellar medium. Recent results were also presented on the supernova 1987 A, although because of their preliminary nature it was decided not to discuss them at length in these Proceedings. The spectra of stellar and galactic sources are rich in emission lines: it was shown that the interpretation of these spectra can give information on temperature, density, elemental abundances and non-equilibrium conditions. Finally, considerable attention was devoted to hot thin plasmas in isolated galaxies and in clusters of galaxies. Phenomena such as galactic fountains and ram pressure stripping were discussed in some depth. The structure and evolution of clusters was discussed at length, and the evidence was reviewed for the presence of cooling flows in the central regions of clusters.

In summary, the single most important aspect of the ASI was trying to avoid a narrow-minded approach to astrophysical problems and the usual subdivision of the study of hot plasmas into separate areas with little relation one to another. These Proceedings reflect such an approach. We felt that this was particularly important for young people, who in this way can obtain a better understanding of the relevant physical processes and may be ready to work on a variety of different problems, thus taking full advantage of the many opportunities offered by future space missions. The presence at the ASI of a large audience of young astronomers who had only recently started their scientific careers also explains why we devoted so much attention to future space opportunities.

I wish to express my gratitude to the Scientific Affairs Division of the North Atlantic Treaty Organization for the generous support given to the Institute, and to the lectures and participants who contributed so much to the success of the ASI. I also thank the European Physical Society for granting its sponsorship and the Osservatorio Astrofisico di Arcetri and its Director Prof. Franco Pacini for contributing to the organization of the meeting by providing personnel and facilities. The National Science Foundation is gratefully acknowledged for providing travel grants to four young American participants.

Special thanks are due to Mrs. Fern Bongianni for her dedicated work during the organization of the ASI and her assistence throughout the meeting, and to Miss Marie-France Hanseler for arranging the great hospitality of the Institut d'Etudes Scientifiques de Cargèse and for taking care of the local organization. I am deeply grateful to her for ensuring a smooth and pleasent running of the meeting throughout its duration. Least but not last, one of the reasons of the success of the ASI was certainly the beautiful setting of Cargèse, and the wonderful experience we had of visiting such an extraordinary place as the island of Corsica.

Roberto Pallavicini

Director of the ASI

LIST OF PARTICIPANTS

Dr. L. Angelini	EXOSAT Observatory, ESTEC, Noordwijk, The Netherlands
Prof. K. M. V. Apparao	Tata Institute of Fundamental Research, Bombay, India
Dr. M. Arnaud	CEN Saclay, Gif sur Yvette, France
Dr. B. Aschenbach	MPE, Garching, Federal Republic of Germany
Dr. J. Ballet	CEN Saclay, Gif sur Yvette, France
Dr. R. Bandiera	Osservatorio di Arcetri, Firenze, Italy
Dr. M. G. Baring	University of Cambridge, United Kingdom
Dr. M. A. Barstow	University of Leicester, United Kingdom
Dr. T. S. Bastian	NRAO, Very Large Array, Socorro, NM, USA
Dr. F. Bely-Dubau	Observatoire de Nice, France
Dr. H. Bohringer	MPE, Garching, Federal Republic of Germany
Dr. J. N. Bregman	NRAO, Charlottesville, VA, USA
Dr. A. Brinkman	Space Research Laboratory, Utrecht, The Netherlands
Dr. C. Cardinali	Università "La Sapienza", Roma, Italy
Prof. A. Cavaliere	II Università di Roma, Italy
Dr. R. Chevalier	University of Virginia, Charlottesville, VA, USA
Prof. C. Chiuderi	Università di Firenze, Italy
Dr. S. Colafrancesco	II Università di Roma, Italy
Dr. A. Collura	Osservatorio di Palermo, Italy
Dr. E. Corbelli	Osservatorio di Arcetri, Italy
Dr. F. Christensen	Danish Space Research Institute, Lyngby, Denmark
Dr. O. Demircan	Middle East Technical University, Ankara, Turkey
Dr. A. Edge	University of Leicester, United Kingdom
Dr. I. N. Evans	Space Telescope Science Institute, Baltimore, MD, USA
Prof. A. Fabian	University of Cambridge, United Kingdom
Dr. F. Fiore	Università "La Sapienza", Roma, Italy
Prof. A. Gabriel	LASP, Verrières Le Buisson, France
Dr. H. S. Ghataure	Nuffield Radio Astronomy Lab., Jodrell Bank, United Kingdom
Dr. F. Gungor	Technical University, Instanbul, Turkey
Dr. F. Haberl	EXOSAT Observatory, ESTEC, Noordwijk, The Netherlands
Dr. J. Heise	Space Research Laboratory, Utrecht, The Netherlands
Dr. A. Hornstrup	Danish Space Research Institute, Lyngby, Denmark
Dr. J. C. Houck	University of Virginia, Charlottesville, VA, USA
Mr. N. Howard	University of Cambridge, United Kingdom
Dr. J. Hughes	Center for Astrophysics, Cambridge, MA, USA
Prof. C. Jordan	University of Oxford, United Kingdom
Dr. M. Kizilyalli	University of Ankara, Turkey
Dr. B. C. Monsignori Fossi	Osservatorio di Arcetri, Firenze, Italy
Dr. F. Moreno Insertis	Instituto de Astrofisica de Canarias, Tenerife, Spain
Dr. R. Mushotzky	Goddard Space Flight Center, Greenbelt, MD, USA
Dr. K. N. Nagendra	Indian Institute of Astrophysics, Bangalore, India
Dr. J. E. Neff	JILA, University of Colorado, Boulder, CO, USA
Dr. R. Pallavicini	Osservatorio di Arcetri, Firenze, Italy
Dr. L. Pasquini	MPE, Garching, Federal Republic of Germany
Dr. A. Peacock	ESTEC, Noordwijk, The Netherlands
Prof. C. Perola	Università "La Sapienza", Roma, Italy
Ms. T. Pulkkinen	University of Helsinki, Finland
Dr. J. C. Raymond	Center for Astrophysics, Cambridge, MA, USA
Dr. R. Rothenflug	CEN Saclay, Gif sur Yvette, France

Dr. S. H. Saar	Center for Astrophysics, Cambridge, MA, USA
Prof. E. E. Salpeter	Cornell University, Ithaca, NY, USA
Dr. J. Sanchez Almeida	Instituto de Astrofisica de Canarias, Tenerife, Spain
Dr. J. Schmitt	MPE, Garching, Federal Republic of Germany
Prof. H. W. Schnopper	Danish Space Research Institute, Lyngby, Denmark
Dr. K. P. Singh	Tata Institute of Fundamental Research, Bombay, India
Dr. J. Schwarz	Center for Astrophysics, Cambridge, MA, USA
Dr. A. Smith	ESTEC, Noordwijk, The Netherlands
Dr. G. Tagliaferri	EXOSAT Observatory, ESTEC, Noordwijk, The Netherlands
Prof. Y. Tanaka	ISAS, Tokyo, Japan
Dr. H. Tananbaum	Center for Astrophysics, Cambridge, MA, USA
Dr. G. Trinchieri	Osservatorio di Arcetri, Firenze, Italy
Prof. J. Truemper	MPE, Garching, Federal Republic of Germany
Dr. G. Umana	Università di Catania, Italy
Dr. C. von Montigny	MPE, Garching, Federal Republic of Germany
Dr. M. V. Zombeck	Center for Astrophysics, Cambridge, MA, USA.

1. RADIATIVE PROCESSES

RADIATION FROM HOT, THIN PLASMAS

John C. Raymond
Harvard-Smithsonian Center for Astrophysics
60 Garden St.
Cambridge, MA 02138 USA

ABSTRACT. We discuss the processes which produce radiation in astrophysical plasmas, emphasizing the accuracy with which the spectrum can be predicted. Detailed comparisons among different computations are made for simple cases, and the effects of non-negligible densities, fields, photon fluxes and time variations are discussed.

1. INTRODUCTION

The emissivity of an astrophysical plasma, P_ν, links the physical conditions in the plasma to an X-ray count rate or UV emission line flux which we measure. In some cases, say a smooth continuum produced by bremsstrahlung, this is quite simple. In other cases, such as the emission in a single spectral line, the abundance and ionization state of the emitting element and an excitation cross section must be known. And in other cases many spectral lines contribute to the observed emission, so that the abundances and ionization states of several elements and hundreds of atomic rate coefficients enter into the predicted emissivity.

Clearly, one cannot infer the physical conditions of the observed plasma more accurately than the level of uncertainty in the emissivity calculation, so I'd like to discuss the level of uncertainty of the currently available rates. This uncertainty may be totally insignificant compared with uncertainties such as detector response, or it may dominate, depending on the measurement being made. It is quite difficult to assess the uncertainty of a theoretical prediction, especially when a great many atomic rates interact in complex and confusing ways to yield the predicted emissivity. I'll try to enrich that confusion by going through the various atomic processes, discussing which ones are important under what circumstances, and guessing at the likely accuracy of the available cross sections. Having done that, we can compare actual emissivity calculations with each other and with observations of the Sun.

To keep the complexities under control, we'll start with the simplest case: a hot (but non-relativistic) low density plasma with weak electric and magnetic fields, no significant photoionization, no time dependence, no diffusion, no noticeable optical depth, and a Maxwellian velocity distribution. Later we can discuss the effects of relaxing these restrictions and where the approximations are valid.

2. IONIZATION STATE

We must compute the ionization state of the gas. There are about a dozen abundant elements, and element number A has A+1 ionization states to consider. Under the assumptions of low density and low photon flux, all the ions can be taken to be in their ground states (the coronal approximation), and the rate of change of the population of ion i (zero in equilibrium) is

$$\frac{dn_i}{n_e dt} = 0 = q_{i-1}n_{i-1} - (q_i + \alpha_i)n_i + \alpha_{i+1}n_{i+1}. \tag{1}$$

The q_i and α_i are the ionization and recombination rates of ion i.

The A+1 equations for the ions of element A can be solved for n_i. The population of stage i clearly depends on the four rates which connect it with the neighboring ionization stages. If ion i-k is most abundant, then all the rates connecting the stages between i-k and i enter. If, for instance, all the ionization rates are systematically high by a factor of two, then the predicted population of i might be off by 2^k. Fortunately, if k is more than 1 or 2, the population is generally to small to matter.

2.1. Ionization Rates

One popular ionization rate is the Lotz (1966) formula,

$$q_i = A \xi F T^{1/2} \left(\frac{I_H}{\chi}\right)^2 e^{-\chi/kT} / (1 + akT/\chi), \tag{2}$$

the ionization rate for an electron shell with ξ electrons at ionization potential χ. F is a "focussing factor", I_H is the ionization potential of hydrogen, and k is Boltzmann's constant. F and a are generally fit to experimental data or extrapolations, so the predicted rates are about as good as the experimental cross sections used.

Another widely used formula is the Exchange Classical Impact Parameter (ECIP) formula of Summers (1974). This is based on more detailed atomic calculations, but it also has been adjusted to fit experimental data (see Burgess et al. 1977 and Burgess and Chidichimo 1983). The predicted rates are typically about half those given by the Lotz formula.

The ionization cross sections now coming into widest use are Distorted Wave calculations (e.g. Younger 1981). These rates include a four parameter fit to the computed cross section for each ion, so they deal with the details of the atomic structure more effectively than can a simpler formula. Unfortunately, DW cross sections have not yet been published for all the important ions, so Coulomb Born calculations and extensive interpolations are employed.

Figure 1 compares the three ionization rates for a simple ion, lithium-like silicon. The parameters for the Lotz formula are taken from Shull and van Steenberg (1982). Figure 1 is typical in that the ECIP rate is about half the Lotz rate, and the Distorted Wave values lie in between. Many ionization cross sections have been measured to 10 or 20% accuracy at Oak Ridge (e.g. Gregory et al. 1987a,b). They generally agree with the DW cross sections to 30% or better. Arnaud and Rothenflug (1985) present a set of ionization rates based on Younger's results and fits to the available experimental cross sections.

An additional ionization process is important for some ions. In cases such as the sodium iso-electronic sequence ($1s^2 2s^2 2p^6 3s$) an inner electron can be excited to a level above the ionization threshold and subsequently autoionize ("Excitation-Autoionization"). Measured cross sections, such as those of Gregory et al., include this process, and Arnaud and Rothenflug present a comprehensive compilation based on Sampson's (1982) calculations where measurements are not available. In many cases, the rates should be good to the 10% uncertainties of the measured cross sections, but the computed cross sections are probably uncertain at a 30% level.

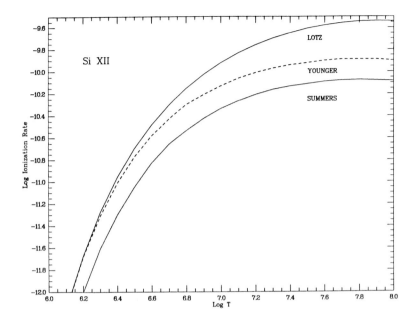

Figure 1. Ionization Rate Coefficients.

2.2. Recombination Rates

In equilibrium, each ionization must be balanced by a recombination. At low densities there are three options.

2.2.1. <u>Radiative recombination.</u> The radiative recombination process

$$A^{+j} + e^- \rightarrow A^{+j-1} + h\nu \qquad (3)$$

is the inverse of photoionization, so the rate is computed from the photoionization cross section with the help of the detailed balance relation. For hydrogenic ions it should be accurate to a few percent. For more complex ions, one uses the photoionization cross section from the ground state (e.g. Reilman and Manson 1979) to compute the rate of recombination directly to the ground state, then adds scaled hydrogenic rates for recombination to excited levels. The hydrogenic approximation is accurate for high n. Typically half of the radiative recombinations go to the ground state or other levels of the lowest shell, however. The accuracy of the theoretical ground state photoionization cross sections is difficult to assess due to the lack of measured photoionization cross sections for ions, but the level of agreement of among the various calculations seems to be around 30%, with a few ions more discrepant. Fortunately, the computed cross sections should be most reliable for photoionization of the simple ions of the H-, He-, Li- and Na-like sequences, and dielectronic recombination generally dominates for other ions under coronal conditions. However, radiative recombination to the 2p level of the Li-like ions and to the 3p and 3d levels of Na-like ions dominates over recombination to the ground states of those sequences, and the hydrogenic approximation for those low-lying excited states is likely to be poor.

2.2.2. Dielectronic recombination. The recombination rates of most coronal ions are dominated by dielectronic recombination. Hahn (1985) reviews the theory. As an example, consider a lithium-like ion in its 2s ground state being struck by an electron of energy ϵ and angular momentum hl relative to the ion. If ϵ is greater than the 2s-2p excitation threshold E, the electron can excite the ion to the 2p state and continue on with energy ϵ-E. If ϵ is less than E, the excitation may still occur, but now the electron has negative energy, meaning that it is bound to the ion in a state $n^2 = I_H z^2/(E-\epsilon)$. The process

$$2s + \epsilon l \rightarrow 2pnl' \qquad (4)$$

can be quite rapid.

Once the doubly excited 2pnl' level has been formed, it has two options. It can autoionize, which is exactly the inverse of the capture process which created the doubly excited level, or else to emit a photon. If it autoionizes, the net result of the whole encounter is just an elastic scattering. If a photon is emitted, it is most likely a 2p $\rightarrow$ 2s transition, with a slight shift to lower frequency due to the presence of the nl' electron (a satellite transition). In that case, the system is now in the 2snl' bound level of the beryllium-like ion.

The cross section for forming the doubly excited 2pnl' state is related by detailed balance to the autoionization rate, A_a. The branching ratio for the dielectronic recombination channel is $A_r/(A_a+A_r)$, where A_r is the radiative decay rate. The dielectronic recombination rate by way of the 2pnl' level is therefore proportional to

$$\frac{A_a A_r}{A_a + A_r}. \qquad (5)$$

The A_a can be found by extrapolating the 2s $\rightarrow$ 2p excitation cross section below threshold. They vary as n^{-3}. For small n, A_a is likely to be much larger than A_r for the values of l' which contribute strongly, and the above expression is then just $\sim A_r$. For lithium-like ions a few times ionized, n levels up to several hundred and l' levels up to about 6 are important.

Burgess (1966) computed dielectronic recombination rates for many ions at many temperatures and fit the results to a General Formula

$$\alpha_{di} = 0.03 \text{ A B f } T^{-3/2} e^{-\bar{E}/kT}, \qquad (6)$$

where f is the absorption oscillator strength of the inner electron transition, A and B are functions of ionic charge and the energy of the inner electron transition, and $\bar{E}$ is the average energy of the doubly excited levels, a value somewhat below the inner electron transition energy. The expected accuracy was around 30%. The Burgess formula is only a fit, however, so there is no reason to expect it to be valid outside its intended range, which was elements up through calcium and temperatures near T_m, the temperature at which the ion concentration peaks. The dielectronic recombination rate at temperatures much above T_m rarely matters, but the rate can be quite important at temperatures far below T_m in photoionized plasma or in rapidly cooling gas. Nussbaumer and Storey (1983) have computed low temperature dielectronic recombination rates for abundant elements through silicon.

The Burgess formula is best for inner electron transitions which involve no change in principal quantum number, $\Delta n = 0$, and modified versions of the formula have been proposed for $\Delta n \neq 0$ (Merts et al. 1976; Hahn 1985). Another modification is the inclusion of other decay channels of the doubly excited state. The denominator of expression 5 becomes $\Sigma A_a + \Sigma A_r$, with the summation over all channels, and the A_r in the numerator is replaced by the sum of radiative rates to bound levels. While radiative decays of the outer electron seem to increase α_{di} by less than 30% for the astrophysically abundant elements (Raymond 1979), autoionization of the doubly excited state excited levels can drastically reduce α_{di} (Jacobs et al. 1977). As an example, the resonance transition of a neon-like ion is $1s^22s^22p^6 \to 1s^22s^22p^53d$, and the $3dnl$ states dominate the dielectronic recombination. However, many of these levels can autoionize to $3p + \epsilon l'$ continuum states an order of magnitude faster than the they autoionize to the ground state. Each of these alternate channels corresponds to an excitation of a $1s^22s^22p^53p$ level of the neon-like ion by way of a resonance in the excitation cross section (at the 3dnl energy), rather than a recombination to the sodium-like ion. Jacobs et al. found that this process reduced the dielectronic recombination rates by way of most $\Delta n \neq 0$ transitions by an order of magnitude. It turns out, though, that the relative importance of this process diminishes quickly with increasing charge. The quantum numbers n of the doubly excited levels which contribute strongly to α_{di} decrease with Z because A_r increases as Z^4 (for $\Delta n \neq 0$ transitions), while A_a stays approximately constant. In the example of the neon-like iron, Smith et al. (1985) found that $3dnl$ levels with n up to about 10 are energetically incapable of autoionizing to the 3p levels, and that these levels account for most of the dielectronic recombination even if autoionization to 3p is ignored. The Jacobs et al. rates should be good for low Z ions, but they seriously underestimate α_{di} for high Z, and there is no complete set of α_{di} available which treats autoionization to excited levels correctly. Smith et al. suggested a simple correction formula based on a very approximate extension of their Fe XVII calculation, but better calculations are needed. They have been done for some iso-electronic sequences, including He-like (Younger 1983b; Bely-Dubau, Gabriel and Volonte 1979; Nasser and Hahn 1983) and Li-, O-, and F-like ions (Roszman 1987), though they tend to focus on very highly charged ions.

Unfortunately, there is less experimental guidance than for ionization rates. Crossed beam experiments have measured dielectronic recombination cross sections, but only in apparatus with electric fields that limit the n of the doubly excited levels to ~ 30. Figure 2 compares the computed rates for a simple case, Li-like iron. The predictions agree at very low

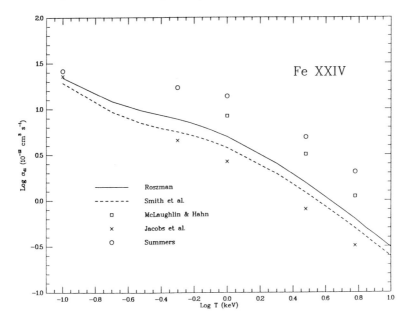

Figure 2. Dielectronic recombination rates.

temperatures, where the $\Delta n=0$ transition dominates, but they scatter over a factor of two in the important temperature range. In this case, the Summers rate is expected to be too large, since autoionization to excited levels was not included in the calculation, while the Jacobs *et al.* rate is expected to be an underestimate, since some of the autoionization channels they included are not energetically allowed. This still leaves an unexplained factor of two discrepancy between the Roszman and the McLaughlin and Hahn (1984) calculations. The level of agreement among various calculations is typically 20% for H-like and He-like ions and for most cases in which $\Delta n=0$ transitions dominate (most low to moderate Z ions), so that 40% may be a reasonable overall estimate of the uncertainty.

2.2.3. Charge transfer. While charge transfer can be important for either ionization or recombination, the most important charge transfer process is generally the capture of an electron from neutral hydrogen by a charged ion, which amounts to a recombination of the ion. The rate coefficient for a charge transfer process can be several orders of magnitude larger than the rate coefficients of the other recombianation processes if a favorable channel is available, so that this process may dominate even if the ratio of neutral hydrogen to electrons is quite small. Reliable charge transfer rates of modest accuracy can be obtained with the Landau-Zener approximation, and more accurate quantal calculations, as well as measurements, are available for many of the astrophysically important cases. A review is given by Butler and Dalgarno (1980).

2.3. Overall Ionization Balance

Having estimated ionization and recombination rate uncertainties, we can now guess at the uncertainties in the overall ionization balance. Given 20 - 40% uncertainties in both ionization and recombination rates, we expect the predicted ratio of adjacent ionization states A_i/A_{i+1} to be off by a factor of about 1.5. Some cases are better; both ionization and recombination rates for H- and He-like ions are known more accurately than most, and X-ray lines of these ions are among the strongest in astrophysical plasmas. In some cases is doesn't matter very much; essentially all the silicon is He-like Si XIII between 3×10^6 and 6×10^6 K, and for many diagnostics it is not important whether 93% or 96% of the silicon in Si XIII. It may matter a lot whether the Si XII ionization fraction is 4% or 7%, but only if the plasma is isothermal. In most astrophysical objects, a wide temperature range exists, and the Si XII emission will be dominated by lower temperature gas having a larger Si XII fraction. Few astrophysical measurements have the dynamic range to detect ions having very low ionization fractions.

In spite of the cases where ionization and recombination rate uncertainties don't matter, they usually do. What is the effect of a systematic error? Let's say that all the ionization rates are overestimated by a factor of two. Since they vary as $e^{-\chi/kT}$, and since ions typically peak at $\chi/kT \approx 7$, the peaks will be shifted to $\chi/kT \approx 8$, so that all the ions will be shifted by about 0.1 in log T. That doesn't sound too serious. The errors are probably not so systematic, however. If the ionization rate of ion i-1 is underestimated and that of i is overestimated, ion i will be squeezed out, and the intensities of its lines underestimated.

How do existing calculations actually compare? Figure 3 shows the ionization fractions of Fe XVI, Fe XVII and Fe XVIII. Fe XVII is the neon-like ion, and one of the ions for which $\Delta n \neq 0$ dielectronic recombination and Excitation-Autoionization of the sodium-like ion are crucial. The solid curve is the current version of the Raymond and Smith (1977) X-ray emission code, which uses Distorted Wave ionization rates (Younger 1983a), Excitation-Autoionization based on an extension of Cowan and Mann (1979), and dielectronic recombination from Smith *et al.* (1985). The dashed curve is the Arnaud and Rothenflug (1985) calculation which uses the Younger ionization rates for these ions, Excitation-Autoionization from

Sampson (1982), and Jacobs et al. (1977) recombination rates.

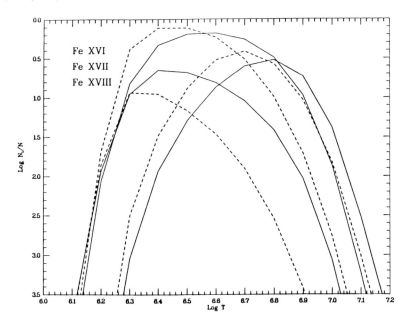

Figure 3. Iron ionization balance.

The overall shapes of the curves are quite similar. The sodium-like Fe XVI ion has a loosely bound 3s electron, while the neon-like Fe XVII ion has a closed shell structure, so the ionization rate of Fe XVI is high. Also, the neon-like ion has no $\Delta n=0$ transition, so that its dielectronic recombination rate is relatively low. These features combine to make the Fe XVII curve stand out above the others, while the Fe XVI never reaches a very large abundance. For the same reasons, the helium-like ions of all the elements are quite prominent, while the lithium-like ions are never very abundant. The Fe XVII and Fe XVIII curves are very similar in shape except that the Arnaud and Rothenflug curves lie at lower temperatures by about 0.1 in log T. Fe XVI is most discrepant, largely due to the different dielectronic recombination rates. Not only does the Raymond and Smith version peak at higher temperatures, but the peak is twice as high. The Jacobs et al. curves for these ions lie in between the two sets of curves, with the Fe XVI curve lying close to the Arnaud and Rothenflug values. The ionization calculations of Shull and van Steenberg (1982) are essentially the same as those of Jacobs et al.. The Summers (1974) ionization balance uses ECIP ionization rates. Those curves are again similar in overall shape, but shifted to higher temperatures by 0.1 to 0.3 in log T.

Depending on how you look at it, the level of agreement indicated by Figure 3 is quite heartening or quite distressing. The discrepancy is only about 0.1 in log T, and that is quite good enough for many purposes. On the other hand, the abundance of a given ion at a given temperature is likely to differ by a factor of two between the two calculations. Which way you look at it depends on what sort of data you wish to analyze. It should be kept in mind that the discrepancies nearly vanish for the simpler and more thoroughly studied ions of the He- and H-like sequences, and that the uncertainties are much worse for the low stages of ionization of iron and nickel having 3d electrons. The ionization state in a plasma of accurately known temperature can be inferred from some satellite line diagnostics applied to solar flare observations (Doschek and Feldman 1981). Only a few ions can be observed this way, but the ratio of Li- and Be-like ions to the He-like ion seems to agree with the current calculations.

3. RADIATION

Now that the ionization state of the abundant elements is known, the emissivity can be computed. In the low field, low photon flux regime, the basic processes are bremsstrahlung, recombination and collisional excitation.

3.1 Collisional Excitation

Spectral lines excited by electron collisions dominate the X-ray emission and the total cooling of astrophysical plasmas at temperatures up to about 10^7 K. In the low density regime, the excitation rate of a spectral line is given by

$$q = 8.63 \times 10^{-6} \frac{\Omega}{\omega \sqrt{T}} \, e^{-E/kT} \text{ cm}^3 \text{ s}^{-1}, \qquad (8)$$

Here Ω is the collision strength for the excitation and ω is the statistical weight of the ground state. The wavelengths and excitation energies of the strong transitions are generally known from laboratory studies, so all that's left is to find Ω. A rough value for optically allowed transitions can be obtained from the gaunt factor formula,

$$\Omega = \frac{8\pi}{\sqrt{3}} \frac{\omega \, f \, \overline{g}}{E/13.6}, \qquad (9)$$

where f is the absorption oscillator strength and $\overline{g}$ is the gaunt factor, which is typically 1.0 for $\Delta n=0$ transitions and 0.2 for $\Delta n \neq 0$. This estimate is generally better than a factor of two, but more accurate values and values for optically forbidden transitions are also needed. There are still cases where nothing better is available, such as most transitions with $\Delta n > 2$, but there are more reliable collision strengths for most of the transitions strong enough to contribute to the total radiative losses significantly or to be very useful as diagnostics. Seaton (1975) reviews the theoretical methods for computing Ω.

Many Coulomb-Born approximation computations are available, in which the wave function of the free electron is represented by that of an electron in the field of a point charge. This gives accurate results if the cross section is dominated by collisions at large impact parameters, since that is where a point charge provides a good approximation to the ion's field, and Coulomb-Born cross sections are best well above the excitation threshold. We can't get by so easily, however. Ions are typically found at $\chi/kT \approx 7$. The $\Delta n=0$ transitions are usually found at $E/\chi \sim 1/4$, while $E/\chi \sim 2/3$ for $\Delta n \neq 0$. Therefore, for the important spectral lines $E/kT \sim 2\text{-}5$, and electrons near threshold dominate the excitation. The Coulomb-Born approximation tends to overestimate the collision strength near threshold, typically by 20 - 50%.

A better result is obtained by using a more realistic model of the ion potential, the Distorted Wave approximation. DW calculations are now available for most $\Delta n=0$ or 1 transitions of ions of the abundant elements or for isoelectronic transitions close enough for interpolation. An accuracy of 20% is expected for uncomplicated transitions, and calculations with different codes usually agree that well. The agreement with laboratory measurements also seems to be that good (e.g. Lafayetis and Kohl 1987), but most of the measurements at that level of accuracy are for fairly simple ions of modest charge. DW calculations are expected to become more accurate with increasing ionic charge.

The most sophisticated theoretical cross section computations include the coupling among all the reaction channels, the Close Coupling method. It naturally handles the resonances in the excitation cross sections mentioned in connection with dielectronic recombination. Its major limitation is that it requires a lot of computer time, and the time needed grows rapidly with the number of states included. It has mostly been used where high accuracy is needed for line ratio diagnostics, such as He-like ions (e.g. Kingston and Tayal 1983) or where resonances make very large contributions to the total excitation rate. In general, the resonances are of modest importance (up to 20%) in the excitation of strong allowed transitions, but they may more than double the effective cross sections of forbidden or intercombination transitions (e.g. Dufton et al. 1978).

Aggarwal et al. (1985) and Gallagher and Pradhan (1985) have recently compiled collision strengths. As a general appraisal, the collision strengths for many of the strongest X-ray lines from astrophysical plasmas, the $\Delta n=1$ lines of H- and He- like ions, are know with better than 20% accuracy. The other very strong X-ray lines are $2l \rightarrow 3l+1$ lines of the Li- to Ne-like ions of iron near 1 keV and of Si and S near 200 eV. These may be known to 20% accuracy from DW calculations, but resonances in the excitation cross sections and cascades among the excited levels have only been examined in detail for a few cases. At lower energies, the $2s \rightarrow 2p$ transitions of these ions of iron and the $3l \rightarrow 3l'$ transitions of lower ions of iron become important. Most of these are probably known to about 20%. In the EUV range (100 Å to 1000 Å) the 2s - 2p transitions of the lighter elements dominate, and DW or CC calculations are available for many of these.

3.2 Two Photon Continuum

Most of the collisional excitations of ions produce emission in spectral lines, but a few produce contina. The $2s^2S$ and $1s2s^1S$ states of H- and He-like ions decay by emitting pairs of photons whose energies sum to the excitation energy of the level. At 10^6 K the two photon continuum dominates over other continuum emission processes between about 35 and 55 Å, but the two photon continuum is typically only 10-20% of the emission of the H-like and He-like ions.

3.3 Bremsstrahlung

When an electron passes close to an ion it is accelerated by the ion's electric field, and so it radiates. The formula

$$P_\nu = 1.6 \times 10^{-23} \ Z^2 \ T^{-1/2} \ e^{-h\nu/kT} \ \overline{g}_{ff} \quad \text{erg cm}^3 \ \text{s}^{-1} \text{eV}^{-1} \tag{10}$$

is classical electrodynamics except for the Gaunt factor $\overline{g}_{ff}$. The Gaunt factor can be computed very accurately for collisions with stripped ions, and in hot astrophysical plasmas that have any significant bremsstrahlung emissivity, collisions with ions other than H^+ and He^{++} contribute only a per cent or so to the total. With normal abundances, bremsstrahlung dominates the X-ray emission above about 10^7 K.

3.4 Recombination

The radiative recombination rates are computed as described earlier. Recombination to a level of ion i of element A having photoionization cross section $\sigma_0(\chi/h\nu)^3$ above its threshold χ gives

$$P_\nu = 1.31 \times 10^{-22} \ \frac{n_A}{n_H} \ \frac{n_i}{n_A} \ \frac{\omega_{i-1}}{\omega_i} \ \chi^3 \ e^{(\chi-h\nu)/kT} \ \sigma_0 \ T^{-3/2} \quad \text{erg cm}^3 \ \text{s}^{-1} \ \text{eV}^{-1}. \tag{11}$$

for $h\nu > \chi$. Recombination can be reliably computed for hydrogenic and helium-like ions, and fortunately these are the important contributors. The recombination continuum of oxygen accounts for only about one quarter of the continuum at 1 keV at a temperature of 10^7 K, and its importance decreases at higher temperatures, but at lower temperatures the recombination continua of C, N and O dominate over bremsstrahlung above the K edges of these elements. The accuracy of the recombination continuum is determined by the accuracy of the photoionization cross section, which is probably better than 10% for the n=1 levels of H- and He-like ions which matter most.

Recombination to excited levels produces continua at longer wavelengths, but these are generally less important than bremsstrahlung. Recombination to excited states also leads to line emission as the recombined ion cascades down to its ground state. This process seldom dominates the intensities of strong lines in collisional ionization equilibrium due to the fact that each recombination to ion i must be balanced by an ionization, and the ionization rate is smaller than the excitation rates of strong permitted lines. The recombination contributions do become important in photoionized or rapidly cooling plasmas, and the contributions are large enough, especially for some intercombination transitions, that they must be included for accurate interpretation of some diagnostic line ratios. Those of helium-like ions have received the most attention (Mewe and Schrijver 1978).

Since α_{di} is larger than the radiative recombination rate for most ions, the radiation produced by dielectronic recombination must also be greater. It arises in two ways. First there is the stabilizing radiative transition, which produces a satellite line to a resonance line of the recombining ion. Then the outer nl electron cascades down to the ground state producing lines of the recombined ion. Each recombination must be balanced by an ionization. The satellite lines are most important for He- and Ne-like ions, because for those sequences the relevant ionization rate is that of the weakly bound electron outside a closed shell, the 2s or 3s electron of a Li- or Na-like ion. The ratio of satellite lines to the resonance line goes as

$$T^{-1} e^{(E-\bar{E})/kT}, \qquad (12)$$

so that they tend to be strong for lines of high excitation potential and temperatures at the lower end of the ion's temperature range. Some satellites can also be formed by innershell excitation of the lower ionization stage. Either way, satellites can be very useful diagnostics, and they can make important contributions to the total line flux measured with low resolution instruments (e.g. Pravdo and Smith 1979; Rothenflug and Arnaud 1985). The lines formed during the cascade haven't gotten much attention, but Mewe and Schrijver have computed their contribution to the excitation of He-like ions, and Nussbaumer and Storey (1984) have computed those produced by low temperature dielectronic recombination.

3.5 Comparison of Calculations

Some care is required in comparing various calculations in that different authors will lump together multiplets in different ways, treat satellite line contributions differently, assume different abundances, and even define the emissivity differently (some being multiplied by n_e^2 and others by $n_e n_H$ to get erg cm^{-3}s^{-1}). I hope that those differences have been properly accounted for in Figure 4, which shows the O VII resonance line at 21.6 Å including its satellite lines as computed by several codes. The Raymond & Smith curve is taken from the current version of that code, which predicts an O VII ionization fraction nearly identical to the Arnaud and Rothenflug (1985) values used by Mewe, Gronenschild and van den Oord (1985). The extremely close agreement between these curves requires not only very similar ionization balances, but also similar excitation rates. Both calculations took collision strengths from Pradhan, Norcross and Hummer (1981), and the RS code uses correction terms for recombination to excited levels and cascades from excitation to higher levels taken from Mewe and Schrijver (1978). Thus the close agreement reflects almost identical choices for the atomic rates important for this line. The Gaetz & Salpeter (1983) and Landini et al. (1985) calculations both use the Shull & van Steenberg (1982) ionization balance. Oxygen is more highly ionized at a given temperature, and this accounts for the difference between this pair of calculations and the other pair.

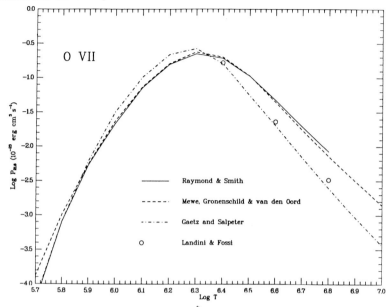

Figure 4. Predicted emissivity of O VII resonance line.

Figure 5 shows a more complex case, the 15.0 Å line of Fe XVII. Here the differences among ionization and dielectronic recombination rates are much larger, and the collision

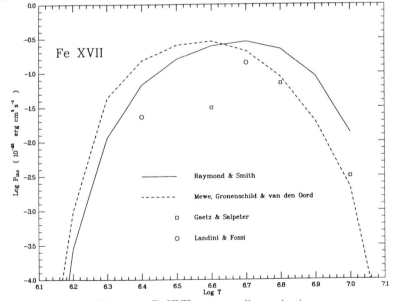

Figure 5. Fe XVII resonance line excitation.

strength is probably less accurately known. The scatter near the peak is somewhat larger, but the different assumed ionization balances account for most of the differences in emissivity. The discrepancies for this line are fairly typical for the strong iron lines. If they are viewed as a shift of 0.1 in log T, they don't seem too bad, but the emissivity at a specific temperature may differ by a factor of 2 or 3 among different calculations.

How do overall spectra compare? Figure 6 shows the emissivity computed with the current Raymond & Smith code for 10^6 K in 0.2 Å bins. Horizontal lines show the emissivities given by Mewe, Gronenschild and van den Oord (1985) for the strong lines. As anticipated from the O VII and Fe XVII comparisons, the C V, C VI and O VII lines (40, 33.7 and 22 Å) agree quite well, while many of the longer wavelength lines differ by factors of two. Some of this can be traced to the ionization balance, in that the Raymond & Smith calculation gives a somewhat higher ionization state for iron. A few of the other differences come from different ways of handling multiplets, but most result from differing estimates of the collision strengths of $n = 2 \rightarrow 3$ excitations of silicon and sulfur. Few calculations are available for these lines. A systematic difference is that the Raymond and Smith calculation used the results of Smith et al. (1985) to estimate the contribution of resonances and cascades to these lines.

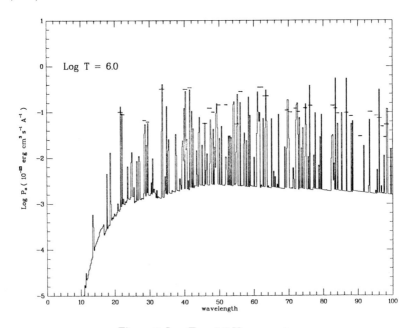

Figure 6. Log T = 6.0 X-ray spectrum.

Overall then, the agreement among the various calculations for the strongest X-ray lines, those of the H- and He-like ions, approaches the 20% anticipated near the temperatures of peak emission. Factor of two discrepancies among the computed intensities of the 1 keV iron lines and the 1/4 keV Si and S lines are common, though the total emission over typical X-ray bands agrees somewhat better.

4. COMPLICATIONS

We began with a list of simplifying assumptions and the promise to examine their validity. Validity is relative, depending on whether we are worried about accuracy at the 1% level or the order of magnitude level. The uncertainty level in the basic atomic rates, $\sim 30\%$, seems most appropriate.

4.1 Density

At high densities, ionization of the highly excited nl levels suppresses α_{di}. The magnitude of this suppression scales roughly as $(n_e/Z^7)^{0.2}$ (Summers 1974). It is greatest for Li-like and Na-like ions, which are dominated by high nl levels, so it is more likely to be important for UV emission lines than for the X-ray emission of higher ionization species. For the $\Delta n=0$ transition of O VI, this reduction reaches 30% at 10^8 cm^{-3}, but because of the fairly slow n_e dependence, 10^{10} cm^{-3} is needed to reach factor of two suppression. Thus models of stellar transition regions should include this effect, but the X-ray spectrum is only affected in stellar flares. High densities also lead to population of metastable levels which can be more easily ionized due to their lower binding energies. In particular, the populations of triplet levels of Be-like and Mg-like ions equal those of the ground states at high densities. This effect is comparable to the suppression of α_{di} (Vernazza and Raymond 1979). At still higher densities, three body recombination becomes important.

Aside from ionization state, the population of metastable levels changes the emission spectrum in that different lines can be excited. This tends not to affect the total radiative loss much, but to shift the excitation from, say singlets to triplets, at somewhat different wavelengths. Fairly high spectral resolution is needed to notice the difference, but if that resolution is available, there are many density diagnostic line ratios. Calculations are available for many individual line ratios, and Mewe *et al.* give the most complete treatment of the density dependence of lines in a general X-ray spectral calculation.

4.2 Electromagnetic fields

Fields can of course Zeeman split or Stark broaden spectral lines, but of the processes discussed above, only dielectronic recombination is very sensitive the the fields, again due to its dependence on very high nl. Li- and Na-like ions are again most affected. E fields of ~ 0.1 keV/cm or B fields of ~ 10000 G cut off the n by field ionization and mix the l for lower n. The complicated effects have been studied only for a few ions studied in the laboratory, as such fields are present in the experimental apparatus (e.g. Müller *et al.* 1987). Similar fields may well be present in solar flares and active regions, and may drastically affect α_{di}.

Magnetic fields on the order of 1500 G exist on the Sun, 10^7 G in magnetic white dwarfs and 10^{12} G in neutron stars. The most important effect is cyclotron or synchrotron radiation at radio (for the Sun), optical (accreting magnetic white dwarfs) or even X-ray (HZ Her cyclotron line) wavelengths. While the emissivities can be computed quite accurately, the emitting regions tend to be optically thick, so that the geometrical structure and radiative transfer are more important than uncertainties in the emissivity.

4.3 Photoionization

Photoionization can easily be included to the accuracy of the photoionization cross sections, provided that the ionizing flux is known. Unfortunately, the geometries and density structures of the emitting regions are usually known only approximately. In the solar case, photoionization of helium by coronal radiation leads to significant recombination radiation. The ionization balance at transition region temperatures (10^5 K) is not much affected by photoionization in the quiet sun. Even in solar flares photoionization effects are modest if the pressure is constant throughout the flare, since the ratio of ionizing flux to electron density doesn't increase much.

In gas at X-ray emitting temperatures, photoionization is probably important in AGN's, X-ray binaries and cataclysmic variables. Only high resolution spectra, such as the OGS spectrum of SCO X-1 (Kahn *et al.* 1984) provide the observational detail to pin down models with this extra degree of freedom.

4.4 Time-dependence

A plasma approaches the equilibrium expressed in Equation 1 over a time comparable to the ionization and recombination times of the relevant ions. If the temperature does not remain constant at least that long, time-dependence must be included. As with photoionization, this can be easily computed, but once the history of the plasma is included it is difficult to define, much less tabulate, a set of models with which to fit observations, and observations of high enough quality to discriminate among such models are rare.

Fortunately, a few simple cases which are commonly encountered. Hot gas cooling radiatively with no other heat input is likely to be found in settings as diverse as post-flare loops in the solar corona and cooling flows in galaxy clusters. Departures from ionization equilibrium are determined by the slowest recombination rates, in particular those of the helium-like ions. The recombination time of O VII is 30% of the radiative cooling time (for constant density cooling and solar abundances) just below 10^6 K. Thus the gas cools in equilibrium at higher temperatures, but it gets far enough from equilibrium that the total radiative cooling rate is reduced by a factor of 2 at $\sim 2 \times 10^5$ K. When the temperature falls below 10^4K, the cooling time again becomes long, and the gas eventually returns to equilibrium.

The other simple case is instant heating in shock waves or solar flares. At flare densities, equilibration times are typically a few seconds, so the hot gas is probably in equilibrium for most of the flare duration. In the shock case, low ionization material is continually swept up. In Tycho's supernova remnant, for instance, neutral hydrogen emission from 5×10^7 K gas is detectable, and the equilibration time for strongly emitting ions such as Si XIII is longer than the age of the remnant. In general, the ionization proceeds quickly up to the He-like ion and much more slowly through He- and H-like species. The two approaches to analyzing SNR spectra are to compute global models, specifying a Sedov blast wave solution, for instance (e.g. Itoh 1977), or to find values of the average temperature and $n_e t$ that can account for the line ratios (Winkler *et al.* 1981).

4.5 Optical Depth

At optical depths of a few to resonance line scattering, the photons in the thickest lines are scattered a few times, but not destroyed, so that the intrinsic emission spectrum is unchanged. However, if the emitting region is not spherical, the radiation will be anisotropic, and line ratios will depend on the observer's position. In principle one could extract geometric information from the spectra, but in practice this scattering confuses interpretation. It is probably a much more common problem than we would like to believe. Acton and Brown (1978) reported attenuation of the O VII resonance line relative to the forbidden line in a solar flare. The optical depth in the O VII resonance line in the northeast section of the Cygnus Loop is about 10 along the line of sight and 1 in the plane of the sky, so the same thing probably happens there, but we are a long way from having good enough data to require elaborate models. Resonant scattering at modest optical depths can also change the effective branching ratios of lines from a common upper level. For instance, $Ly\beta \rightarrow H\alpha + Ly\alpha$.

At higher optical depths and densities ($n_e \tau$ on the order of spontaneous radiative decay lifetimes) photons are destroyed and the intrinsic spectrum is modified. Even before that point is reached, the excited levels may attain large enough populations to dominate the ionization rates and the excitation of some transitions. This is likely to be the case in the accretion disk coronae and in the winds of Wolf-Rayet stars and Of stars. The computation of these effects requires simultaneous solution for level populations, radiative transfer in many lines, and gas temperature.

Finally, at very large column densities the continuum optical depths become large. Photoelectric absorption can be relatively easily included. Once the electron scattering and free-free optical depths become large the gas reverts to local thermodynamic equilibrium.

4.6 Diffusion

Steep temperature and density gradients are common in astrophysics, and diffusion of ions may become important. Observations show abundance changes in the solar wind which may arise from diffusion in the solar transition region. While the calculations are non-local, the major problem again lies in knowing the geometrical structure well enough to choose a model. Calculations have been done for the sun (Rousel-Dupre and Beerman 1981), but not for other coronae.

4.7 Non-Maxwellian Velocity Distributions

Nearly all astrophysical model computations assume either a power-law non-thermal electron velocity distribution or Maxwellian thermal distribution. Departures from a Maxwellian are most likely to arise at high velocities, since the Coulomb collision time is much longer for electrons at several kT. As mentioned earlier, though, these are just the electrons which excite and ionize the ions. A deficit of high energy electrons could arise if the fast electrons lose their energy by exciting ions more rapidly than Coulomb collisions shuffle ions with energy near kT to higher energies. This might occur in mostly neutral hydrogen gas or in the highly enriched gas seen in some supernova remnants. In a steep temperature gradient, fast electrons from the high temperature regions may penetrate into the lower temperature plasma, creating a high energy excess, or non-thermal tail. Such distributions are observed in the solar wind. They most strongly affect the ionization rates, since those rely on electrons farthest above the Maxwellian peak (Owocki and Scudder 1983).

5. OBSERVATIONAL TEST

The real test of the spectral model is not comparison with other models, but comparison with observed spectra. This isn't as straightforward as it sounds. The observed spectrum must include many more resolved spectral features than the number of model parameters if the test is to mean anything. Still worse, there is no way to discriminate between errors in the model and observational errors.

X-ray observations of the required quality exist only for the Sun. Solar flares, in particular, are bright enough for high resolution observations. McKenzie *et al.* (1980) present the 8 - 22 Å spectrum of a flare observed by the SOLEX experiment, a Bragg crystal spectrometer aboard the P78-1 satellite. Acton *et al.* (1985) present a longer wavelength spectrum of a different flare observed during a rocket flight. That spectrum covers the 15-100 Å range. A composite spectrum can be obtained by scaling the Acton *et al.* spectrum to the O VII λ21.6 intensity and using these scaled intensities for all the longer wavelength lines.

The entire temperature range from 10^4 to $10^{7.5}$ K is present in solar flares. The model spectrum is obtained by specifying the emission measure as a function of temperature, computing the X-ray emission at each temperature, and adding up the total intensity in each line.

The model is compared with observation by plotting ratios of observed to predicted line intensities as a function of wavelength. The emission measure distribution is adjusted to minimize the scatter on this plot. The composite spectrum includes lines formed between about $10^{5.8}$ and $10^{7.2}$ K. The lines are typically formed over temperature intervals about 0.3 wide in log T, so in effect there are about 5 independent temperature intervals. Elemental abundances can also be adjusted to reduce the scatter. The model presented here takes the Allen (1973) cosmic abundance set, but doubles the Ca and Ni abundances, increases the Mg abundance by a factor of 1.6.

The ratios of observed to predicted line intensities are shown in Figure 7. The scatter is about a factor of two. There are several contributions to this scatter. One is the errors in the atomic rates assumed in the model. Another is inappropriateness of the model. Some of this is uninteresting, such as incorrect choice of the emission measure distribution as a function of temperature or incorrect abundance choices. Some of it may result from a breakdown of the standard model assumptions listed in section 4. Of all the effects listed there, only density

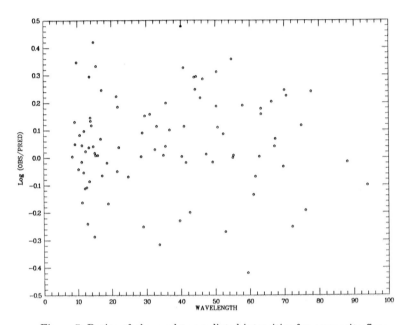

Figure 7. Ratios of observed to predicted intensities for composite flare.

dependence ($n_e = 10^{11}$ cm^{-3}) was included in the model. Observational errors also contribute to the scatter. The relative calibration uncertainty was estimated at less than 25% for the stron unblended lines in the SOLEX spectrum. However, the intensities were derived from the counts at the peaks of the lines, rather than counts integrated over the line profiles. This tends to underestimate the intensities of blended lines, and most of the lines are blended. The uncertainty in the intensities of the longer wavelength lines seems to be around 50% (Acton et al. 1985). Finally, there are errors which are neither model nor observation errors, but errors in the comparison itself. The joining of the two spectra at the O VII line is only strictly valid if the two flares have similar distributions of emission measure with temperature. Also, the model predicts intensities for entire multiplets, while in many cases some members of a multiplet are not observed, and their intensities must be inferred from those multiplet members which are seen. In the more complicated multiplets this can introduce considerable error.

All in all, it seems likely that errors in the model, errors in the observation, and errors in the comparison contribute roughly equally to the scatter in Figure 7, and that 50% is a likely estimate for the model errors.

6. SUMMARY

The uncertainties in the emissivities of high temperature plasmas range from a few percent (bremsstrahlung, recombination continua) to tens of per cent (lines of H- and He-like ions) to a factor of two for some more complex species. A great many processes can affect the emissivity at this level, including density, optical depth, time dependence and non-Maxwellian velocity distributions. Many of these processes can be taken into account if the structure of the emitting gas is understood. Comparison of a model calculation with the composite X-ray spectrum of a solar flare suggests that inaccuracies in atomic rates (or breakdown of simplifying model assumptions) lead to errors on the order of 50% in the predicted strengths of individual lines.

This work has been supported by NASA grant NAG-528

REFERENCES

Acton, L.W., and Brown, W.A. 1978, *Ap. J.*, **225**, 1065.
Acton, L.W., Bruner, M.E., Brown, W.A., Fawcett, B.C., Schweizer, W., and Speer, R.J. 1985, *Ap. J.*, **291**, 865.
Aggarwal, K., Berrington, K., Eissner, W., and Kingston, A. 1986, "Report on Recommended Data", Atomic Data Workshop, Daresbury Laboratory.
Arnaud, M., and Rothenflug, R. 1985, *Astr. Ap. Suppl.*, **60**, 425.
Bely-Dubau, F,, Gabriel, A.H., and Volonte, S. 1979, *MNRAS*, **189**, 801.
Burgess, A. 1965, *Ap. J.*, **141**, 1588.
Burgess, A., and Chidichimo, M.C. 1983, *MNRAS*, **203**, 1269.
Burgess, A., Summers, H.P., Cochrane, D.M., and McWhirter, R.W.P. 1977, *MNRAS*, **179**, 275.
Butler, S.E., and Dalgarno, A. 1980, *Ap. J.*, **241**, 838.
Cowan, R.D., and Mann, J.B., Jr. 1979, *Ap. J.*, **232**, 940.
Doschek, G.A., and Feldman, U. 1981, *Ap. J.*, **251**, 792.
Dufton, P.L., Berrington, K.A., Burke, P.G., and Kingston, A.E. 1978, *Astr. Ap.*, **62**, 111.
Gaetz, T.E., and Salpeter, E.E. 1983, *Ap. J. Suppl.*, **52**, 55.
Gallagher, J.H., and Pradhan, A.K. 1985, JILA Data Center Report No. 30, JILA, University of Colorado, Boulder.
Gregory, D.C., Meywe, F.W., Müller, A., and Defrance, P. 1987a, *Phys. Rev. A*, **34**, 3657.
Gregory, D.C., Wang, L.J., Meyer, F.W., and Rinn, K. 1987b, *Phys. Rev. A*, **35**, 3256.
Hahn, Y. 1985, *Adv. Atomic and Mol. Phys.*, **21**, 123.
Itoh, H. 1977, *Pub. Ast. Soc. Japan*, **29**, 813.
Jacobs, V.L., Davis, J., Kepple, P.C., and Blaha, M. 1977, *Ap. J.*, **211**, 605.
Kahn, S.M., Seward, F.D., and Chlebowski, T. 1984, *Ap. J.*, **283**, 286.
Kingston, A.E., and Tayal, S.S. 1983, *J. Phys. B*, **16**, 3465.
Lafayetis, G., and Kohl, J. 1987, *Phys. Rev. A*, **36**, 59.
Landini, M., Monsigniori Fossi, B.C., Paresce, F., and Stern, R.A. 1985, *Ap. J.*, **289**, 709.
Lotz, W. 1967, *Ap. J., Suppl.*, **14**, 207.
McKenzie, D.L., Landecker, P.B., Broussard, R.M., Rugge, H.R., Young, R.M., Feldman, U., and Doschek, G.A. 1980, *Ap. J.*, **241**, 409.

McLaughlin, D.J., and Hahn, Y. 1984, *Phys. Rev. A.*, **29**, 712.
Merts, A.L., Cowan, R.D., Magee, N.H. 1976, LA-6220-MS, Los Alamos.
Mewe, R., and Schrijver, J. 1978, *Astr. Ap.*, **65**, 99.
Mewe, R., Gronenshcild, E.H.B.M., and van den Oord, G.H.J. 1985, *Astr. Ap. Suppl.*, **62**, 197.
Müller, A., Belic, D.S., DePaola, B.D., Djuric, N., Dunn, G.H., Mueller, D.W., and Timmer, C. 1987, *Phys. Rev. A*, **36**, 599.
Nasser, I., and Hahn, Y. 1983, *JQSRT*, **29**, 1.
Nussbaumer, H., and Storey, P.J. 1983, *Astr. Ap.*, **126**, 75.
Nussbaumer, H., and Storey, P.J. 1984, *Astr. Ap. Suppl.*, **56**, 293.
Owocki, S.P., and Scudder, J.D. 1983, *Ap. J.*, **270**, 758.
Pradhan, A.K., Norcross,D.W., and Hummer, D.G. 1981, *Ap. J.*, **246**, 1031.
Pravdo, S.H., and Smith, B.W. 1979, *Ap. J.*, **234**, L195.
Raymond, J.C. 1978, *Ap. J.*, **222**, 1114.
Raymond, J.C., and Smith, B.W. 1977, *Ap. J., Suppl.*, **35**, 419.
Reilman, R.F., and Manson, S.T. 1979, *Ap. J. Suppl.*, **40**, 815.
Roszman, L.J. 1987, *Phys. Rev. A.*, **35**, 3368.
Rothenflug, R., and Arnaud, M. 1985, *Astr. Ap.*, **144**, 431.
Rousel-Dupre, R. ,and Beerman, C. 1981, *Ap. J.*, **250**, 408.
Seaton, M.J. 1975, *Adv. Atomic and Mol. Phys.*, **11**, 83.
Shull, M.J., and van Steenberg, 1982, *Ap. J. Suppl.*, **48**, 95.
Smith, B.W., Raymond, J.C., Mann, J.B., Jr., and Cowan, R.D. 1985, *Ap. J.*, **298**, 898.
Summers, H.P. 1974, *Culham Laboratory Internal Memo* IM-367, Culham Laboratory, Ditton Park, Slough, England.
Vernazza, J.E., and Raymond, J.C. 1979, *Ap. J.*, **228**, L29.
Younger, S.E. 1981, *JQSRT*, **26**, 329.
Younger, S.E. 1983a, *JQSRT*, **29**, 61.
Younger, S.E. 1983b, *JQSRT*, **29**, 67.

DIAGNOSTIC TECHNIQUES FOR HOT THIN ASTROPHYSICAL PLASMAS

F. Bely-Dubau
Observatoire de Nice
BP 139
06003 Nice Cedex
France

ABSTRACT. Soft X-ray spectra with high resolution represent a rich source of information on the nature of hot thin plasmas. These emission spectra correspond to multiply ionized atoms and for their interpretation in terms of physical parameters it is necessary to understand the atomic processes involved and to compute the corresponding parameters. These X-ray diagnostic techniques are used mainly for solar active regions and flare plasmas. Spectra from Ca and Fe have been observed and fitted with the best available theoretical spectra giving the temperature, density, ionization state, abundances, velocity, non thermal electron effects... Diagnostics for other optically thin plasmas as supernova remnants and galaxy clusters are also briefly described.

I. INTRODUCTION

There has been a close association between astrophysical studies and the development of the atomic collision physics. This interaction took a major step when some fourty years ago the nature of the solar corona was understood as a very tenuous plasma. The formulation of the "coronal equilibrium" equations pointed out the importance of various atomic processes for determining the state of the plasma and for understanding its emitted spectra. In the major part of this paper the "coronal model" will be the starting point for analysing spectra in order to obtain more information on the observed plasmas.

Observations made above the earth's absorbing layers have greatly enriched our knowledge of the Sun. Its radiation can be observed over a very extended range of wavelengths far out into the UV and X-ray regions. The greater part of the solar energy output is hovewer in the optical and infra-red range coming from the photosphere ($T \approx 6000$ K). The corona extends out several times the photospheric radius and has a high temperature $T \geq 10^6$ K. Between the photosphere and the corona are situated the chromosphere ($T \approx 10^4$) and the transition region ($T \approx 2.10^5$), this last one being a very thin layer of the sun atmosphere with a high temperature gradient. The richest parts of the chromospheric and coronal spectra occur at UV and X-ray ranges which correspond to hot plasma emissions. In recent years chromospheres and coronæ have also been observed in stars other than the sun.

In the more readily accessible UV region $\lambda \geq 912$ Å observed spectra contain many resonance and intercombination lines of ionized atoms such as CIII, CIV, OIII, OIV, OV, NIV, SiIII...For the shorter wavelengths the emitting atoms are more highly ionized and the identification of some of them by Edlen was the first definite evidence of the high temperature of the corona.

The soft X-ray spectrum (23-1Å) corresponds to multiply ionized atoms belonging to plasmas characterized by temperature in the range of 2 to 50 10^6 K as active regions and flares, the most energetic components of the solar corona. Solar flares are explosive events which occur in active regions. They are connected with the magnetic fields of the sunand they result in the release of large amounts of energy. Two main phases can be identified: (i) the rise phase which can include a short impulsive phase (secs) characterized by hard X-ray emission with an acceleration of electrons to very high energies, (ii) the quasi- thermal phase which is characterized by a rapid rise (1-5 mins) in the soft X-ray emission followed by a decay phase ($\geq$ 1hr). The soft X-ray emission enables us to follow the thermal component of the events during all their manifestation. During the maximun solar activity many such soft X-ray high resolution spectra were obtained from spectrometers aboard spacecrafts: Solflex, flown by the Naval Research Laboratory (Doschek *et al* ,1979) , XRP on the Solar Maximun Mission (Acton *et al* ,1981) and later SOX on Hinotori (Tanaka *et al* ,1982).

The only high resolution X-ray non solar object is reported by Winckler *et al* ,1981 of the supernova remnant Puppis A taken from the Einstein observatory. Morever such spectra have been observed from Tokamak plasma devices or laser experiments.

In this paper we shall mainly discuss the X-ray observations made on solar flares and the physical parameters deduced from these observations. We shall examine in particular the data obtained by the Bent Crystal Spectrometer (BCS) of the XRP and indicate some progress towards understanding the flare phenomenon permitted with the X-ray spectrum analysis. At last two other types of investigations concerning supernova remnants and clusters of galaxies will be presented.

1. X-RAY SPECTRA

1.1 Definition

The solar soft X-ray spectrum (1-23Å) is rich in emission lines from the $n = 2 \to 1$ transition in ions of Ni, Fe, Ca, Ar, S, Si, Mg, Ne and O, and its important features are due to:

(1) H-like ions: Lyman series lines;

(2) He-like ions (Fig. 1):

- resonance line   (w) $1s^2\,{}^1S_0 - 1s2p^1\,P_1^0$

- forbidden line   (z) $1s^2\,{}^1S_0 - 1s2p^3\,S_1$

- intercombination lines   (x) $1s^2\,{}^1S_0 - 1s2p^3\,P_2^0$

                                        (y) $1s^2\,{}^1S_0 - 1s2p^3\,P_1^0$

and also the transitions such as $1s^2 - 1snp$ with $3 \leq n \leq 5$ for the most abundant elements;

(3) Ne-like ions: transitions such as $1s^2 2s^2 2p^6 - 1s^2 2s^2 2p^5 nl$ with $n \leq 5$;

(4) Other types of lines which are important: the so-called satellite lines. They form systems of lines appearing close to or blended with the above-mentioned resonance lines. They arise from doubly excited levels above the first limit of ionisation and correspond to transitions such as $1s_i^2 l_i - 1s2p_i l_i$ with $n_i \geq 2$, which can be interpreted as resonance transitions in the presense of additional bound perturbing electrons $n_i l_i$ (Fig.1). The excited level can be formed either by dielectronic recombination (line j) or by an inner-shell excitation process (lines q and β). The stronger satellites arise from states with n=2 for the perturbing electrons and appear on the long wavalength side of the parent line (line w). The group of satellites arising from n=3 appears closer or blended with the resonance line. Those arising from n=4,5... are practically indistinguishable and blended with the resonance line (Bhalla et al ,1975, Bely-Dubau et al ,1979a,b). It is possible to separate some of the spectral line blends into discrete components using a high spectral resolution ($\Delta\lambda/\lambda \sim 10^4$ for $\lambda = 1 - 10$Å);

(5) All ionization stages of Fe between Fe XXI -Fe XVII in the range 1.85 - 1.91 Åwith lines such as $1s^2 2s^2 2p^q - 1s2s^2 2p^{q+1}$ and the $K\alpha$ lines at 1.936 and 1.940Åwhich are excited normally by fluorescence.

2.2 Line emission intensity

The ions responsible for these emissions are mainly excited by energetic electron impacts. For this reason their spectra are sensitive to the electron temperature and density and represent a rich source of information on the emitting plasma properties. Their interpretation requires an understanding of the atomic processes involved in the emission mechanisms and the precise determination of the corresponding atomic parameters.

The total emitted intensity for an optically thin spectrum line with a wavelength λ is

$$I(\lambda) = \frac{1}{4\pi D^2} \int_V \epsilon(\lambda) dV$$

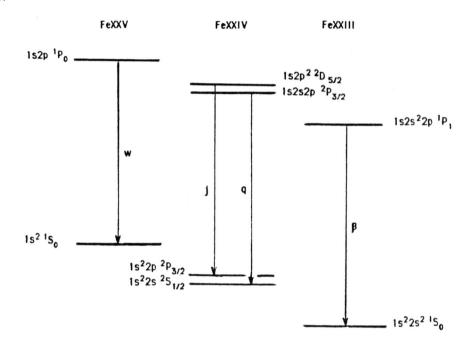

Fig. 1. Energy level diagram of some Fe XXV, Fe XXIV and Fe XXIII states, showing the resonance line w and associated satellite transitions.

where D is the Sun-Earth distance, V the volume in the approximate physical state for the emission of that line and $\epsilon(\lambda)$ the theoretical emissivity given by

$$\epsilon(\lambda) = h_i(X^{+m})A_r$$

with $h\nu$ the emitted photon energy in erg, A_r the spontaneous transition probability in s^{-1} and $N_i(X^{+m})$ the number density in cm^{-3} of the emitting level of the ion X^{+m} (in th m th ionization stage of the element X). In general, the quantity $N(X^{+m})$ is calculated by solving the time-dependent statistical equilibrium equations. Its determination begins with the analysis of all important population processes involved in the emission including ionization and recombination and is followed by the calculation of the required atomic data. It is possible to reduce the number of equations if the time scales for changes in the physical parameters are much longer than the time scales for atomic processes. Also, the number of the levels can be limited to the low lying ones which can be populated. Moreover, if collisional and radiative processes occur more quickly between low lying levels within the ion than ionization and recombination between ions, the calculation of level populations and ionization equilibria can be performed separately.

In the plasma model of interest here, the excited states of allowed transitions are populated predominantly by electron collision from the ground state and their emissivity becomes

$$\epsilon(\lambda) = h\nu N_f(X^{+m})CN_e$$

where C is the effective excitation rate coefficient from the ground state population density. $N_f(X^{+m})$ can be expressed as

$$N_f(X^{+m}) = \frac{N_f(X^{+m})}{N(X)}\frac{N(X)}{N(H)}\frac{N(H)}{N_e}N_e$$

with $N_f(X^{+m})/N(X)$ the ionization ratio, $N(X)/N(H) = A$ the element abundance relative to hydrogen and $N(H)/N_e$ the abundance of hydrogen relative to electrons (=0.8 in the higher temperature region of the Sun). Then the total intensity can be written as

$$I(\lambda) = \frac{0.8Ah\nu}{4\pi D^2}\int_V \frac{N_f(X^{+m})}{N(X)}CN_e^2 dV \tag{1}$$

$$I(\lambda) = \frac{0.8Ah\nu}{4\pi D^2}\int_V G(T_e)Y(T_e)\frac{dV}{dT_e}dT_e \tag{2}$$

The differential emission measure is defined as

$$Y(T_e) = N_e^2(dV/d(logT_e))$$

and $G(T_e)$ depends only on the line and is normally assumed to have a gaussian form.

For many of the diagnostic methods which follow we use measurements of spectral line ratios. However, when using relative data from different elements or instruments channels, or when deriving absolute emission measures, the full equations become important.

2.3 Atomic data

Before interpreting the BCS spectra it is necessary to understand the physical processes which are involved in the emissions and to compute the corresponding atomic parameters. The first problem is the determination of the wavefunction and energies from which the wavelengths λ and the spontaneous transition probabilities A_r are deduced. Then the other atomic parameters have to be calculated as for the case of Fe ions (Fig.1) those corresponding to :
  - collisional excitation from Fe XXV (ground level) ;
  - inner-shell collisional excitation from either Fe XXIII or Fe XXIV;
  - collisional ionization from Fe XXIV;
  - radiative cascades from higher Fe XXV excited levels;
  - dielectronic capture from Fe XXV.

For this purpose, the computation must be reliable and the computer code has to provide large amounts of atomic data, as for instance, does the computer program package of the University College of London in a modified version (Bely-Dubau et al ,1979a,1981). For a heavy element as Fe a relativistic Hamiltonian has to be used.

3. ANALYSIS OF OBSERVED SPECTRA AND DIAGNOSTICS

The XRP experiment has been prepared to cover the soft X-ray range for the Solar Maximun Mission and one of its components the BCS was more devoted to observe fixed wavelength intervals around strong lines of Fe and Ca with a high spatial and time resolution. The diagnostic possibilities that arise from the measurement of ratios of satellite to resonance lines involve only a single BCS channel, so that many constant terms cancel in the interpretation (e.g. channel sensitivity, element abundance). There are two processes in general for the production of the satellite lines(see 2.1): the first by dielectronic recombination from the parent ion (Fe XXV) , and the second by excitation of the inner-shell electron of the ion stage which has one more electron (Fe XXIV). Since the importance of each one depends on the magnitude of the atomic decay rates, A_a by autoionization, and A_r by spontaneous radiation, some satellites are produced predominantly by the first process, some by the second, and others by a combination of the two.

3.1 Electron temperature

The intensity ratio of a satellite line arising only from dielectronic recombination (for example j of Fe XXIV on Fig.1) to the Fe XXV resonance line w can be written as

$I_j/I_w = F_1(T_e)F_2(j)$

where $F_2(j)$ depends only on atomic parameters of the j-line as A_a and A_r, and $F_1(T_e)$ cannot be expressed simply since it includes the effective excitation rate coefficient C_w for exciting the resonance line. However, it is approximately of the form $\approx 1/T_e$. Thus the above ratio for a line formed predominantly by dielectronic recombination is an excellent diagnostic for T_e in the region of formation.

3.2 Relative abundance of ions from the same element

When T_e is known, the intensity ratio of a satellite line arising by direct excitation of an inner-shell electron (for example q of Fe XXIV and β of FeXXIII on Fig. 1) to the resonance line w can be written as

$I_q/I_w = F_2(q)N(FeXXIV)/N(FeXXV)$

$I_\beta/I_w = F_2(\beta)N(FeXXIII)/N(FeXXV)$

The F_2 functions depend on the atomic parameters of their respective lines and more particularly of their corresponding inner-shell excitation rates. The relative

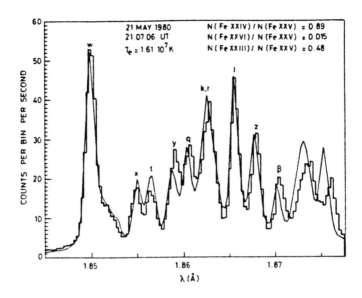

Fig. 2. A BCS iron spectrum from the flare of 21 May 1980, showing the fitted theoretical curve (smooth) and the parameters used to obtain this fit.

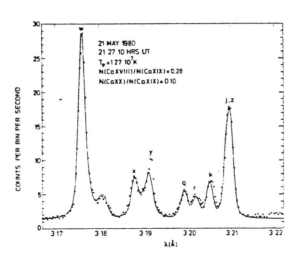

Fig. 3. A BCS calcium spectrum from the flare of 21 May 1980, together with the fitted curve.

abundance could be either smaller or greater than the value derived for an ionization equilibrium plasma. From these estimates, one can determine if the plasma is in an ionizing or a recombining phase.

For accurate application of the above it is necessary to correct the resonance line for its blending with unresolved high n satellites (see 2.1). The detailed application together with the essential atomic parameters and the merged high n satellites, can be found in a series of papers by Bely- Dubau *et al* (1979a, b;1982a) for Fe and Bely -Dubau*et al* (1982b) for Ca.

The flare spectra showing the helium-like ion lines and their satellites are shown in Fig.2 and 3 for Fe and Ca from the BCS observations of the event of 21 May 1980. The smooth line indicates the spectrum computed using the above theory, with the parameters of electron temperature effective doppler temperaturefor the width, and relative ion stage populations adjusted to give the best fit to the observations. In practice the electron temperature is determined primarily by the intensity of the j-line for Fe and the k-line for Ca and the Li-like to He-like ion ratios and the Be-like to the He-like ion ratios by the relative intensity of the q-line and β -line respectively. However, all parameters are iterated to provide the best overall fit, taking account of line blending effects.

The electron temperature and the abundance ratios can be determined for different moments of one flare. In this way it is possible to follow its temporal evolution. Fig. 4 shows for Fe, the way in which these parameters vary with time throughout the flare. The fact that the minima of Be/He and Li/He ion ratios occur at the maximun of T_e lends support to the viewpoint that the ionization balance follows closely its limiting steady-state value throughout most of the flare duration.

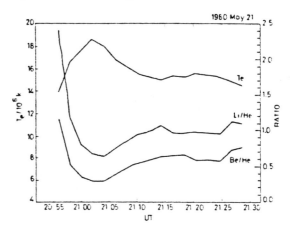

Fig. 4. The derived time variation of temperature and ionization ratios for iron ions, from the 21 May 1980 flare.

The ratios of ion stage populations can be compared with the ratios obtained from theoretical calculations of ionization balance. Whenever we can be sure of the steady-state condition of the flare plasma, these observations can be used to verify or choose between conflicting theoretical values. An analysis has been carried out over 368 spectra of active regions and flares: Ca XVIII/CaXIX abundance ratio measurements were compared with predictions from several theories (Antonucci et al.,1984). By avoiding the onset phase of the flares, the points are found to be on a single curve, verifying the steady-state assumption. A similar analysis was made in mixing 300 spectra of flares from both Fe observations of BCS and SOX (Antonucci et al. 1987a) leading to the same conclusion.

3.3 Relative abundance of different elements

Starting from the above analysis on Ca spectra which used the q-line relative intensity, Antonucci et al.(1987b) have pointed out that this line blended with a line of ArXVII ($1s^2 - 1s4p$),was unsuitable for determining the CaXVIII/CaXIX ratios. Using another "inner-shell excitation" line, less intense than q, they have corrected these ratios and deduced the relative abundance of Ar to Ca in flares, less abundant than expected.

3.4 Turbulent motions and plasma motions

Using Ca XIX and FeXXV resonance lines and associated satellite lines, Antonucci et al.(1982) have shown the existence of features of the line profiles characteristic of the impulsive phase of flares. The profiles indicate the dynamic processes of the plasma during this phase. Turbulent motions reach velocities higher than 100 km s^{-1} and upward motions of plasmas reach vertical velocities of a few hundred km s^{-1} corresponding to a well defined blue shift on the resonance lines.

3.5 Electron density

Different techniques can be used depending on the emission conditions. The total emission line from a particular line is given by eq.1. If both the size of the emitting region and T_e can be estimated, then $I \approx N_e^2 V$ and the average N_e is roughly deduced.

Another determination can be obtained from line intensity ratios independently of any geometrical assumption: for some "forbidden" or "intercombination" transitions, the radiative decay is so small that the electron collisional deexcitation competes as a depopulation mechanism. Then the population of the emitting or "metastable" level can be comparable with the population of the ground level and the ratio of two differently excited lines enables one to determine N_e but only in an intermediate range depending on the ion considered. Gabriel and Mason (1982) have reviewed many possible situations depending on the lines and the density conditions.

Finally N_e can be obtained when the ground configuration of the emitting ion is complex (FeXIX-FeXXII). The population of levels within the ground configuration depends on N_e which gives rise to a density dependence of the intensities of the lines due to these levels. Phillips et al.(1983) have used some satellite lines of FeXX which have confirmed this density dependence.

3.6 Differential emission measure and non thermal electron detection

Using a set of well resolved allowed lines (either for one element or for several), it is possible to invert a set of eq.2 (one for each line) and obtain the differential emission measure $Y(T_e)$. This quantity represents the emitting material distribution as a function of T_e. The limit in the T_e range covered by the BCS leads to a substantial statistical uncertainty in the inverted results. In order to resolve the above problem, a method has been developed (Gabriel et al. 1984) to broaden the temperature or energy range of the observations used, by extending the data to include BCS lines plus broad-band channels from the Hard X-ray Spectrometer (HXIS) and the Hard X-ray Burst Spectrometer (HXRBS) on SMM. In extending to these higher energy continuum spectra it is also necessary to include in the model, not only the thermal differential emission measure, but also the Bremsstrahlung produced by fast electron streams. The interpretation of the observations for the 29 June 1980 flare is based upon a single broad maximum in the differential emission measure, plus a non-thermal component consisting of two power-law electron distributions with 5.3 and 7.5 as indexes respectively . This analysis has given the following results: (1) departures from ionization equilibrium are rare, but exist detectably during the first 30 secs of the flare; (2) the energy content of the thermal component is consistent with the time integral of energy deposited by non-thermal electrons; (3) many channels of HXIS and HXRBS include contributions from both thermal and non-thermal components; (4) a non-thermal component persists always.

4. THE PUPPIS-A SUPERNOVA REMNANT

The highest quality X-ray line spectrum of a supernova remnant is reported by Winkler et al.(1981). among other features, this shows the three lines of OVII w, y and z defined in 2.1. The ratio of these lines can be used for detection of non-thermal electrons. This possibility arises because non-thermal high-energy electrons excite only "normal" transitions as w, and not intercombination and forbidden lines as y and z. In order to explain the strange ratio observed from Puppis A, previous interpretations (Winkler et al. 1981) were based upon high temperatures $> 510^6$ K and departure from ionization equilibrium. Gabriel et al. (1984) have developed an alternative model. The conditions prevailing in supernova remnants of a collisionless shock propagating through the interstellar medium are ideal for the acceleration of some electrons to high energies. The existence of such electrons (for instance few per cents at 20 keV) imbedded in an otherwise thermal plasma at 1 or 2 10^6 K, would have just the effect required to produce the observed spectrum.

5. GALAXY CLUSTERS

X-ray spectoscopy data for 22 galaxy clusters revealed the existence of Fe line features around 7 keV. These emission lines must originate in a hot plasma ($T \approx 10^7 - 10^8$ K). Due to low resolution of proportional counters used, obserations give the complex set of lines corresponding to Fig.2 as a blend. To derive reliable iron mass and abundance estimates from these raw observations, Rothenflug and Arnaud (1985) have deduced the variation of Fe line equivalent width with temperature from the same set of atomic data presented in 2.3, continuum emission and ionic equilibrium. The cluster Fe abundance has been obtained from the comparison between measurements and computations. Its value appears to be universal, i.e. 0.53 ± 0.03 solar values, whatever the cluster characteristics are: the sample includes clusters with temperatures from 1.8 keV to 10 keV and as shown in 3.1 the dielectronic satellite lines can be very strong for low temperatures and thus should be included in the line equivalent width calculations.

6. CONCLUSION

High resolution spectroscopy allows us to study the physics of "coronal" type plasmas which are present in many sources both inside or outside our Galaxy. Measurements of emission line intensities enable us to estimate gas temperature and density, ionization stages, element abundance, gaz velocity for various objects and to detect non thermal electron streams. The combination of the study of variability and energy spectra leads to advance in our understanding of the physics of the sources and their evolution. The diagnostic accuracy depends strongly on the atomic parameter data which characterize the physical processes involved in the emission.

REFERENCES

Acton,L.W.,Culhane,J.L.,Gabriel,A.H.,Wolfson,C.J.,Rapley,C.G.,Phillips,J.H., Antonucci,E.,Parma,A.N.,Strong,K.T.andVeck,N.J.:1981,*Astrophys.J.Lett.* **244**,L137.
Antonucci,E.,Gabriel,A.H.,Acton,L.W.,Culhane,J.L.,Doyle,J.G.,Leibacher,J.W., Machado,M.E.,Orwig,L.E. and Rapley,C.G.:1982,*Solar Phys.* **78**,107.
Antonucci,E.,Gabriel,A.H.,Doyle,J.G.,Dubau,J.,Faucher,P.,Jordan,C.andVeck,N. 1984, *Astron.Astrophys.* **133**,239.
Antonucci,E.,Dodero,M.A.,Gabriel,A.H.,Tanaka,K.,Dubau,J.:1987a,*Astron. Astrophys.* **180**.263.
Antonucci,E.,Marocchi,D.,Gabriel,A.H.,Doschek,G.A.:1987b,*Astron.Astrophys.*

in press.
Bely-Dubau,F.,Gabriel,A.H.,Volonté,S.:1979a,*Mon.Not.R.astr.Soc.* **186**,405.
Bely-Dubau,F.,Gabriel,A.H.,Volonté,S.:1979b,*Mon.Not.R.astr.Soc.* **189**,801.
Bely-Dubau,F.,Dubau,J.,Faucher,P.,Gabriel,A.H.:1982a,*Mon.Not.R.astr.Soc.* **198**,239.
Bely-Dubau,F.,Dubau,J.,Faucher,P.,Gabriel,A.H.,Loulergue,M.,Steenman-Clark,L. Volonté,S.,Antonucci,E ,Rapley,C.G.:1982b,*Mon.Not.R.astr.Soc.* **201**, 1155.
Bhalla,C.P.,Gabriel and A.H.,Presnyakov,L.P.:1975,*Mon.Not.R.astr.Soc.* **172**, 359.
Gabriel,A.H.,Mason,H.E.:1982,"Solar physics",*Applied Atomic Collision Physics*, Vol.1,ed. H.S.W. Massey, B. Bederson and E. W. McDaniel, Academic Press, p.345.
Gabriel,A.H.,Bely-Dubau,F.,Sherman,J.C.,Orwig,L.E. and Schrijver,J.:1984, *Adv.Space Rev.* **4**,221.
Gabriel,A.H.,Acton,L.W.,Bely-Dubau,F.,Faucher,P.:1985,*ESA SP* -239.
Phillips,K.J.H.,Lemen, J.R.,Cowan,R.D.,Doschek,G.A. and Leibacher,J.W.:1983, *Astrophys.J.***265**,1120.
Rothenflug,R.,Arnaud,M.:1985, *Astron.Astrophys.* **144**,431.
Winkler,P.F. *et al.*:1981,*Astrophys..J.***246**,L27.

2. X-RAY INSTRUMENTATION

INSTRUMENTATION FOR THE STUDY OF COSMIC X-RAY PLASMAS

Herbert W. Schnopper
Danish Space Research Institute
Lundtoftevej 7
DK-2800 Lyngby
Denmark

ABSTRACT. The capacity for broad and narrow band spectrophotometry of point sources and spectral imaging of extended sources over a broad energy range are prominent features of X-ray missions now being developed. X-ray concentrators, energy sensitive broad-band imaging detectors and wavelength dispersive gratings and crystals can be combined to provide powerful tools for performing diagnostic measurements on cosmic X-ray plasmas. A selection of these devices is described.

1. INTRODUCTION

Parameters which lead to the understanding of the physical conditions in hot cosmic plasmas are to be found from studies of the temporal and positional variation of the spectra emitted by the sources. Electron and ion temperatures, density, mass motion and the presence or lack of equilibrium are typical of the parameters which can be obtained from the detailed analysis of spectral data. The purpose of this discussion is to describe in an outline form some of the instruments which are proposed or have been used to obtain X-ray spectra from cosmic sources. Several approaches to the design of X-ray concentrators are presented first. The properties of imaging- and non-imaging, broad- and narrow band X-ray spectrometers follow the sections on X-ray optics.

The problem to be investigated will determine the required angular and spectral resolutions and energy range. Energy sensitive spectrometers can have resolutions which range from a few parts in 10 to a few parts in 10^4 over an energy range from several hundred to more than ten thousand eV. Optical systems can have resolutions as good as 1 arcsec or less for the very best optical systems and a few arcmin for large area concentrators. Combinations of detectors and optics exist which cover the full matrix of possible combinations of the parameters, some with greater usefulness than others.

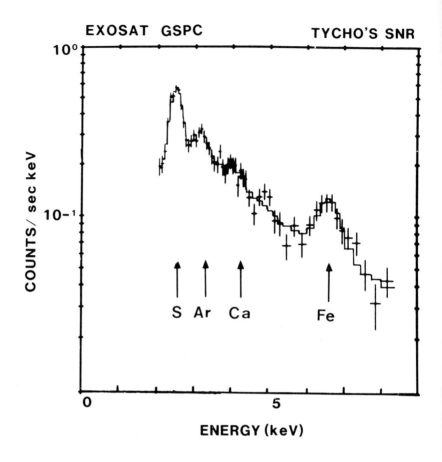

Figure 1a. (Above). The X-ray spectrum of Tycho's supernova remnant obtained from the GSPC on EXOSAT. (Courtesy A. Peacock, ESTEC).

Figure 1b.(Facing page) High resolution helium like calcium and iron X-ray spectra of a solar flare obtained from the Bent Crystal Spectrometer on SMM (Culhane et al., 1981).

A typical broad band X-ray spectrometer is the gas scintillation proportional counter (GSPC). ESA's X-ray satellite, EXOSAT, had one on board and it was used to obtain the spectrum of Tycho's SNR which is shown in Figure 1a. In addition to the underlying thermal continuum component the line emission from highly excited, cosmically abundant ions between sulfur and nickel contribute strongly. There are other features as well coming from line of sight absorption in the source and the intervening interstellar medium but they cannot be seen in the data.

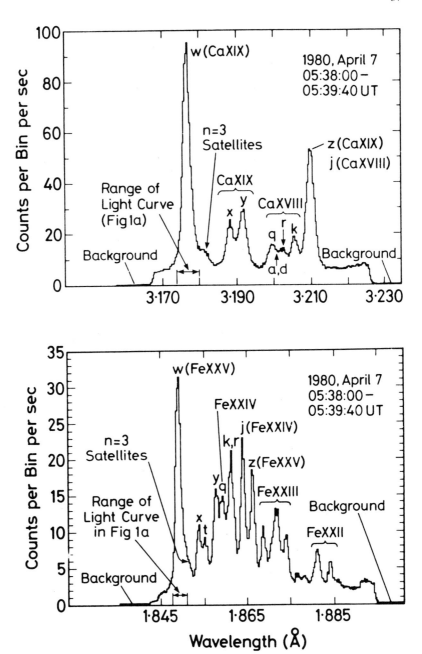

The small diameter of the source (~ 8 arcmin) fits well within the field of view of the mechanically collimated detector and the data, therefore, represent some weighted average over position, abundance and temperature of the plasma. The interpretation of this data is discussed in the literature and by others in this volume.

Also shown in Figure 1b are two solar spectra obtained from the Solar Maximum Mission (SMM). A high resolution spectrometer was used to reveal the richness of the line structure arising from multiply ionized states of calcium and iron (Culhane et al., 1981). Similar structures can also be expected to be seen in high resolution spectra from Tycho's SNR which will be obtained by future X-ray missions.

Figure 2 is an X-ray image of Tycho's SNR obtained with a telescope and an imaging detector on board NASA's X-ray satellite EINSTEIN (Seward, Gorenstein, and Tucker, 1983). The good angular resolution shows details in the hot plasma shell which is the interaction of the expanding remnant and the quiescent interstellar medium. The distribution of intensity is far from uniform and numerous hot spots are seen. The detector is a channel plate which has very good positional resolution but almost no spectral resolution at all. This instrument was also used to study the SNR N132D in the LMC and to obtain the image of the Puppis A SNR shown in Figure 3 (Petre, Canizares, Kriss, and Winkler, 1982).

The data shown in Figures 1-3 are a good start and point the way to requirements for the next generation of X-ray satellites. Future studies of SNRs should have the benefit of improved spatial and spectral resolution and detailed studies should not be limited to objects in the Galaxy but be extended to nearby galaxies such as the Magellanic clouds and Andromeda (M31). These goals apply to other objects as well and the methods of achieving them are discussed below.

2. CONCENTRATORS

Specular reflection for X-rays is possible only at small incident angles under conditions of total external reflection. The critical angle, beyond which reflection does not occur, depends on the density of electrons in the reflector and, therefore, on the reflecting material as well as the energy of the incident photon. These requirements have lead to the development of optical systems which reflect at shallow angles of incidence.

Most past and proposed satellite X-ray optics have two reflections and are either true or approximate Wolter I systems (see Figure 4) (Wolter, 1952 a,b). The concentrator together with its detection system is generally referred to as a telescope. In either case, the concentrators are generally nested systems and the reflecting surfaces are elements which have been machined and polished [EINSTEIN (Giacconi et al., 1979); ROSAT (Aschenbach, Bräuninger, and Kettenring, 1983; AXAF (van Speybroeck, 1988)], replicas taken from mandrels which have been machined and polished [EXOSAT (de Korte et al., 1981); SAX (Citterio et al., 1985; XMM (de Korte, 1988)], or assemblies of thin foils [BBXRT (Serlemitsos, Petre, Glasser, and Birsa, 1984; XSPECT (Schnopper et al., 1988); ASTRO-D (Tanaka and Makino, 1988)]. The special case of Baez

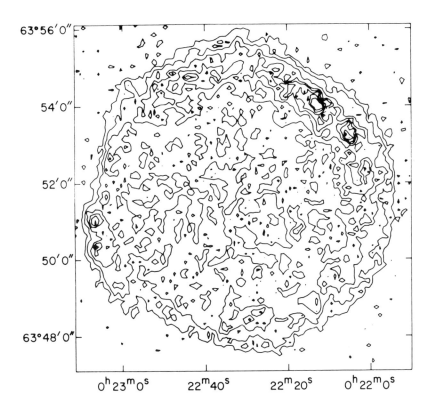

Figure 2. A contour map obtained from the EINSTEIN image of the CAS A supernova remnant. The high resolution imager (HRI), a channel plate, was used to record the image (Seward, Gorenstein, and Tucker, 1983).

optics falls into the last mentioned case (Gorenstein, Cohen, and Fabricant, 1985).

No surface is perfectly smooth and roughness, both short and long range, on scales as small as atomic sizes will introduce small angle scattering halos around the geometrically perfect image of a point source (the concentrator diffraction limit is not important at X-ray wavelengths). There are also deviations from the exact figure of the concentrator which add to the blur in an image. The angular diameter of the circle which encloses half of the photons in the image is defined as the Half Power Width (HPW) of the concentrator and is very useful in defining the optical performance of the system. It is usually referred to when the question of source confusion arises. For a recent review of these topics see Aschenbach, 1985.

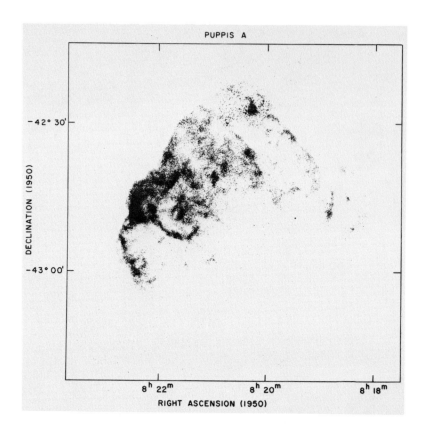

Figure 3. A high resolution image of the Puppis A supernova remnant in the energy range 0.1-4 keV obtained with the EINSTEIN HRI (Petri, Canizares, Kriss, and Winkler, 1982).

2.1. Exact Wolter I Ground and Polished Optics

The most exacting optical techniques are required to produce mirror surfaces which follow exactly the Wolter I figure with tolerances in the order of microns and which have less than ~ 5-10Å surface roughness. The mirror material is usually an optical quality low expansion material such as glass or quartz-glass. The mirrors must be thick enough to provide the strength to retain their figure during manufacturing, test and launch activities. Shell thicknesses of up to several centimeters are not uncommon. The effective collecting area of the concentrator is determined by the fraction of the projected geometrical area of the system for which the incident angle is less than the energy dependent critical angle. The projected area is a fraction (usually less than

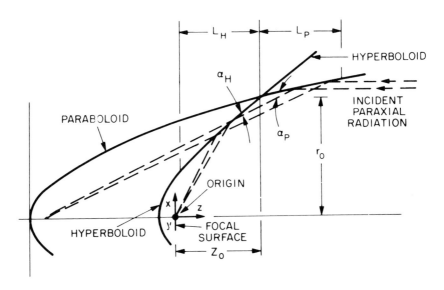

Figure 4. Wolter I X-ray optics (Giacconi et al., 1979).

half) of the area defined by the diameter of the outermost shell since the shell thickness and mirror supporting structures limit the maximum geometrical reflecting area. In addition, the minimum incident angle on each shell defines an upper limit for photon energies which can be reflected at angles below the critical angle and, therefore, only the innermost shells contribute to the high energy response of the concentrator. Table I and Figure 5 give details of previously flown and proposed optical systems.

2.2. Exact Wolter I Replicated Optics

Replicas of well figured and polished mandrels can also be assembled into concentrators. This technique is particularly useful for missions where a cluster of telescopes is to be used, for example XMM. The two small concentrators flown on EXOSAT were prepared by replicating glass mandrels. In this case, accurately machined beryllium carriers are coated with a thin epoxy layer which makes contact with the mandrel. During the setting process gold which has been evaporated onto the mandrel is transferred to the epoxy which also conforms to the shape of the mandrel. This technique also serves as the basis for the development of mirrors for the XMM concentrators. Thin, carbon fiber reinforced epoxy shells are used as the carrier in place of beryllium.

Variations of this method include the electrodeposition of the reflecting material onto the mandrel followed by the application of a thermoreactive plastic coating (Hudec and Valnicek, 1985). When set the plastic shell together with the reflecting material is separated from the mandrel. Typical thicknesses of the reflecting layer are from 0.1 to 1.0 mm and the plastic is from 0.5 to 1.5 cm. Small, unbacked, replicated mirrors are being developed for the SAX mission (Citterio et al., 1985).

Replicated optics will, in general, make more efficient use of a given aperture than will ground and polished optics because the mirror shells can be made thinner and more of them can be nested. Various studies have shown that a mandrel can be replicated many times before its surface deteriorates. Residual stresses in the shell, distortions introduced during separation and errors accumulated during assembly will generally limit the angular resolution and HPW when compared with ground and polished optics. The difficulties associated with separating replicas with small grazing angles from the mandrel limit the high energy response of such systems.

2.3. Approximate Wolter I Optics

If the length of a conical surface is made sufficiently short, then the deviation from the ideal paraboloidal and hyperboloidal Wolter I surfaces shown in Figure 4 can be made extremely small. Ray tracing studies have shown that the contribution to the HPW from these deviations can be held to ≤ 20 arcsec. The object of this approach is to make the most efficient use of the aperture by using a very dense nest of thin shells. This not only produces a very large effective area, but also extends the energy range of the telescope to perhaps 20 keV since the mirrors with shallow incident angles can be placed close to the telescope axis.

Commercially available aluminium foils with thicknesses between 0.1 and 0.15 mm can be obtained with relatively smooth surfaces. Sectors of the conical surfaces are cut and rolled to approximately the correct shape. The surfaces of the conical section are then coated with a thin layer of acrylic resin which tends to smooth out the surface. Finally, a layer of gold is evaporated to form the reflecting surface. The surface roughness achieved this way appears to be ~ 3Å RMS, an indication that small angle scattering is not a problem. The conical sectors are assembled into a mechanical structure which provides support and alignment (Kunieda and Serlemitsos, 1988).

Plastic sheets have also been investigated as potential foil material. In this case the complete conical surface is cut from a single sheet, rolled and glued in a fixture and assembled into the structure after gold evaporation (Tanaka and Makino, 1987).

Both approaches have yielded concentrators with HPWs between 3 and 4 arcmin with contributions coming from surface defects (called orange peel) and residual alignment errors in manufacturing the structure. Improved mounting structures could lead to a much better angular resolution and HPWs ~ 2 arcmin. The use of replication techniques in which both conical surfaces are produced with the same mandrel is an alternative approach (Citterio et al., 1985).

2.4. Kirkpatric-Baez Optics

In its simplest form this concentrator consists of two confocal parabolic cylindrical sheets, one following the other but with the second sheet turned to have its reflecting surface orthogonal to the first. The first sheet forms a line image in the focal plane; the second reduces the line to a point. Large collecting areas can be obtained by using nested stacks of reflectors in each direction but it is then not possible to maintain the excact focusing condition for on axis rays. Since both reflections do not take place in the same plane, the high energy performance is less efficient than that of a Wolter I telescope with the same aperture and focal length.

Specially selected float glass plates which have been gold coated have been used to produce Kirkpatric-Baez concentrators for rocket flights. Recent developments have produced such concentrators with a HPW just over 0.5 arcmin (Fabricant, Cohen, and Gorenstein, 1988).

2.5. The Question of "Confusion"

The EINSTEIN satellite was used to make a number of medium and deep surveys of the X-ray sky. These data, when combined with results from earlier satellites, can be used to predict the number of sources N which can be expected to be seen per unit solid angle at a given flux level S in a particular energy range. A discussion of this distribution, usually expressed as the logN(>S) vs. logS curve is well beyond the present subject matter, but there are practical considerations which merit consideration, especially when extragalactic X-ray sources at large distances are to be investigated. Ignoring relativistic effects for the moment, it is expected that increasingly larger numbers of faint sources should be found as the distance from the observer increases. At some distance the distribution of sources on the celestial sphere will become dense enough that the telescope may not be able to resolve two neighbouring objects in the field of view. At this point, the telescope is said to be "confused". The definition which is usually accepted is that the confusion limit is reached when there is more than 1 source in a region containing 40 pixels, each of which corresponds to the area of the HPW circle.

Future missions will have telescopes with large effective collecting areas and, therefore, fainter limiting source sensitivities for a given observing time. Provided that the sources can be resolved, this rule works. Thus, for very high resolution telescopes such as AXAF confusion will not be a factor for any practical observing time but, for low resolution telescopes, such as XSPECT confusion will be a factor for observing times on the order of 10^3 s. These considerations apply to broad band studies over the full energy range of each system. The XSPECT system has greater collecting area than AXAF and will, therefore, be a "faster" system.

It makes a difference if there is a priori knowledge of the source fluxes and their positions in the field of view. Data from confused regions can be ignored and there is always the statistical possibility

TABLE I. Concentrators for previous or proposed X-ray missions.[1]

	EINSTEIN	EXOSAT	ROSAT	AXAF	XMM	XSPECT
Focal Length, m	3.44	1.09	2.4	10	7.5	8
Max. diameter, cm	58	27.8	83.5	120	70	60
Collecting area, geometric, cm^2	350	90	1150	1400	2450	2000
at 1 keV	200	35	400	1000	2260	1800
at 8 keV	-	-	-	100	770	1100
Energy range, keV	0.2-4	0.04-2	0.1-2	0.1-10	0.2-12	0.2-15
HEW, arcsec at 1 keV	10	30	5	1.0	30	120
at 8 keV	-	-	-	?	?	120
Construction ground and polished		fused quartz	zerodur	quartz or zerodur		
replicated			epoxy on beryllium		epoxy on carbon fiber	
foil						epoxy on aluminium
Reflector(s)	nickel	gold	gold	nickel gold	nickel gold iridium	gold plus?

(1) Adapted in part from B. Aschenbach (1985)

of having very faint sources in unconfused regions. High resolution spectroscopic studies require bright sources and confusion is not a problem.

When equipped with suitable imaging detectors, high resolution Wolter I systems are particularly well suited to studying distant point sources in crowded regions of the sky and for broad-band imaging spectrophotometry of extended sources with sufficient surface brightness. Optical systems with poorer angular resolution lose the ability to perform deep surveys in confused regions and cannot produce detailed images. These losses are generally offset by greater collecting areas and an energy range which extends to higher energies. Broadband detectors of various types can be coupled to the optics to form spectrometers which take advantage of unique properties of the optics and detector.

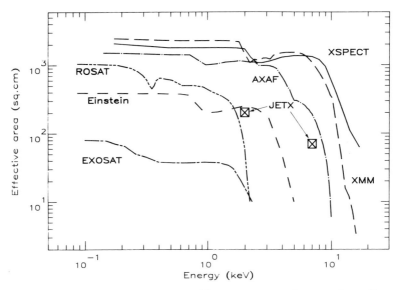

Figure 5. A comparison of the on axis effective area for various X-ray concentrators. Detector efficiency is not included.

For each kind of optics, suitable Bragg crystal and grating spectrometers can be designed to give moderate ($E/\Delta E\sim 10^2$) to high ($E/\Delta E\sim 10^3$) energy resolution for bright point sources and, in some cases, extended ones. The various spectroscopic instruments are discussed next.

3. NON-DISPERSIVE, BROAD-BAND, ENERGY-SENSITIVE, X-RAY SPECTROMETERS

The conversion of an X-ray photon's energy E into a cloud of ionized particles, a burst of low energy photons or heat (phonons) are the most commonly used methods for detection. The conversion can take place in a gas or a solid. Since the creation of a secondary particle requires, on average, a constant energy, the number of secondary particles is proportional to the energy of the incoming photon. In each case, it is the statistical fluctuations in the creation of secondary particles during the conversion process which determines the uncertainty in the energy measurement. This uncertainty is proportional to $E^{1/2}$. In each case, the small signal is amplified internally and/or externally to the detector, a process which can add additional fluctuations to the current pulse.

Gas detectors are made imaging either by introducing arrays of position sensing electrodes or by detecting the light flash produced in in the primary absorption event with an array of photo detectors. Arrays of solid state detectors (either made up of discrete elements or monolithic devices) are used to form images. A summary of the properties of the various detectors discussed is given in Table II.

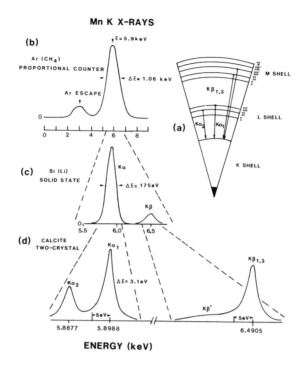

Figure 6. K X-ray emission spectra from ^{55}Mn as recorded with spectrometers having different resolving power. Increasing the resolving power reveals more details concerning the electronic structure of the system (Schnopper, 1984).

3.1. Gas Filled Detectors

3.1.1. <u>Position sensitive proportional counter (PSPC)</u>. The simplest proportional counter consists of a detector box with a single anode wire. The box is normally grounded and a suitable high voltage is applied to the anode. X-rays enter through an appropriately thin window and are absorbed in the gas creating an energetic photoelectron and an excited ion. The photoelectron dissipates its energy creating further ionization. The excited ion decays either by emitting photons and/or electrons which also create further ionization. In some cases the ion emits a characteristic K shell X-rays which may or may not be reabsorbed in the gas.

Under the influence of the field between the anode and the wall, the cloud of positive ions drifts towards the wall and the electrons drift towards the anode wire. As the electrons gain energy in the strong potential near the wire further ionization is produced and this process is called gas amplification. Typical values of gas gain are from 10^4 to 10^5. A low noise preamplifier connected to the anode produces additional gain. Further signal processing and amplification produces a voltage pulse whose height is proportional to the energy of the X-ray. In the case where the initial ion emitted a K X-ray which was not reabsorbed, the voltage pulse is proportional to the difference between the two energies which is called an escape peak in the distribution. Further electronic processing is used to produce a spectrum of pulse heights and Figure 6 is an example of such a distribution.

The useful energy range for a gas filled detector is defined by the transmission properties of the window at low energies and by the absorbing properties of the gas at high energies. Very thin and, therefore, fragile windows plus a gas reservoir are required if gas detectors are to be able to detect X-rays with energies below about 1 keV. Absolute cleanliness must be maintained during manufacture to prevent the counting gas from being contaminated by outgassing from the walls of the detector.

Non-X-ray events can also be detected and they must be discriminated against. Typical events are those produced by very energetic, minimum ionization charged particles traversing the detector or by Compton recoil electrons produced by gamma ray interactions in the walls of the detector. Pulse height discriminators can be set to reject events whose energy falls outside the range of interest. Events caused by ionizing particles which take time to traverse the detector have a pulse shape which differs from true X-ray events. Additional anodes can be placed between the walls and the detecting anode. Events caused by ionizing particles will be detected on more than one anode and the close coincidence between these events can be used to reject them. Modern techniques will generally reject > 99 percent of the unwanted events while allowing > 90 percent of the true events to pass.

Imaging is made possible by building multi-electrode arrays. Two approaches are in common use. The first has an array of closely spaced anode wires with two cathode planes also closely spaced, one above and one below the anode plane. The cathode planes have their wires running orthogonal to each other. They are placed very close to the anode creating a high field in the vicinity of the anode. Additional electrodes are used for rejecting non-X-ray events.

For a single event, the sum of all signals detected on the anode wires is used to determine the energy of the X-ray photon. The drift of electrons towards the anode wires will also induce signals on several wires in each of the crossed cathode planes. By measuring the amplitude of the signal in each wire the centroid of the event in each plane can be determined. The intersection of the centroids determines the position of the event. All the anode wires can be connected in parallel and the sum signal sent to a single amplifier. In principle, each cathode wire requires a separate amplifier chain. The best energy resolution can be obtained only when the spacing between the wires is very uniform. This produces a uniform field distribution and, therefore, a constant gain from place to place in the detector (Pfefferman and Briel, 1985).

The second approach has a single cathode wire plane just above the anode plane with both sets of wires parallel to each other. A second cathode plane just below the energy sensing anode plane is divided into strips. Each strip is divided into 4 different segments (Figure 7) and all of the segments of one kind are interconnected. The relative areas of the segments on each strip are varied in such a way as to allow the position of the induced signal to be calculated from the four signals by charge division: $x = (q_1+q_2)/(q_1+q_2+q_3+q_4)$; $y = (q_1+q_4)/(q_1+q_2+q_3+q_4)$. This approach requires only five amplification chains (Madsen et al., 1985).

Apart from the influence of electronic noise the spatial resolution of PSPCs is limited by the spreading caused by diffusion of the electrons as they drift towards the multiplication region and by the track of the photoelectron.

3.1.2. <u>Position sensitive gas scintillator (PSGS)</u>. The primary operation of this detector is similar to the PSPC. The electron cloud produced in the initial event drifts towards a region of high field. The field, however, is higher than in the PSPC causing the gas to scintillate. Ultraviolet (UV) photons are emitted which can be detected with high efficiency by the photocathode of an appropriate detector. On average, less energy is required to liberate one UV photon in a PSGS than is required to liberate one electron in a PSPC and, therefore, the energy resolution is improved.

If an array of UV detectors is used then the centroid of the light detection will give positional information while the total signal yields energy information. When the UV detectors are an array of photomultiplier tubes the device is called an Anger camera (Davelaar, Peacock, and Taylor, 1982). An alternative approach is to introduce a gas mixture in which the scintillation will cause photoionization (Hailey, Ku, and Vartanian, 1983; Simons and de Korte, 1987).

3.2. Solid State Detectors

3.2.1. <u>Lithium Drifted Silicon [Si(Li)]</u>. Semiconductor diodes offer the possibility of producing good X-ray detectors. In a way analogous to gas filled detectors, electron-hole pairs are created when the primary X-ray is absorbed. The significant difference is that much less energy is required to create a free electron in the diode than in the gas and the energy resolution of a Si(Li) detector is much better than that of either a PSPS or a PSGS. There are some practical matters which must be considered. Reverse current in the diode produces noise and the only way to reduce it is to lower the operating temperature of the diode. In the laboratory liquid nitrogen (77°K) is typically used but successful operation of the detector can be obtained with somewhat higher temperatures. This helps for space applications where cooling is provided by stored solid cryogens, mechanical refrigerators or passive radiators (only in deep orbits). Solid state detectors have no equivalent to gas amplification and the primary charge must be detected by extremely low noise preamplifiers which also must be cooled although not to as low a temperature as the detector. The dominant source of noise is the capacitance of the detector which is seen by the preamplifier.

The combination of layers of gold which acts as the charge collector and of non-active ("dead") silicon on the front surface of the detector limit the low energy response of the detector. In practice, additional low energy cutoff is introduced by filters which are required to reduce contributions from infrared, visible and ultraviolet light to which the silicon is very sensitive. The thickness of the detector determines the high energy limit. Solid state detectors have been flown on a rocket (Rocchia et al., 1984) and on Einstein (Joyce et al., 1978).

Imaging detectors can be produced only by making arrays of discrete elements. There is no real analog to imaging gas systems. There is no practical way to match the requirements of a very high resolution telescope like AXAF, but it is within the realm of possibility to consider Si(Li) arrays for XMM and XSPECT. A formidable electronics problem must be faced even with these lower resolution systems since up to several thousand detectors and their preamplifiers could be required depending upon how much of the field of view is to be covered. Cooling such an array might also be a problem.

3.2.2. Charge Coupled Devices (CCD). These detectors are also made out of silicon and offer all the advantages of improved energy resolution that apply to the Si(Li) detector. In addition, they also can be used as imagers with spatial resolution that matches the requirements demanded by high resolution concentrators such as AXAF. Although originally developed as the imaging devices for portable video systems, CCDs have seen wide application in astronomy over a broad range of wavelengths.

CCDs are produced by modern techniques of microlithography. Arrays of capacitors "grown" in silicon are charged when a voltage is applied. The charge created when electron-hole pairs liberated by an X-ray (or any other wavelength photon) is trapped and partially discharges the capacitor. Other electrodes, activated by a series of clocking pulses, shift, in turn, the contents of each row of the CCD to the next and, finally, into a common output bus. The end result of this readout is a serial string of pulses, each one corresponding to a particular pixel in the detector according to its position in the string. The amplitude of the pulse is proportional to the energy liberated by the photon in that pixel. The entire "frame" must be scanned in a time which is less than the time it takes for a second photon to arrive in one pixel.

The same low energy considerations apply to the CCD as to the Si(Li) but achieving a thin entrance window presents more difficulties to the CCD because of the way they are constructed. In addition, the thickness of the active or "depleted" region of the CCDs designed for optical wavelengths is too small to meet the high energy requirements in most cases. CCDs specially designed for the needs of X-ray astronomy are now being evaluated (Chowanietz et al., 1986; Janesick et al., 1985). Because the volume of the sensitive region of each pixel is very small it is possible to reduce the amplifier noise to a level where the intrinsic statistical contribution from the silicon becomes the factor which limits the energy resolution. CCDs should, therefore, give somewhat better resolution than Si(Li) detectors.

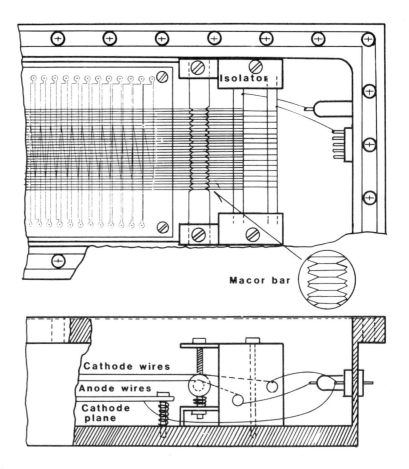

Figure 7a (Top). A typical PSPC showing multiwire anode and cathode planes which produce the electron avalanche. The imaging cathode plane is also shown.

Figure 7b (Bottom). A side view showing the relative orientation of the various electrodes.

Figure 7c (Facing page). The geometry of the four segment imaging cathode plane (Madsen et al., 1985).

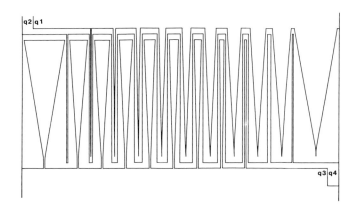

CCD chips are small and a mosaic is required to cover a substantial portion of the field of view. CCD systems are being studied for both AXAF and XMM. Cooling is, of course, required but the temperature need not be as low as for Si(Li) detectors.

3.2.3. <u>Calorimeter</u>. After an X-ray is absorbed and all the excitation created in the absorber has decayed the net result is a small temperature rise which depends upon the energy of the photon and the thermal properties of the absorbing material. The temperature change can be converted to an electrical signal if the absorber is a temperature sensitive resistor called a thermistor. At room temperature the small temperature change caused by an X-ray event is not noticeable against the thermal noise (phonons) in the detector. The temperature change is small enough to require operating temperatures less than 0.1 to 0.2°K before being detectable. The energy resolution for this device is determined by the thermal fluctuations in the detector and can be extremely good, a ΔE of less than ~ 10 eV based on theoretical considerations, and is independent of energy. Most recent devices consist of an absorber chosen because of its X-ray and thermal properties in close contact with the thermistor. A silicon thermistor-bismuth absorber and a similar germanium-diamond combination are being studied for AXAF (Moseley, Mather, and McCammon, 1984) and XMM (Coron et al., 1985), respectively.

Following an event, the detector should reach the bath temperature before a second event occurs. The cooling time depends on the thermal conductivity of the calorimeter which typically limits the count rate to < 50 s^{-1}. The detector size is small, about 1x1x1 mm, and a small array could be used at the center of the field of view. The cooling problem is formidable. Not only is it required to have ~ 0.1°K but fluctuations should not exceed 1 part in 10^4. Laboratory versions of these systems are based on cooling which is achieved by adiabatic demagnetization in a helium refrigerator.

TABLE II. A comparison of broad band spectrometers.[1]

	PSPC	PSGS	Si(Li)	CCD	Calorimeter
Energy range, keV	0.2-20[2]	0.2-20[2]	0.1-30	0.35-7.5	0.2-10
Energy resolution, FWHM eV					
at 1 keV	500	170	100	100	<20
at 8 keV	1150	470	150	150	<20
Functional form	$400E^{1/2}$	$170E^{1/2}$	[3]	[3]	
Resolving power, $E/\Delta E$					
at 1 keV	2	5	10	10	50
at 8 keV	7	15	50	50	400
Position resolution, μm					
at 1 keV	500	2500	4000[4]	25	500[4]
at 8 keV	300	1000	4000[4]	25	500[4]
Time resolution,					
μs	5	5	5		
ms				1-30[5]	<20
Non-X-ray Background[6] $cm^{-2} s^{-1} keV^{-1}$	2×10^{-3}	3×10^{-3}	3×10^{-3}	6×10^{-3}	?
Detector size, cm×cm	15x15	15x15	0.4x0.4	1x1[7]	0.05x0.05

(1) Adapted in part from XMM, Report of the Instrument Working Group, ESA SP, to be published in 1988.
(2) Gas filled detectors require a thin window (~1μm stretched polypropylene) which introduces a severe loss of efficiency just above the carbon K-absorption edge.
(3) Energy resolution is dominated by electronic noise.
(4) Si(Li) and calorimeter detectors are discrete elements (typically 4 mm and 0.5 mm, respectively). Arrays can be made.
(5) Better time resolution can be achieved by reducing the portion of the CCD which is scanned.
(6) Orbit dependent. The numbers quoted are for a high earth orbit.
(7) Arrays of CCD chips can be made. Larger size chips are under development.

4. WAVELENGTH - DISPERSIVE SPECTROMETERS - MEDIUM TO HIGH RESOLUTION

Spectral measurements with broad band detectors are most effective when the shape of a featureless continuum is to be studied. When many lines and/or absorption features are present it is generally true that broad band instruments cannot resolve the individual lines and, for cool plasmas, the blending of the many closely spaced features will override the continuum completely. Spectra obtained from broad band instruments must be interpreted with models which include all the plasma parameters together with parameters relating to chemical abundance and equilibrium. A best fit set of parameters obtained from such an analysis may or may not point uniquely towards the best interpretation of the data. If, for each class of source, a reliable model could be found from some other set of measurements, then the broad band data would not be so difficult to interpret. Data obtained with higher spectral resolution than is attainable with broad band spectrometers is required and, until the calorimeter spectrometer reaches its ultimate potential, this resolution can be provided only by wavelength dispersive spectrometers such as diffraction gratings or Bragg crystals.

In selected regions of the spectrum, detailed information concerning the plasma under study is available from medium- to high resolution measurements of a carefully selected set of emission and absorption lines such as those shown in Figure 1b. The positions, shapes and ratios of these lines lead most directly to independent values of electron and ion temperatures, electron density, relative chemical and ion abundances and their distribution, mass motion within the source and redshift. The details of the analysis of high resolution spectra are discussed elsewhere in this volume by F. Bely-Dubau. Practical difficulties limit the number of cosmic X-ray sources for which good high resolution studies can be obtained for the brightest members of each class. Although the number is small - several hundreds of candidates out of a potential sample of millions which are within reach of the large X-ray telescopes - the quality of the data obtained from them and the uniqueness of its interpretation when selecting from among the many possible models allow this approach to serve as the benchmark against which data from broad band measurements can be interpreted more reliably. Dispersive spectrometers are compared in Table III.

4.1. Diffraction Gratings

4.1.1. Objective grating spectrometer (OGS). High resolution microlithographic techniques have made it possible to construct very accurate, free standing transmission gratings with line densities of up to several thousand per mm. These can be assembled into an array which can fill the aperture of an X-ray concentrator. The array could be placed either in front of or behind the concentrator. The latter position is chosen since the rays diffracted by a grating in front of the concentrator would no longer be parallel to the optic axis and degraded optical performance would result. When placed behind the concentrator, the array should follow a toroidal Rowland surface (Trümper,1984;Brinkman et al.,

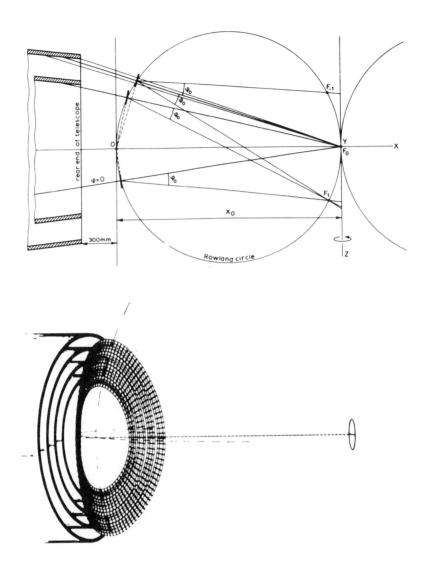

Figure 8a (Top). The Rowland circle geometry for an OGS. It is important that the detector surface also follows the circle (Brinkman et al., 1985).

Figure 8b (Bottom). The mounting surface for the grating facets is generated by rotation of the Rowland circle about the z-axis (Trümper, 1984).

1985) to compensate for the shape of the wavefront which emerges from the concentrator (see Figure 8).

The grating array forms a spectrum in the focal surface of the concentrator. The spectrum is multiplexed over a broad energy range. For a point source the spectrum is a line. For an extended source, however, an image of the source is formed at each point along the line and the spectrum is difficult to interpret.

For wavelength λ the dispersion in the focal plane for a gratingwith a line spacing d and a concentrator with focal length f is $x \sim (\lambda/d)f$. The dimension of the focal plane blur from the HPW is $\Delta x \sim \theta_{HPW} f$ which is a constant and the resolution $\Delta\lambda/\lambda = \Delta E/E = \Delta x/x \sim \theta_{HPW}(d/\lambda)$ is, therefore, best at long wavelengths. Low dispersion combined with telescope blur and the detector pixel size limit the resolution at short wavelengths (high energy) while telescope aberrations limit the low energy resolution. Resolutions better than $\Delta E/E \sim 10^{-2}$ are difficult to achieve. Typical combined first order efficiencies are $\sim 10\%$ but, in special cases, the efficiency can be increased to as much as 40-50% by using partially transmitting gratings (phased gratings, Schnopper et al., 1977).

An OGS requires a high resolution detector and can be used effectively only with a high resolution concentrator such as those flown on EINSTEIN and proposed for AXAF.

4.1.2. Reflection grating spectrometer (RGS). An OGS on XMM is ruled out by the large contribution from θ_{HPW} to the focal plane blur. To overcome the blur requires a large dispersion which implies transmission line densities which simply cannot be manufactured. Diffraction by ruled gratings at grazing incidence is an attractive alternative since the effective line spacing can be made very small. The effective line spacing $d_{eff} \sim d\sin\theta_i$ where d is the ruled spacing and θ_i is the incident angle. The grating should be constructed with the reflecting line facets tilted slightly (blazed) to maintain conditions of total reflection over a broad energy range.

Optical arguments again require that the RGS be located behind the concentrator and, in this case, the compensation for the non-plane wave front is to vary the line spacing (Hettrick and Kahn, 1985). Instead of a single, very long grating a cheveron array is used to cover the telescope aperture (see Figure 9). One complication, however, is that the average surface of the grating is now part of the image forming system (which is not the case for the transmission grating) and must not introduce significant additional contributions to the HPW which could come from small angle scattering or deviations from the ideal plane surface. The RGS has the same difficulties with extended sources as does the OGS.

4.2. Bragg Crystals

Gratings play a role intermediate between broad band and Bragg spectrometers. Their increased resolution allows some strong, well separated lines to be resolved and the continuum will be seen for bright sources. Grating spectrometers work best below 1 - 2 keV, where they have high

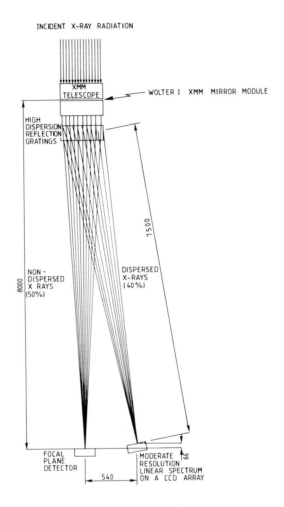

Figure 9. RGS requires an auxiliary high resolution detector (Hettrick and Kahn, 1985).

dispersion, good multiplexed efficiency and are a good match to the concentrator and detector resolutions. But gratings do not work well for extended sources and high resolution $\Delta E/E \sim 10^{-3}$> is difficult to achieve.

Bragg crystal instruments can provide high resolution but the multiplex advantage inherent in broad band and grating spectrometers is lost. The 3-dimensional nature of the process of X-ray diffraction in crystals restricts their efficiency to a very small energy range for each setting of the spectrometer. These instruments have, however, very high dispersion since the scattering centers have spacings which are characteristic of atomic dimensions. A direct benefit of the high dispersion is that the effects of telescope blur do not dominate the spectrometer resolution and it is possible to design instruments which will operate at high energy resolution in conjunction with low resolution concentrators such as XMM and XSPECT. Imaging spectrometers are also possible since the bright spectral lines are far enough apart in energy to be imaged without the confusion of overlapping which occurs for gratings with insufficient dispersion. The high spectral resolution will not only separate the substructure within the helium like emission features of most elements, but also resolve the diagnostically important satellite features that accompany them. In some cases, it will be possible to study the shape of strong, well isolated lines to establish the degree of line broadening under various plasma conditions and, in addition, to measure the red shift of bright, not too distant sources.

4.2.1. Focal plane crystal spectrometers (FPXS). The most general form of an FPXS is based on a doubly curved crystal positioned behind the focal plane of a grazing incidence concentrator. X-rays reflected on the crystal's concave side are refocused on a position sensitive detector. Ray paths for a typical FPXS geometry (Byrnak et al., 1985) are illustrated in Figure 10. The radius of curvature in the plane of dispersion together with the distance from the crystal to the concentrator focus determine the range of energy ΔE dispersed by the crystal. Curvature in the orthogonal direction refocuses the divergent beam. The whole range of energy is recorded simultaneously and the spectral resolution is nearly that defined by the crystal properties alone. In addition, the spectrometer provides nearly stigmatic imaging of extended or composite sources.

Bragg spectrometers are generally signal limited devices i.e., the integration time to obtain a meaningful signal is determined mostly by the fluctuations in the signal itself and there are only small contributions from other background sources. Of course, this assumes that a detector with good background rejection and imaging properties has been coupled to the instrument. A well designed PSPC is a good choice.

4.2.2. Objective crystal spectrometer (OXS). Spectrometers of the type already developed for use in the focal plane of a concentrator are intended primarily for point sources. An appreciable loss of spectral resolution and efficiency occurs when these instruments are used for sources with large angular extent. One way around these difficulties is to separate the spectroscopy portion of the instrument from the imaging portion.

A large flat crystal would by itself be a fine spectrometer for a point source or with suitable mechanical collimation for an extended source. The beam from the source is finely collimated compared with the rocking curve width of the crystal and a simple scan over the desired spectral range would produce the required spectrum. This simple approach has produced excellent solar X-ray spectra. For cosmic X-ray sources the system does not work well at all since the sources are weak and the noise introduced from the necessarily large active detector volume over

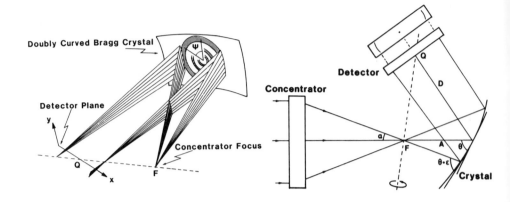

Figure 10a (Left). The geometry of a typical FPXS. A small portion of the spectrum is dispersed along the x-axis according to the curvature in the x-z plane. The curvature in the y-z plane focuses the beam into a line.

Figure 10b (Right). The shape of the crystal is obtained by a rotation about the line FQ (Byrnak, Christensen, Westergaard, and Schnopper, 1985).

whelms the signal from the source. The solution to this problem is to add a concentrator following the crystal (Schnopper and Byrnak, 1987) (see Figure 11). Two benefits are provided: The detecting pixel volume shrinks dramatically and detector noise can be almost totally suppressed. The combination of the crystal and concentrator also provides spectral imaging of extended sources. The angular range of the crystal rocking curve is typically around 30 arcsec and, therefore, this concept is ideally suited to low resolution concentrators such as XMM and XSPECT.

Data is recorded as the concentrator axis, as reflected by the crystal, is scanned across the source by rocking the crystal and/or the concentrator. The results of a computer simulation of an extended source are shown in Figure 12.

5. SENSITIVITIES

A useful way of characterizing the sensitivity of a particular instrument is to ask how many sources are available to be investigated if the length of observation is fixed at some predetermined value t. Sources are distributed according to the logN-logS distribution and the weaker the source which can be observed in a time t the greater the number of

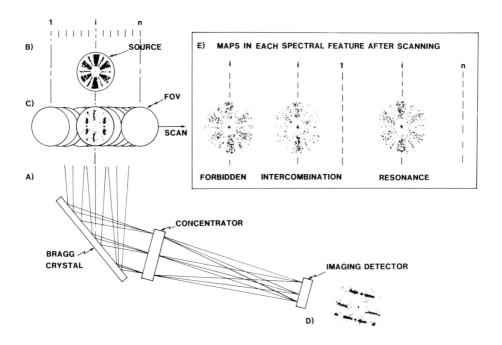

Figure 11. The OXS concept together with a simulation of an observation of a hypothetical supernova remnant. Separate images in the principal heliumlike oxygen lines are shown. (A) Geometry of the OXS system showing the Bragg reflection and imaging of a heliumlike emission system (F,I,R). (B) Surface brightness distribution of the hypothetical supernova remnant. (C) Portion of the source which satisfies the Bragg conditions for the F, I, and R lines at the ith step of the scan. (D) Recorded image of the source at the ith step of the scan. (E) Complete scan of the source provides images in each of the spectral lines. The contribution from the ith step is indicated in each case. The first and last steps for the R line are shown by 1 and and n, respectively (Schnopper and Byrnak, 1987).

sources available at the fainter end of the distribution. It is useful to calculate the minimum number of source photons $n_{s,m}$ collected in one pixel from either the continuum or an emission line which is required to produce a statistically significant feature above the background count n_b. For continuum detection, n_b includes contributions from charged particle induced non-X-ray events and from galactic and extragalactic diffuse X-ray background. For line detection, the continuum under the line should also be included as background in n_b. A feature is considered to have a positive detection with n_σ standard deviations

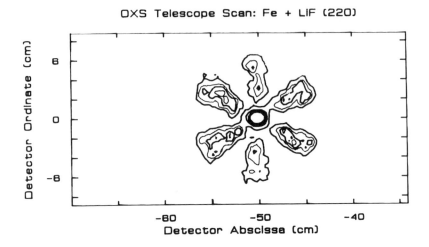

Figure 12. A simulation of the image of an extended source in one of the bright spectral lines emitted by the hot plasma. The source is one degree in diameter, has six spokes and a point source at the center (as indicated in Figure 11b).

above the background when $n_{s,m} = (n_\sigma^2 n_b^2 + n_s^2)^{1/2}$ where n_s is the number of source counts in the pixel which are required for a positive detection in the absence of background. A number of authors have made calculations based on $n_s = 10$ and $n_\sigma = 5$.

6. ACKNOWLEDGMENTS

The staff at DSRI including: B.P. Byrnak, F.E. Christensen, A. Hornstrup, P. Jonasson, M.M. Madsen, H.U. Nørgaard-Nielsen and N.J. Westergaard have made significant contributions to the work reported here. I thank them warmly for their assistance in preparing the manuscript.

TABLE III. A comparison of wavelength dispersive spectrometers.[1]

Grating	Transmission							Reflection	
Mission	EINSTEIN		EXOSAT		AXAF			XMM	
Lines per mm	500	1000	500	1000	1000	1700	5000	500	
Resolving power, $E/\Delta E$(FWHM)								order	
0.2 keV	50	50	20	40	1000			1	2
1.0	50	200	10	10	400	400		230	400
2.0						200	500	115	230
7.0							130		
Effective area, cm²									
0.2 keV	10		10		20				
1.0	1		0.5		20	60		130	90
2.0						200	30	20	60
7.0							15		

Crystal	Focal Plane		Objective
Mission	EINSTEIN	AXAF	XSPECT
Resolving Power, $E/\Delta E$(FWHM)			
0.2 keV		60	
1.0	100	200	880
2.0	90	1300	4000
7.0		600	1900
Effective area, cm²			
0.2 keV		20	
1.0	0.5	10	45
2.0	0.2	20	90
7.0		5	40

[1] Data are from various sources. Values given for AXAF, XMM and XSPECT are preliminary and subject to change during the final design phases of these missions.

REFERENCES

B. Aschenbach, H. Bräuninger, and G. Kettenring, 1983, 'Design and construction of the ROSAT 5 arcsec mirror assembly', Adv Space Res **2**, 251-254.

B. Aschenbach, 1985, 'X-ray telescopes', Rep Prog Phys **48**, 579-629.

A.C. Brinkman, J.J. van Rooijen, J.A.M. Bleeker, J.H. Dijkstra, J. Heise, P.A.J. de Korte, R. Mewe, F. Paerels, 1985, 'Low energy X-ray transmission grating spectrometer for AXAF', SPIE **597**, 232-240,

B.P. Byrnak, F.E. Christensen, N.J. Westergaard, and H.W. Schnopper, 1985, 'Doubly curved imaging Bragg crystal spectrometer for X-ray astronomy', Appl Opt **24**, 2543-2547.

E.G. Chowanietz, D.H. Lumb, and A.A. Wells, 1985, 'Charge coupled devices (CCDs) for X-ray spectroscopy applications', SPIE **597**, 381-388.

O. Citterio, G. Conti, E. Mattaini, B. Sacco, E. Santambrogio, 1985, 'Optics for X-ray concentrators on board of the Astronomy Satellite SAX', SPIE **597**, 102-110.

N. Coron, G. Artzner, G. Dambier, G. Jegoudez, J. Leblanc, J.P. Lepeltier, J.Y. Deschamps, R. Rocchia, A. Tarrius, O. Testard, P.G. Hansen, B. Jonson, H.L. Ravn, H.H. Stroke, and E. Turlot, 1985, 'Composite bolometers as spectrometers for X-ray astronomy', SPIE **597**, 389-396.

J.L. Culhane, A.H. Gabriel, L.W. Acton, C.G. Rapley, K.J. Phillips, C.J. Wolfson, E. Antonucci, R.D. Bentley, R.C. Catura, C. Jordan, M.A. Kayat, B.J. Kent, J.W. Leibacher, A.N. Parmar, J.C. Sherman, L.A. Springer, K.T. Strong, and N.J. Veck. 1981, 'X-ray spectra of solar flares obtained with a high resolution bent crystal spectrometer', Ap J **244**, L141-L145.

J. Davelaar, A. Peacock, and B.G. Taylor, 1982, 'A gas scintillation camera for X-ray astronomy', IEEE Trans Nucl Sci **NS 29**, 142-145.

P.G. Fabricant, L.M. Cohen, and P. Gorenstein, 1988, 'X-ray performance of the LAMAR proto-flight mirror', SPIE **830**, in press.

R. Giacconi, G. Branduardi, U. Briel, A. Epstein, D. Fabricant, E. Feigelson, W. Forman, P. Gorenstein, J. Grindlay, H. Gursky, F.R. Harnden Jr., J.P. Henry, C. Jones, E. Kellogg, D. Koch, S. Murray, E. Schreier, F. Seward, H. Tananbaum, K. Topka, L. Van Speybroeck, S.S. Holt, R.H. Becker. E.A. Boldt, P.J. Serlemitsos, G. Clark, C. Canizares, T. Markert, R. Novick, D. Helfand, and K. Long, 1979, 'The EINSTEIN (HEAO 2) X-ray Observatory', Ap J **230**, 540-550.

P. Gorenstein, L. Cohen, and D. Fabricant, 1985, 'X-ray telescope module for the LAMAR Space Experiment', SPIE **597**, 128-134.

C. Hettrick, and M. Kahn, 1985, 'A reflection grating spectrometer for the X-ray Multiple-Mirror (XMM) Space Observatory: Design and Calculated Performance' SPIE **597**, 291-300.

R. Hudec and B. Valnicek, 1985, 'Czechoslovak replica X-ray mirrors for astronomical applications', SPIE **597**, 111-118.

J.R. Janesick, S.T. Elliot, J.K. McCarthy, H.H. March, and S.A. Collins, 1985, 'Present and future CCDs for UV and X-ray scientific measurements', IEEE Trans Nucl Sci **NS-32**, 409-416.

R. Joyce, R. Becker, F. Birsa, S. Holt, and M. Noordzy, 1978, 'The Goddard Space Flight Center solid state spectrometer for the HEAO-B mission', IEEE Trans Nucl Sci **NS-25**, 453-458.

P.A.J. de Korte, R. Giralt, J.N. Coste, C. Ernu, S. Frindel, J. Flamand, and J.J. Contet, 1981, 'EXOSAT X-ray imaging optics', Appl Opt **20**, 1080-1088.

P.A.J. de Korte, 1988, 'High throughput replica optics', SPIE **830**, in press.

W. H-M. Ku, C.J. Hailey, and M.H. Vartanian, 1982, 'Properties of an imaging gas scintillation proportional counter', Nuc Inst Meth **196**, 63-67.

K. Kunieda and P.J. Serlemitsos, 1988, 'Optical assessment of grazing incidence X-ray mirrors', SPIE **30**, in press.

M.M. Madsen, P. Jonasson, P.L. Jensen, H.E. Rasmussen, P. Ørup, and H.W. Schnopper, 1985, 'A multiwire proportional counter for low energy X-ray imaging', SPIE **597**, 199-205.

S.H. Moseley, J.C. Mather, and D. McCammon, 1984, 'Thermal detectors as X-ray spectrometers', J Appl Phys **56**, 1257-1262.

R. Petre, C.R. Canizares, G.A. Kriss, and P.F. Winkler, 1982, 'A high-resolution X-ray image of Puppis A: Inhomogeneities in the interstellar medium', Ap J **258**, 22-30.

E. Pfefferman and U. Briel, 1985, 'Performance of the position sensitive proportional counter of the ROSAT telescope', SPIE **597**, 208-212.

R. Rocchia, M. Arnaud, C. Blondel, C. Cheron, J.C. Christy, R. Rothenflug, 1984, 'Spectral observations of the soft X-ray background with solid state detectors: Evidence for line emissions', Astron Astrophys **130**, 53-61.

Schnopper, H.W., L.P. Van Speybroeck, J.P. Delvaille, P. Epstein, E. Källne, R.L. Bachrach, J. Dijkstra, and L. Lantward, 1977, 'Diffraction grating transmission efficiencies for XUV and soft X rays', Appl Opt **16**, 1088-1091.

H.W. Schnopper and B. Byrnak, 1987, 'Bragg imaging of extended cosmic X-ray sources: the objective crystal spectrometer', Appl Opt **26**, 2871-2876.

P.J. Serlemitsos, R. Petre, C. Glasser, and F. Birsa, 1984, 'Broad band X-ray astronomical spectroscopy', IEEE Trans on Nuc Sci **NS-31**, 786-790.
S. Seward, P. Gorenstein, and W. Tucker, 1983, 'The mass of Tycho's supernova remnant as determined from a high-resolution X-ray map', Ap J **266**, 287-297.

D.G. Simmons, P.A.J. de Korte, and H. Heppener, 1987, 'An X-ray imaging gas scintillation spectrometer', Lab. for Space Research, Leiden, preprint.

L. van Speybroeck, 1988, 'Grazing incidence optics for the United States high resolution X-ray astronomy program', SPIE **830**, in press.

Y. Tanaka and F. Makino, 1988, 'Grazing incidence optics for the X-ray astronomy mission SXO', SPIE **830**, in press.

J. Trümper, 1984, 'ROSAT', Physica Scripta **T7**, 209-215.

N.J. Westergaard and P.L. Jensen, 1985, 'A thin foil high throughput X-ray telescope', SPIE **597**, 68-73.

H. Wolter, 1952a, Ann Phys NY **10**, 94.

H. Wolter, 1952b, Ann Phys NY **10**, 286.

SYSTEMATIC ERRORS IN THE DETERMINATION OF X-RAY SPECTRUM PARAMETERS AND THE CALIBRATION OF SPACE EXPERIMENTS

Martin V. Zombeck
Harvard/Smithsonian Center for Astrophysics
Cambridge, MA 02138
U.S.A.

ABSTRACT. As statistical errors in X-ray observations are reduced by means of high-throughput observatories, it will be necessary to take more careful account of systematic errors that result from the imprecision of instrument calibrations, time varying background and temporal variations in the instruments' response. Future space experiments will have to be calibrated more accurately than in the past and the calibration will have to be planned with closer attention to the the scientific goals of the observations. The lecture presents a review of parameter estimation in X-ray astronomy and, using examples from X-ray observations, the significance of systematic errors in determining best-fit parameters. The calibration of X-ray observatory instruments is also discussed.

Introduction

As the statistical precision of X-ray spectral data improves through the availability of present and future high sensitivity X-ray observatories such as GINGA, ROSAT, AXAF and XMM, systematic errors will be significant contributors to the overall error in an observational result and will have to be taken into account with greater care than in the past to correctly interpret the data. The sources of systematic errors are, for example, the imprecision of the ground calibration of the instrument providing the data, temporal variations in the instrument's response, time variations in the background and inaccurate parameters (e.g., assumed interstellar abundances) used in the physical interpretation of the data. Future space experiments will have to be calibrated more accurately than they have been in the past and the calibration will have to be planned with closer attention to the scientific goals and realistic data collecting times of the observations. The aim is to reduce both systematic and random contributions to the total error in an efficient manner because of the expense of both observations in space and calibration on the ground. It will do no good to reduce one without a concomitant reduction in the other.

The purpose of this lecture is to encourage in the student a healthy skepticism of the scientific interpretation of space observations, especially his own interpretations. Because of the nature of the enterprise, large groups at various instititutions and long development times, the interpreter of the data is often far removed, both in space and time, from the instrument builders and operators and is, therefore, often not aware of the subtle idiosyncracies of the instument's behavior or the limitations of its calibration. He may be using "canned" data reduction and analysis software packages of certain origin but not of universal applicability to his particular scientific problem.

Estimating Spectrum Parameters

Figure 1 illustrates schematically a "typical" X-ray telescope consisting of a mirror assembly, a filter and an imaging focal plane detector. For the purpose of this discussion we will consider this detector to be a non-dispersive imaging spectrometer such as an imaging proportional counter (IPC) or a charge coupled device (CCD). For a point source imaged by the telescope, we will obtain a finite-sized image due to the combined imaging performance of the mirror and focal plane detector. The data we are interested in is the pulse height spectrum of the events detected within the image.

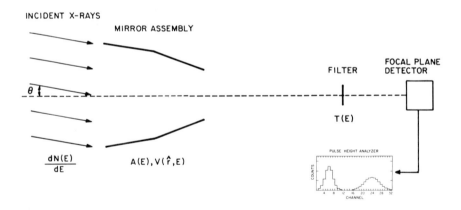

Fig. 1. Schematic diagram of an X-ray telescope.

The total number of counts in the image, N_i, in detector pulse height channel i in the observation time $\hat{T}$ is given by a convolution of the telescope response function $R(E,E')$ with the incident spectrum dN/dE:

$$N_i = \int dt \left[\int dE \int dE' (dN/dE') R(E,E') + B_i \right]$$

In general, we wish to solve for dN/dE knowing R(E,E') and the background rate B_i. When calibrating the telescope, we wish to solve for R(E,E') and B_i knowing the calibration source spectrum dN/dE.

To be more expicit we can write

$$N_i = \int_0^T dt \int_0^R 2\pi\rho\, d\rho \int_{E_{il}}^{E_{iu}} dE \int_0^\infty dE'\, (dN/dE')\, P(\hat{r},E,E')\, PSF(\rho;E',\hat{r})\, A(E')\, V(\hat{r},E')\, T(E')\, v(\hat{r}) +$$

$$+ \pi R^2 \int_0^T dt\, b_i(\hat{r},t)$$

where,

ρ = the radial coordinate of the image

R = the radius of the image detection cell

E_{il} = the lower energy limit of pulse height channel i

E_{iu} = the upper energy limit of pulse height channel i

dN/dE' = the incident photon number spectrum at energy E'

$P(\hat{r},E,E')$ = the probability that a photon of energy E' incident on the detector at the field position $\hat{r}$ results in a pulse height signal corresponding to an energy between E and E + dE

PSF(ρ;E',$\hat{r}$) = the point spread function of the combined mirror and detector

A(E') = the effective area of the mirror assembly

$V(\hat{r},E')$ = the vignetting function of the mirror assembly

T(E') = the transmission of the filter

$v(\hat{r})$ = the "obstruction" function of the detector

$b_i(\hat{r},t)$ = the background count rate per unit detector area in channel i

The instrumental response functions, P, PSF, A, V, T and v are obtained as a result of a series of calibrations. The nature of the detector response function P complicates an observed spectrum. A given pulse height does no correspond to a unique incident energy since the detector has a finite energy resolution and a fraction of the incident energy has a non-zero probability of escaping from the counter. The "obstruction" function v describes opaque or semi-transparent structures that may, for example, be used to support a thin window in a gas counter (Fig. 2). The calibration procedures used to determine the instrumental functions will be discussed in the next section. The background b_i is made up of two components, a detector internal background and a background from external sources such as charged particles and the diffuse X-ray background.

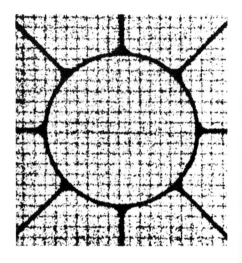

Fig. 2. X-ray shadowgraph of the windowof ROSAT's position sensitive proportional counter (PSPC). The structure seen is the thin window's aluminum support struts and 100 um grid wires. The window material itself has transmission variations of about 5% at 0.9 keV. (Private communication).

I have written the integral equation above to demonstrate the complexity of the detection process and to explicitly show all the functions that must be known in order to properly interpret the data. The problem is to infer the incident source spectrum from the data which consists of the counts in each channel of the pulse height spectrum. A lack of precise knowledge of the instrumental functions will necessarily lead to systematic errors in the determination of the source spectrum upon inverting the above equation.

In principle it is possible to invert the above integral equation directly. Dolan (1) has shown that the problem can be reduced to the solution of a matrix equation. In practice, the matrix equation cannot be solved because the matrix representing the response function of the detector is singular and an inverse matrix cannot be defined. Dolan avoids the problem of inverting the detector matrix by the technique of apodization (2). Blissett and Cruse (3) apply another direct method to infer the incident spectrum. They express both the incident spectrum and the resulting data in terms of complete set of orthonormal eigenfunctions and obtain a discrete form of the Fredholm integral equation of the first kind. By applying a stabilizing filter to handle the Poisson noise they are able to obtain an expression for the restored source spectrum directly from the data. Both methods obtain the incident source spectrum independent of any physical model.

However, the present method of choice for obtaining an incident spectrum is to model the spectrum based on a plausible physical interpretation of the source production mechanism. The model spectrum with adjustable parameters is convolved with the instrumental response functions and the difference between the predicted counts and observed counts is tested using the χ^2 statistic. The model spectrum and the value of its parameters that produce the lowest value of χ^2 is usually then taken as the "true" spectrum. This method has its pitfalls since it does not report the reduced spectrum directly but in terms of a possibly misleading physical interpretation. The χ^2 method can only allow us to reject a model but not select a model out of a set, all of which are permitted by the χ^2 test. One can also approach modeling from a strictly utilitarian viewpoint and look upon it as a means for condensing and summarizing data with a convenient functional form and a limited number of parameters without any physical content. With these considerations in mind, I wish to discuss this method in greater detail since it appears to be universally applied.

For the problem posed χ^2 is given by

$$\chi^2 = \sum_{i=1}^{M} (N_i - C_i)^2 / \sigma_i^2$$

where

C_i = the predicted counts in pulse height channel i

N_i = the observed counts in pulse height channel i

σ_i^2 = the predicted variance in pulse height channel i

M = the total number of pulse height channels

The variance σ_i^2 should include both statistical and systematic errors (4).

For a given model spectrum, the parameters are varied until a minimum χ^2 is obtained. The probability of obtaining a value of χ^2 greater than or equal to that obtained by random chance alone is given by

$$P(>\chi^2) = \left(\frac{1}{2^{n/2}\Gamma(n/2)}\right)\int_{\chi^2}^{\infty} e^{-\mu/2}\mu^{(n/2-1)}\,d\mu$$

n is the number of degrees of freedom and is equal to M minus the number of parameters in the model spectrum. The model spectrum is usually rejected if P is less than 0.05. Thus we follow Judaeo-Christian jurisprudence in declaring a model innocent until proven guilty and possibility accept many incorrect models but reject those that have demonstrated a low probability of being correct.

Spectra are usually represented by models with three parameters (4):

$$dN/dE = Cf(S,E)\exp(-\sigma_H(E)N_H)$$

where

C = a normalization constant

N_H = the hydrogen column density to the source

σ_H = the interstellar photoelectric absorption cross-section per hydrogen atom (5,6)

S = a parameter in the intrinsic spectral shape given by the functional form f(S,E)

Power law: $S = \alpha$, $f(S,E) = E^{-\alpha}$

Exponential law: $S = kT$, $f(S,E) = (1/E)\exp(-E/kT)$

Blackbody radiation: $S = kT$, $f(S,E) = E^2/[\exp(-E/kT)-1]$

Thermal bremsstrahlung: $S = kT$, $f(S,E) = \overline{g(T,E)}\exp(-E/kT)/E(kT)^{1/2}$

where, $\overline{g(T,E)}$ is the temperature averaged Gaunt factor

Often a more complicated model, motivated by the physical nature of the source, is used in which the emission from a hot, optically thin plasma is computed from fundamental atomic interaction coefficients (7). This is especially useful in the interpretation of data from dispersive spectrometers of moderate resolution such as CCD's. Temperature, elemental abundances in the plasma, flux (or normalization), and hydrogen column density are taken as the parameters to be varied for a best fit.

Figure 3 illustrates the overall technique of parameter estimation. The set of parameters $\hat{a}$ that is obtained represents the best estimate of the values of the model parameters based on the input data. The input data is subject to random errors from photon noise and the usual procedure is to estimate the uncertainties in the best fit parameters in terms of confidence limits. If the best fit parameters $\hat{a}$ are perturbed, χ^2 increases by some amount $\Delta\chi^2$ greater than its minimum value. A given $\Delta\chi^2$ defines a confidence region in the multi-dimensional parameter space (8,9). For the case of two parameters, the confidence region is in a plane (Figure 4). The contour that is defined by $\Delta\chi^2 = 2.3$ encloses the true range of possible values of the parameters with 68% probability; the contour defined by $\Delta\chi^2 = 4.6$ encloses the range with 90% probability (Table 1).

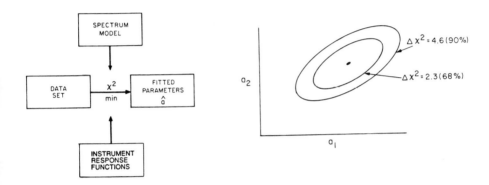

Fig. 3 Procedure for determining best fit parameters.

Fig. 4. Confidence region contours for 68% and 90% confidence for estimated parameters a_1 and a_2.

p	$\Delta\chi^2$ as a Function of Confidence Level and Degrees of Freedom					
	\multicolumn{6}{c}{ν}					
	1	2	3	4	5	6
68.3%	1.00	2.30	3.53	4.72	5.89	7.04
90%	2.71	4.61	6.25	7.78	9.24	10.6
95.4%	4.00	6.17	8.02	9.70	11.3	12.8
99%	6.63	9.21	11.3	13.3	15.1	16.8
99.73%	9.00	11.8	14.2	16.3	18.2	20.1
99.99%	15.1	18.4	21.1	23.5	25.7	27.8

Table 1. (From Ref. 9)

The validity of this prescription for determining confidence regions, particularly for parameter estimation in X-ray astronomy, has been investigated and verified by Avni (10) and Lampton, et al. (11). They generated many synthetic data sets by Monte Carlo simulation. The true spectrum was assumed to be generated by a particular set of parameter values for a given spectral function, convolved with the instrument response function, and the predicted counts were then perturbed using Poisson statistics. This was repeated many times. Each of the new data sets were then analyzed by χ^2 minimization to produce new estimates of the parameters. The statistical distribution of parameter values then yielded the confidence regions for various confidence levels. Figure 5 illustrates the simulation technique.

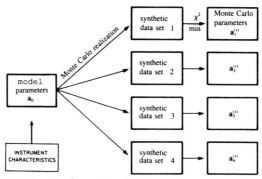

Fig. 5. Monte Carlo simulation of an experiment. Using this technique one can study the statistical distribution of fitted parameters and determine confidence regions for various confidence levels. (Adapted from Ref. 9).

Mark Twain has said that there are lies, damn lies and statistics. Figure 3 illustrates that a knowledge of the instrument characteristics is an integral part of the modeling procedure. Systematic errors in the instrument response function not taken into account can lead to misleading interpretations of the observational data. Madejski and Schwartz (11) have recently analyzed spectral data for two BL Lac objects observed by the IPC (imaging proportional counter) on the Einstein Observatory and have emphasized the role of systematic errors on the derivation of the objects' spectral parameters. The Einstein IPC gain varied with time and position of a detected event in the detector. The gain was mapped within the central 4 x 4 arc minutes of the detector but outside this area the gain could vary by 0.4 pulse height channels out of a total of 16. Madejski and Schwartz (12) found that power laws give reasonable fits to the observed spectra. For an object observed outside the central calibrated region they estimated that the 90% confidence level systematic errors are 0.25 for the power law index, 0.25 for the logarithm (base 10) of the hydrogen column density to the source, and 12% for source fluxes. These errors are larger than the statistical errors for the one object observed outside the central region.

Hughes (13) has recently studied the gas temperature and iron abundance distribution in the Coma Cluster. Using spectral data from EXOSAT's medium energy proportional counters (ME), he derived temperatures and iron abundances by fitting the observational data to the optically-thin equilibrium ionization plasma emission model of Raymond and Smith (7). The parameters of the fit were temperature, elemental abundance, flux, and hydrogen column density. For example, he found that the temperature of the center of the cluster is 8.50 keV (+0.30 -0.30) and the iron abundance 0.20 (+0.038 -0.037) (fraction of cosmic value) considering statistical errors only. Based on the knowledge of inflight adjustments, Hughes concludes that the ME detectors introduce a 0.5% fractional gain uncertainty in the observations. Varying the gain by -0.5% and +0.5%, he found that the best fit kT changed by +0.34 keV and -0.37 keV, and the iron abundance changed by -0.010 and +0.016. Hughes also estimates that the background uncertainty during his observations was 1.5% of the overall rate. With the background increased or decreased by 1.5%, the best fit kT changed by -0.22 keV and +0.09 keV and the iron abundance changed by -0.017 and +0.004.

Calibration

The calibration of a space experiment is usually the last major task before launch. Because major schedule slips and cost overruns have already occurred, there is always tremendous pressure by management to minimize the calibration effort and get on with the launch. Given this situation, it becomes necessary to plan the ground calibration carefully. Most often calibration, on large programs, has been stated vaguely in top level documents and there has been little direct comparison between the planned science observations and the calibration accuracy required to meet the scientific objectives. What is needed is a closer coupling between the observational objectives and the calibration effort. Extensive simulations of the interaction between instrumental uncertainties and scientific results appears to be an excellent approach to providing an optimum calibration. The procedure illustrated in Fig. 3 where one now also perturbs the input instrument characteristics and uses the simulation technique of Fig. 5 to produce synthetic data sets for a variety of spectral models provides a program description for determining the effect of instrument uncertainties on scientific results and for guiding the planning of the calibration effort.

In the calibration of an X-ray telescope, all the components, mirror, filters, gratings, and focal plane instruments, will have to be calibrated both spatially and spectrally. Because of the nature of laboratory X-ray sources and also because of time limitations, instrument responses will only be sampled at discrete energy and spatial points. Therefore, the overall behaviour will have to be modeled. Analogous to the previous discussion of spectral fitting, we have the situation illustrated in Fig. 6. where we now determine the fitted parameters of the instrument model rather than that of the incident spectrum. The choice of a particular model representation of the response of the instrument, uncertainties in the physical coefficents of the model (e.g., atomic interaction coefficients), and statistical errors will result in the instrumental systematic uncertainty. Gorenstein, Gursky, and Garmire (14) and Ref. 4 give excellent treatments of the modeling of the response of gas proportional counters. Hink, Scheit, and Ziegler (15) give various useful forms for the resolution of proportional counters: Gaussian, Prescott function, s-fold interval distribution, Poisson distribution, and the Weibull function. One is now left with the decision of finding the best representation of the calibration data. Since our model function is being used for interpolation, that is, to extend the discrete set of data points into a continuous function, and not to postulate some detailed physical mechanism for the counter behaviour, we don't have to worry about the non-exclusivity of the fitting procedure and just select the model which gives us the lowest χ^2 value.

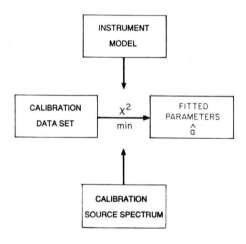

Fig. 6. Procedure for finding best fit calibration parameters.

References

1. Dolan, J.F., 1972, Astrophys. Space Sci., **17**, 472.
2. Lloyd, K.H., 1969, Am. J. Phys., **37**, 329.
3. Blisset, R.J. and A.M. Cruise, 1979, MNRAS, **186**, 45.
4. Gursky, H. and D. Schwartz, in <u>X-ray Astronomy</u>, R. Giacconni and H. Gursky, eds., 1974, D. Reidel Publishing Company.
5. Brown, R.L. and R.J. Gould, 1970, Phys. Rev., **D, 1**, 2252.
6. Morrison, R. and McCammon, D., 1983, Ap. J., **270**, 119.
7. Raymond, J., these proceedings.
8. Meyer, S.L., 1986, <u>Data Analysis for Scientists and Engineers</u>, Peer Management Consultants, Ltd., Evanston, Il.
9. Press, W.H., B. Flannery, S. Teukolsky and W. Vetterling, 1986, <u>Numerical Recipes, The Art of Scientific Computing</u>, Cambridge University Press.
10. Avni, Y., 1976, Ap. J., **210**, 642.
11. Lampton, M., B. Margon and S. Bowyer, 1976, **208**, 177.
12. Madejski, G. and D. Schwartz, 1987, Sub. Ap. J..
13. Hughes, J., P. Gorenstein and D. Fabricant, 1987, sub. Ap. J..
14. Gorenstein, P., Gursky, H., and Garmire, G., 1968, Ap.J., **153**, 885.
15. Hink, W., Scheit, A.N., and Ziegler, A., 1970, Nuc. Inst. Meth., **84**, 244.

3. SOLAR AND STELLAR CORONAE

THE SOLAR CORONA

A. H. Gabriel
Institut d'Astrophysique Spatiale
LPSP, B.P. No 10
91371 Verrière-le-Buisson
France

ABSTRACT. A review is presented covering several selected aspects of the physics of the solar corona, and including a personal viewpoint on a number of outstanding questions. Consideration of the three-dimensional heliosphere provides a link between the evolution of open field structures at the sun and the interplanetary solar wind. Attention is drawn to a number of problems that will be addressed by the proposed ESA mission, SOHO. Studies of the sun and its corona are of critical importance for the understanding of stellar winds, mass loss, activity and evolution.

1. INTRODUCTION

The corona can be loosely defined as that region of the outer solar atmosphere having a temperature of the order of one million degrees Kelvin or higher. Such a definition is bound to include not only the so-called quiet corona, but also those smaller regions within it which are much hotter; the active and flaring regions. Figure 1 shows a simplified model of the quiet solar atmosphere, assuming sherical symetry. In such a model, the corona starts at the top of the steep transition region gradient of temperature, and extends outwards into interplanetary space. The Second Law of Thermodynamics precludes that the corona be heated by thermal energy transport from below, since the photosphere of the sun is at a temperature of only 6000 °K. It is widely assumed that the energy source for the corona must lie at deeper levels, and it therefore follows that the transport mechanism cannot be thermal, and must involve some mode of mechanical or collective energy propagation.

2. CORONAL REGIONS

The spherically symetric model is a poor approximation for the real corona, which has a complex structure dominated by the magnetic field configuration and its evolution. Nevertheless, it is possible to characterise certain typical conditions which can be found to exist in particular coronal regions.

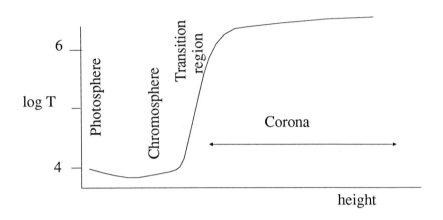

Figure 1. Simplified spherically symetric model of the solar atmosphere

2.1 The "quiet" corona

The most common condition, which occurs typically over some two-thirds of the surface of the sun, occurs in a mean magnetic field strength of between 1 and 10 gauss, and field structures which close onto other parts of the surface with loop sizes of the order of or less than a solar radius. Such closed fields inhibit the flow of the solar wind, and lead to temperatures of the order 1.7 10^6 °K, with densities at the base of the corona of 3 10^8 cm^{-3}. This is the corona that dominates when the sun is observed without spatial resolution.

2.2 Coronal hole regions

When the magnetic field in quiet coronal regions fails to close back onto the sun, but connects to the interplanetary field, there is no inhibition to flow, and the coronal plasma escapes steadily in the form of the solar wind. This flow results in a notable change in the coronal conditions. The temperature is lower, of the order 1 10^6 °K, and the mean density is reduced by a factor of around 4. Such regions have a much lower intesity of emission, which leads to their appearance as "holes" in coronal images. They have well-defined though often irregular boundaries, and can cover up to one third of the sun's surface. A good review of the observations, together with some earlier ideas on interpretation can be found in Zirker (1977).

2.3 Active regions

As part of the mechanism of cyclic evolution of the solar magnetic field, intense flux tubes below the photospheric surface rise and break through the surface locally. This results in a local enhanced and often bipolar field which has a mean value of the order 100 to 3000 gauss. Such intense fields modify greatly the physics of the corona. The resulting complex structures have temperatures of 3 to 7 10^6 °K, with densities of over 10^9 cm^{-3}. An active region can have a life-time varying from 10 to 100 days. There exist from zero to 15 such regions on the surface of the sun, depending on the phase of the solar cycle.

2.4 Flares

Associated with the enhanced field active regions, transient phenomena are observed, in which the energy source is believed to derive from the re-organisation of the magnetic field. These regions have scales of 5000 km or less and a duration at maximum of the order of 10 minutes. Temperatures reach 30 10^6 °K, with densities of over 10^{11} cm^{-3}. At the peak of the solar cycle, several large flares can be observed in one day, accompanied by a large number of smaller flares.

3. OBSERVING THE CORONA

3.1 Impact excited emission spectra

At temperatures of 1 to 2 10^6 °K, intense emission lines from the corona fall in two wavelength regions. The strongest, arising from ions having their resonance lines due to $\Delta n = 0$ transitions, occur between 170 Å and 350 Å. For a variety of reasons, there have been no observatory-class satellites which observed the corona in this important wavelength range with reasonable spatial resolution. The planned ESA mission SOHO will therefore be the first to make effective diagnostic measurements using these intense coronal emission lines.

For ions with $\Delta n = 1$ resonance lines (H- and He-like), the effective wavelength range at coronal temperatures is around 20 Å. However the lower sensitivity of these lines leads to their principal application being limited to active region and flare spectra (Culhane et al 1982).

3.2 Thermal continuum radiation

Free-free and free-bound continuum radiation for quiet coronal temperatures fall in the region 10 to 40 Å. Many observers (eg: Rosner et al 1978) have photographed the corona in this region using X-ray telescopes with broad band filters. The spatial information from such images is good. However the diagnostic information obtained from the broad band spectra is very limited. The temperature descrimination obtained from continuum bands is much poorer than that from line soectra. In addition, these continuum bands include several strong lines. The lines have a different temperature dependence from the neighbouring continuum, this contributing further to the poor descrimination of such bands.

3.3 Resonance scattered lines

Some of the photospheric or chromospheric intense emission lines can be resonce scattered from ions in the corona. This produces good images of the coronal plasma as seen outside the solar limb. Intense images can be observed in the Lyman alpha radiation of H I and He II (Gabriel et al 1986), while weaker signals can be expected from some other ions. Such signals are strongly dependent on the local density variations in the corona, but only weakly dependent on temperature. Proposals to measure the profiles and shifts of coronal scattered lines from SOHO could in principal yield several additional important parameters.

3.4 Thomson scattered radiation

Scattering of the photospheric continuum from free electrons is the principal source of the white light corona seen outside the limb during eclipses, or using coronographs. Such observations can give important data on density distributions, but no information on the temperature. Thomson scattering from isolated emission lines would give temperature data from the spectral profile. However, even the strongest emission line (H I Lyman alpha) has so far proved impossible to observe in this way.

3.5 Planck radiation

The free-free continuum emission of the corona becomes optically thick somewhere in the sub-mm to radio wavelength range, depending on the line density of the plasma. At wavelengths longer than this point, the absolute intensity is related to the temperature by the Planck formula. To exploit this effect, it is necessary to observe the wavelength dependence of the intensity, so as to be sure of working in the Planck regime.

3.6 Forbidden lines

Some of the ions which exist at coronal temperatures are capable of emitting, in addition to the XUV resonance lines, spectra at much longer wavelengths due to forbidden transitions in the ground configuration. Those transitions that fall in the visible produce the classic coronal lines that are observed from limb structures using coronographs or during eclipses. Some others fall in the normal incidence vacuum UV region. They enable normal-incidence spectrometers, normally sensitive only to lower temperature chromospheric plasma, to observe the corona. However, these forbidden lines are always very weak, and can only ever be seen outside the limb or exceptional conditions.

4. THREE-DIMENSIONAL STRUCTURE OF THE CORONA

4.1 The solar cycle

It has long been known that the activity of the sun varies according to an 11 year cycle. At the solar minimum, the magnetic configuration approximates to a classical dipole with the poles corresponding to the rotation poles. As the cycle advances, differential rotation of the solar sphere leads to a winding-up of the field lines that penetrate the interior of the sun. This effect which greatly amplifies the internal field strength leads to the ascent by bouyancy of intense flux ropes. During the period of solar maximum, these break through the surface leading to the well known active regions and solar flares. The net result of the emergence of these local bipolar fields and their reconnection to the interplanetary field structure is to lead to a simplification, and ultimately to a reversal of the solar dipole. At the next minimum we return to a simple dipole, but with the polarity reversed. The true solar cycle thus includes two maximima and two minima, and lasts for 22 years.

A full quantitative understanding of the evolution of fields during the cycle is still lacking. However there is a moderate consensus for the broad causes and phenomena, as outlined above. The complex pattern of field emergence and reconnection during the maximum involves an understanding of the stochastic three-dimensional evolution, for which there is today only a superficial grasp. For the present review, we will confine our attention to the solar minimum, and to limited departures from this condition, leaving the more complex situations to the imagination of the reader.

4.2 The solar minimum configuration

The vacuum or "potential" field of a classical dipole is shown in Figure 2. All of the field lines close onto the sun, although some traverse large distances before returning. This vacuum situation is profoundly modified by (a) the outward thermal pressure of the hot corona, and (b) the cylindrical symmetry imposed by the axial rotation of the sun. The field lines that would have closed at large distances from the sun need very little pressure to be broken open by the flow. Only the smaller loops, which represent higher field strengths are capable of resisting the pressure and thereby inhibiting the flow. The overall result is that shown in Figure 3, which follows generally a model proposed by Pneuman and Kopp (1971). The field lines which have been forced open result in a region near the ecliptic plane in which the field on each side is parallel but reversed in direction. The reversed fields are separated by a current sheet, shown shaded in the figure. There is now a sharp differentiation between the open and closed field regions. In this solar minimum configuration, the open field can be traced back to

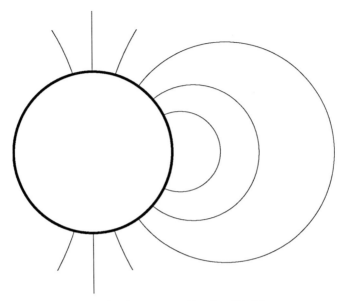

Figure 2. Classical vacuum dipole field.

the sun where it forms the coronal holes as two large polar caps. If we view the field lines in a plane parallel to the ecliptic, we obtain the result of Figure 3(b) for the open field at high solar latitutes, or Figure 3(c) for the closed field at lower latitudes. As there is no flow in the closed field, it can rotate rigidly with the sun. However, the open field will form a rotating spiral, of such a form that the flow can move in a direction approximately radial. At a large distance from the sun, all of the field lines are open, and they can all be traced back to the polar coronal holes at the sun.

The distinction between open and closed field is imposed by the configuration in interplanetary space at some distance from the sun. This picture leads to the conclusion that there is no direct local influence from below the photosphere separating the hole from the non-hole regions, although of course it is the physics of the solar interior which leads to the overall phenomenon.

The earth, positioned close to the ecliptic plane, views the interplanetary medium from a highly unique position. Small deviations between the axis of solar rotation and the magnetic axis result in the field in the vicinity of the earth changing sign dramatically twice per solar revolution. The earliest satellite probes to fly outside the earth's protective magnetosphere observed these changes in field direction, which were at that time attributed falsely to a sector pattern on the sun. This was conceived of as a field direction boundary

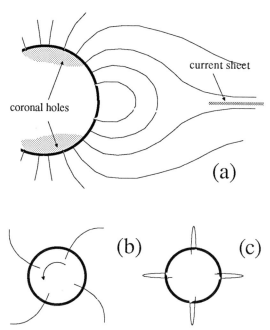

Figure 3. Solar dipole field in the presence of wind pressure and rotation. Section through poles, (a). Sections parallel to the ecliptic for (b) open field region and (c) closed field region.

perpendicular to the ecliptic, co-rotating with the sun. When the dipole field became rather more distorted, the sector pattern was observed to change from 2-sectors to 4-sectors or even more. The model described here and shown in Figure 4 has no difficulty in explaining such features. They can be understood as small undulations in a rotating current sheet, whos principal orientation is parallel to the ecliptic.

4.3 Geometry of the solar wind

The solar wind viewed just outside the earth's magnetosphere is non-uniform both in intensity and velocity. In the case of the velocity, one recognises a "slow" wind of 300 to 400 km/s, and at times a "fast" wind of 600 to 700 km/s. It is claimed that these two components are quite distinct. However all observers are not agreed on this point, and we must keep an open mind on whether there are two distinct separate components, or merely a large spread of observed velocities.

The relation between solar wind velocity and the coronal hole geometry was first understood as a result of studies made by Skylab. These

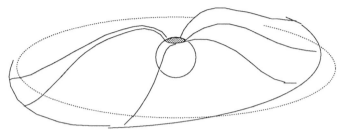

Figure 4. Typical solar minimum configuration, showing a small tilt between the current sheet (solid circumferencial line) and the ecliptic (dotted line).

observations were made in 1973 during the rising phase of solar activity. At this time the holes were not confined to the poles, but as a result of the increasing complexity were frequently observed with extensions that reached the solar equatorial regions. It was possible to observe a clear correlation between the occurence of fast wind streams and the passage of these equatorial holes across the disk. When the transit time of the streams was taken into account, it appeared that the fast streams corresponded with the centre of the coronal hole. At other times, fast wind streams were observed when there were no equatorial holes. A careful analysis of these occurences showed a correlation with a quite small dip towards the equator in the boundary of an otherwise well-behaved polar hole.

For an understanding of these phenomena, it is useful to study the model shown in Figure 4. In the normal nearly-symetrical dipole configuration, the wind observed near the earth can be traced back to the region just inside the very edge of a polar hole. These field lines that connect to the edge of a hole have the greatest length and the largest change of direction. When the field configuration becomes more complex, as shown in Figure 5, the earth can be connected by field lines

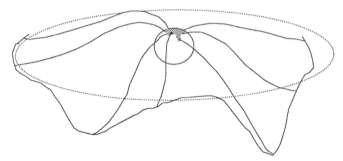

Figure 5. Typical condition some time after the minimum, showing the effects in the ecliptic plane associated with deviations in the boundary of the polar coronal hole.

that come from the interior of the hole. Even a small modulation in the hole boundary can produce a large relative change. We can thus offer a hypothesis that the wind velocity increases dramatically as we move from field lines connected to the boundary towards the interior. A study of Figure 5 opens the possibility that in the ecliptic plane we never see the short direct field line connection. It is possible that the wind experienced over the solar poles is more intense and faster than anything we see in the ecliptic plane. The forthcoming ESA mission Ulyses will traverse this region, and may reveal crucial information concerning the total solar wind and the solar mass-loss rate.

5. HEATING OF THE CORONA

Observations of the inner corona show that the temperature generally increases as a function of height, at least for the first 5 to 10 10^4 km. Such a temperature structure indicates that an important fraction of the energy input must be dissipated or converted into thermal energy at heights in excess of this value. Two problems stand in the way of understanding the physics of the heating process. The fist concerns the process by which mechanical energy is propagated upwards through the inner corona. Earlier ideas involving sound or shock-waves can be dismissed as inadequate. The majority of the energy from such modes would be reflected back downwards by the sharp density gradient in the transition region. It is now usually assumed that some form of magnetic or Alfven mode is responsible. The second problem concerns the manner of dissipation of such waves. This is a particular problem for Alfven waves, which are in general not absorbed. A variety of solutions has been proposed, involving either standing wave resonant interactions or extreme non-linear processes. There is no problem associated with the source of energy. Convective motions observed in the photosphere as granulation can be shown to contain many times the mechanical energy necessary to heat the corona. A review in this Advanced Study Institute discusses the overall problem of heating (Chiuderi, this publication).

Irrespective of the mode, we can identify three limiting concepts for the heating process. The first involves the dissipation of some form of wave motion propagating upwards. The second involves the agency of violent repetitive mechanical motions randomly generated in the chromosphere. This could be identified for example with the "bullets" observed by Brueckner and Bartoe (1983) from the HRTS experiment. These are small regions of the order 10^3 km which are found to move with line-of-sight velocities of up to 400 km/s. The third concept is an extension of the solar flare mechanism. For this it is supposed that the impulsive release of magnetically stored energy occurs repetedly at a very low level in all regions of the corona. These three concepts are of course not totally different, but can be considered as extreme descriptions of a common range of phenomena.

6. ENERGY AND MOMENTUM BALANCE OF THE SOLAR ATMOSPHERE

Logically, as well as historically, it is sensible to consider the sun as a spherically symetrical object for a first approximation, and to see how far this will suffice to explain the observed phenomena. For the relatively small vertical scales involved in the inner atmosphere, this is equivalent to a plane parallel model. Subsequently, one can investigate in turn the various known inhomogeneities, to see whether the averaging implied is a valid approximation, or whether the inhomogeneities themselves significantly alter the average physics.

Figure 1 shows a simplified and schematic diagram of the height variation of temperature in such an atmosphere. It is a synthesis of the various models derived by a variety of methods. The basic energy balance equation essential to all modelling can be presented as a statement that the divergence of the sum of all relevant modes of energy transport is equal to zero (Gabriel 1976a):

$$\mathrm{div}\,(\,F_C + F_R + F_M + F_K + F_E + F_G\,) = 0$$

Here the terms refer respectively to conduction, radiation, mechanical waves, kinetic energy, enthalpy, and gravitational energy. The divergence of radiative energy transport, $\mathrm{div}(F_r)$, is equivalent to the radiative energy loss per unit volume. The last three terms in the equation relate to the solar wind flow, and would be zero in the absence of such a flow. In a plane parallel atmosphere, the divergence operator can be replaced by d/dh.

In addition to the energy equation, we can write the equation for momentum, or pressure balance as

$$-\frac{G\,M_o\,\rho}{R_o^2} + \mathrm{div}(p) + \mathrm{div}(p_M) = \rho.\mathrm{div}(v^2)$$

gravity — gas pressure — wave pressure — wind momentum

whrere ρ is the density, p the pressure, and v the velocity. The right hand side of this equation becomes zero in the absence of a solar wind.

The manner in which the above equations are combined with observations in order to arrive at a model varies according to the region of the atmosphere under study. This is due to the variation throughout the atmosphere of the relative magnitudes of the above terms, as well as the difference in capacity for interpretation of the observed parameters.

In the lower atmosphere, the photosphere and chromosphere, up to the beginning of the steep transition region (Figure 1), one observes spectral lines which are optically thick. The detailed profiles of these lines contain important observational constraints on the gradients in the atmosphere. Against this, several terms in the energy equation are quite poorly known, in particular, the mechanical energy transport and dissipation. The technique in this region therefore consists normally of constructing an empirical atmosphere, which satisfies the observed data, including line profiles and intensities, continuum distributions, etc. Little effort is made to satisfy the energy equation, and only the first two terms of the momentum equation are imposed as constraints, ie hydrostatic equilibrium. A typical result from such an analysis is a model of the chromosphere derived by Avrett (1985).

6.1 The plane parallel inner corona

For the upper part of the atmosphere, above a temperature of $3\ 10^5$ K, the technique is quite different. This results from the fact that the emission lines are now optically thin, thereby enabling a differential emission measure technique to be applied. Furthermore, the second third and fourth terms in the energy equation can be shown to be small, as is also the wave pressure term in the momentum equation. The mechanical energy transport through this region is of course large, being responsible for the heating of the corona. However, its divergence, which appears in the energy equation is small in the region under study, and is often neglected. Thus, the mechanical flux is carried to heights well above those covered by Figure 1, where it is partially dissipated, by some mechanism not fully understood. The resulting heat is then conducted back down the temperature gradient towards the chromosphere, except for any portion used to drive the solar wind. If we examine the dominant terms of the energy equation, ignoring for a moment the radiation loss, we find very approximately:

$$\text{div}(F_C) = -\text{div}(F_E + F_G) \qquad \text{with solar wind}$$

$$\text{or}$$
$$= 0 \qquad \text{without solar wind}$$

The second of the above equations implies that the flux is conducted back through the region at a constant rate for all heights, wheras in the wind case, it is dissipated partially in providing the enthalpy and gratitational energy for the early acceleration of the solar wind.

The redundany of information between interpretation of observations and modelling through the above equations results in a choice of approaches for solution of the problem. One can start with a differential emission measure analysis, in which each spectral line intesity I can be related to the differential emission measure Y(T), through an integral equation of the kind:

$$I = A \int Y(T) \cdot g(T) \cdot d(\log T)$$

where g(T) is the emission function for the line, and A includes geometric and abundance factors. The function Y(T) can be found by inverting the set of such equations, one for each spectral line intensity observed. In order to proceed from Y(T) vs. T, to derive an atmospheric model, it is necessary to use an additional relationship between physical and geometric parameters (usually the momentum equation) plus a boundary condition (normally the pressure at a point high in the corona). Models derived in this way can be evaluated by comparison with *ab initio* models derived using the energy balance and momentum equations alone. It is found at once that the agreement is quite poor. The energy balance models show a transition region too steep by an order of magnitude, and quite different gradients in the region of $1 \; 10^6$ K.

The discrepancies in our modelling can be shown in yet another way. By measurement of emission line profiles for ions formed in the transition region, we can determine the non-thermal broadening (or turbulance) of these ions (Boland et al 1975). There can be many interpretations for this broadening, but by assuming it to be due entirely to the propagating mechanical waves which provide the coronal heating, one can derive an upper limit to the energy carried by such waves. A limit, valid for either accoustic or magnetic waves at reasonable field strengths turns out to be of the order $3 \; 10^5$ ergs cm^{-2} s^{-1}. The conductive fluxes required by either of the models derived as above are at least an order of magnitude greater. We must therefore question our basic assumption on the validity of the plane parellel approximation.

6.2. The supergranulation network

One of the principal modes of convection in the sun is that known as supergranulation. This has a typical scale size, or diameter of cell, of 40 arcsecs, as compared with the smaller granulation, which is of the order 1 to 2 arcsecs. Unlike the granulation, which is clearly visible at photospheric levels, the supergranulation can only be seen as an intensity modulation when the sun is viewed in the radiation formed at upper chromospheric or transition region heights. Since the principal convection flow fields responsible must occur at much lower levels, one is left with the conclusion that one is observing in the transition region some other effect of the motion, ie the magnetic field distortion. This effect, the principal systematic departure from the plane parallel approximation, is therefore the first candidate for explaining the discrepancies noted in the above section.

A model incorporating the effect of this magnetic field configuration has been proposed (Gabriel, 1976b), and is shown in Figure 6 . It can be shown that the effect of the magnetic field in ducting the conductive energy flux to the network boundaries is to produce a brighter network in these regions, as viewed in transition region lines. Moreover, the converging field lines lead to a change in the distribution of the spatially averaged spectral intensities (or emission measure) with temperature, which effectively removes the discrepancy between the theoretical model predictions and the observations. This improvement in agreement for the observed and modelled differential emission measure can be seen clearly in Figure 7. The effect of the network is to reduce the mean conductive flux by an order of magnitude, consistent with the limits placed by line broadening observations.

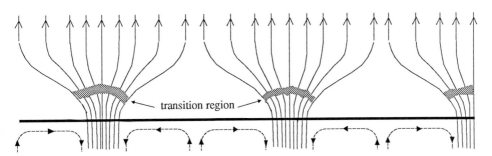

Figure 6. Model for the magnetic field configuration of the chromospheric network (Gabriel 1976b). The field lines are concentrated by the convection at the boundaries of the cells. The principal transition region is shown shaded.

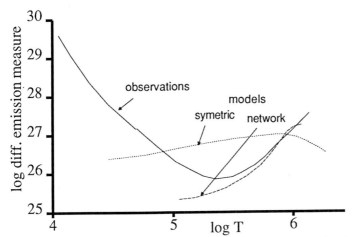

Figure 7. Observation of the differential emission measure compared with models based upon spherical symetry, and on network structure.

It can be seen in Figure 7 that even the network model deviates from the observations at temperatures less than $3\ 10^5$ °K. This will be due to the breakdown of one of the assumptions made in our theoretical model. It probably shows that below this temperature there is direct local absorption of the mechanical energy flux.

6.3 Open and closed field regions

We have already discussed the effects of field closure on the density and temperature of the corona and the existence of the solar wind. We now try to relate this to the network energy-balance modelling discussed here. Models for the network have been derived, both with and without wind flows of typical magnitude (Gabriel 1976a). These can be used to compare with the visibilty of coronal hole structures in the radiation from lines formed at different temperatures. Skylab data showed that although transition region spectra show strong contrast in the supergranular network, the coronal hole structure is almost impossible to detect in these lines. On the other hand, the modelling of such regions shows dramatic changes between open and closed regions. On closer examination, we can resolve this discrepancy. The network models (Gabriel 1976a) show that in hole regions the transition region is lower in density by factors of the order 3. However, since the energy conducted back from the corona is progressively used to accelerate the solar wind (providing enthalpy), the gradient required in the transition region is much smaller. The resulting transition region is up to an order of magnitude thicker than in closed field regions. This greater line-of-sight (when viewed normal to the solar surface) exactly compensates for the lower density and emissivity, eliminating the hole/non-hole contrast. Some of the Skylab observations made at the limb did claim to be able to resolve directly the increased thickness of the transition region in coronal holes.

There is one notable exception to the above remarks concerning the visibilty of holes; the case of helium. It is well-known that the He II 304 Å line shows both the network and the hole structures. Similarly, though at lower contrast, the He I "D-line" visible from the ground can reveal the hole structure. The explanation lies in a complex excitation/ionisation process for the helium ions, which is still much disputed, and not fully understood. One explanation involves the absorption of coronal XUV radiation by the transition region helium continua, thereby superposing coronal patterns on the lower regions. Until this is better understood, interpretation of the line at 304 Å, the strongest of all the XUV emission lines, must be treated with caution.

7. ACCELERATION OF THE SOLAR WIND

We have discussed the momentum and enthalpy of the wind in the region accessible to XUV observations, ie the first 10^5 km of height. We have also discussed the fast and slow solar winds observed near 1 AU in interplanetary space. However, even the slow wind requires the existence of an important acceleration mechanism to explain its velocity. Unfortunately we have no solr wind velocity measurements between 50 10^3 km and 0.3 AU. The SOHO mission will attempt to remedy this by making measurements out to the order of a solar radius (???) above the limb. However, the region most likely to be responsible for the acceleration lies between 5 and 10 solar radii, and remains for the present unobservable.

The early Parker models of the solar wind supposed that it was driven by the high temperature of the corona. It can now be shown that in order to produce the observed wind velocities, coronal temperatures of $4\ 10^6$ °K or higher would be necessary, and these presumably in the coronal hole regions. Our knowledge of the temperature of these regions is limited to the lower one or two scale-heights (10^5 km), where the temperture does not exceed 10^6 °K. Classical thinking supposes that a temperture maximum is reached at around this level, after which the temperture falls as random energy is converted into directed motion. However, since the heating mechanism remains largely unknown, there exists the possibility that the temperature may continue to rise until it reaches the values required for a thermally driven wind. Until XUV spectroscopy of this higher corona is available (hopefully from SOHO), this essntial temperature gradient at the base of the solar wind will remain unavailable.

If we accept that the high temperatures required are not available in coronal holes, then it is necessary to look for another explanation. Skylab observations of the variation with height of the transverse boundary of a polar coronal hole showed an expansion rate of the cross-section which would give rise to a nozzle effect. This would contribute additional terms to the acceleration of the wind. Although somewhat effective, this remains inadequate to explain the high velocity winds observed.

The most promising process considered in recent years concerns the transfer of momentum directly from waves to the wind. For a wave travelling at Alfven velocity one can associate a momentum equal to its energy transport rate divided by the velocity. As such a wave moves outwards from the sun, the fall-off in plasma density leads to an increase in the Alfven velocity. This in turn results in a decrease in the momentum of the wave. Conservation of momentum demands that this lost momentum is transfered to the plasma, leading to an acceleration. This process has several attractions. There is no dissipation or "irreversable" transfer, in the thermodynamic sense. As has already been discussed, true dissipation of Alfven waves presents difficult theoretical problems. In the present case, the plasma is not heated, and high temperatures are not required.

THE SOLAR / STELLAR CONNECTION

Studying the solar corona can be a powerful aid to the interpretation of observations made from other stellar sources. It is of course dangerous to assume that the sun is a typical star, or even that it is typical of stars of the same size and type. However, the ability to view the sun with spatial resolution presents clearly the uncertainties that arise from the point source observation of stars.

Of particular interest is the question of open and closed field lines. Observations of the sun without spatial resolution would result in a corona dominated by the closed field regions in the XUV range. Indeed the coronal holes contribute a completely negligible intensity to the spectra observed. And yet if we consider the origin of the solar wind, it is these regions which are the most important. One can speculate with profit on the application of stellar atmospheric observations and their implications on stellar wind flows and mass loss; quantites of critical importance for the recyling of stellar material within galaxies.

Observations made by the satellites Einstein and Exosat have shown that X-ray emission is far more common from a wider variety of stars than hitherto anticipated. These broad-band X-ray data are not well identified spectroscopically. The tendency is to refer to such data as stellar coronae. There is a body of statistical information correlating their intesties with stellar rotation rate. This raises several important questions. Are these really observations of quiet coronae, or are they more analogous to the solar active regions; features for which the relation to rotation is better understood? Are the observational sensitivities such that we are still not seeing the true "quiet coronae" for the stars? Is the existence of the solar quiet corona tied to the solar rotation in a way that we do not yet understand? Is our concept of the solar quiet corona illusory, in that it is really only an overall average of solar activity? It is only through detailed observations of the sun that we are in a position to pose these more general questions.

REFERENCES

Avrett E H, 1985, in "Chromospheric Diagnostics and Modeling", ed: B W Lites, National Solar Observatory, Sunspot, New Mexico, USA.

Boland B C, Dyer E P, Firth J G, Gabriel A H, Jones B B, Jordan C, McWhirter R W P, Monk P and Turner R F, 1975, Mon Not R astr Soc, **171**, 697-724.

Brueckner G E and Bartoe J D, 1983, Astrophys J, 272, 329.

Culhane J L, Acton L W and Gabriel A H, 1984, Mem Soc Astronomica Italiana, **55**, 673.

Gabriel A H, 1976a, Proc of IAU Colloqu. No 36, Nice 1976, ed Bonnet and Delache, 375-399.

Gabriel A H, 1976b, Phil Trans Roy Soc Lond A, **281**, 339-352.

Gabriel A H, Patchett B E, Lang J, Culhane J L, Norman K and Parkinson J H, 1986, J Brit Inerplanetary Soc, **39**, 207-210.
Pneuman G W and Kopp R A, 1971, Solar Phys, **18**, 258.

Rosner R, Tucker W H and Vaiana G S, 1978, Astrophys J, **220**, 643.

Zirker J B, 1977, ed: "Coronal Holes and High Speed Wind Streams", Colorado Associated University Press.

ULTRAVIOLET STELLAR SPECTROSCOPY

C. JORDAN
Department of Theoretical Physics, University of Oxford,
1, Keble Road, OXFORD, OX1 3NP, U.K.

ABSTRACT. Many objects which emit X-rays are also sources of uv emission from plasmas at temperatures between 10^4 and 2×10^5 K. To obtain a full understanding of, for example, stellar coronae, simultaneous studies in the uv, far uv and X-ray regions are ideally required. The present review summarises the methods by which uv spectra can be analysed to give plasma conditions such as the density and outlines how the uv spectra can be used to investigate the physical processes controlling energy balance in chromospheres and coronae.

1. INTRODUCTION

The aim of ultraviolet and X-ray studies of stellar chromospheres and coronae is to understand the processes which heat these regions and thus, together with the energy loss and transfer processes, control the structure of the atmosphere. The investigations to date have followed two lines of approach; studies of the correlations between selected line fluxes (or broad-band X-ray fluxes) for a large sample of stars; studies of emission lines from spectra obtained with IUE, to establish the physical conditions within the atmospheres of individual stars, and thus examine the energy requirements at different temperatures. In recent work these two approaches have been brought together to try to understand the flux-flux correlations in terms of the physical conditions.
 Here we concentrate on what information can be obtained from uv spectra, but the physics dictates that the X-ray emitting coronae must also be considered at the same time. A recent review (1) gives details of research in this area.
 Figures 1 and 2 illustrate IUE high dispersion SWP (2,3).

2. METHODS OF ANALYSING LINE FLUXES

2.1 Absolute Intensities.

The methods by which absolute line intensities are analysed to determine the amount of emitting material are now routine. The

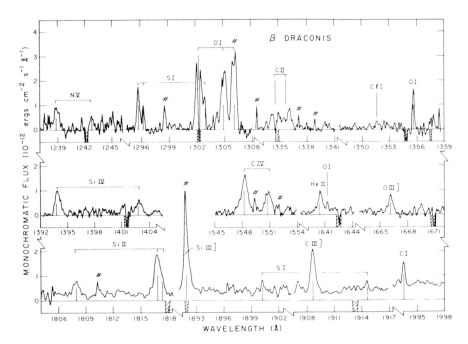

Figure 1. Sections of a high resolution IUE SWP spectrum of β Dra (G2 II), a bright giant with C IV and X-ray emission (2).

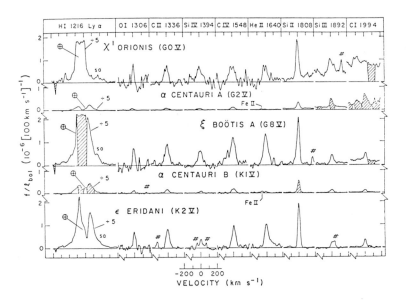

Figure 2. Sections of high resolution IUE SWP spectra of five main-sequence stars (3).

discussion is restricted to the region where $T_e \geq 10^4$ K. The stellar surface flux, F, in an effectively optically thin emission line can be found assuming that the emission comes from the full stellar hemisphere. Then, in a transition between the ground state, 1, and an excited state, 2, the flux is

$$F_{12} = hc \int N_2 A_{21} dh / 2 \lambda_{12} \qquad (1)$$

where N_2 is the population density of level 2 and A_{21} is the spontaneous transition probability. The factor ½ allows for radiation emitted back towards the star.

For a two-level atom, - in practice multi-level calculations are made,

$$N_2 A_{21} = N_1 C_{12} N_e \qquad (2)$$

since in the conditions in stellar transition regions and coronae, ion-electron collisions are the dominant process exciting level 2 and for permitted transitions $A_{21} >> C_{21} N_e$. The population density N_1 can be rewritten as

$$N_1 = (N_1/N_{ion}) (N_{ion}/N_E) (N_E/N_H) \; 0.8 \; N_e \qquad (3)$$

where for this simple case, $N_1/N_{ion} \approx 1.0$. N_{ion}/N_E is the relative ion abundance, which can be calculated as a function of temperature from ionization and recombination rates, N_E/N_H is the element abundance, N_e is the electron density. The collisional excitation rate, C_{12} is given by

$$C_{12} = 8.6 \times 10^{-6} \; \Omega_{12} \; T_e^{-\frac{1}{2}} \; \exp(-W_{12}/kT_e) \; /\omega_1 \qquad cm^3 s^{-1} \quad (4)$$

where Ω_{12} is the average collision strength, ω_1 the statistical weight of the lower level and W_{12} the excitation energy. The line flux can then be re-written as

$$F_{12} = 6.8 \times 10^{-22} \Omega_{12} (N_E/N_H) \int N_e^2 g(T_e) dh / \lambda_{12} \omega_1 \quad erg \; cm^{-2} s^{-1} \; (5)$$

where

$$g(T_e) = T_e^{-\frac{1}{2}} \exp(-W_{12}/kT_e) \; (N_{ion}/N_E) \qquad (6)$$

For most ultraviolet lines the function $g(T_e)$ peaks quite sharply at a particular temperature, say, T_m, and $g(T_e)$ can be removed from the integral and replaced by an average value $\bar{g}(T_e)$, normalised over a fixed range of temperature, $\Delta \log T = \log T_m \pm 0.15$. This procedure is less appropriate for ions which have a broad distribution with T_e, e.g. Si III and the He I-like ions which produce strong X-ray lines. It is important to fix the range of $\Delta \log T$ and use the appropriate normalisation so that two lines with the same T_m but different shapes of $g(T_e)$ yield the same value of the emission measure, which is defined as

$$Em(0.3) = \int_{\Delta h} N_e^2 \, dh. \tag{7}$$

Thus each line flux can be used to calculate a value of $Em(0.3)$ at a particular temperature, over the region Δh where $\log T_e = \log T_m \pm 0.15$. The emission measure distribution can then be built up as a function of T_e by using many lines. In the solar spectrum there are sufficient lines observed to give a good definition of the mean distribution, but only the stronger lines are observable with IUE. In this case it is useful to also consider the emission measure calculated as if all the line were formed at each T_e in turn, giving a locus which is the upper limit to the true emission measure as a function of T_e.

Figures 3a and 3b show two emission measure distributions, for ξ Boo A (G8 V) and α Cen A (G2 V), from a recent study of five main-sequence stars (4), which also included x^1 Ori (G0 V), ϵ Eri (K2V) and α Cen B (K0 V), referred to hereafter as the five dwarfs. The Si III] loci are density sensitive and are discussed below. The distributions have a broadly similar shape in spite of differing in absolute level by about an order of magnitude.

2.2 Relative Intensities used to Measure the Electron Density.

When a transition has a low A-value, as for an electric dipole, spin-forbidden line, collisional de-excitation can compete with or exceed the spontaneous decay rate. Then, labelling such a level, 3,

$$N_3 A_{31} = N_1 C_{13} N_e \cdot A_{31} / (A_{31} + C_{31} N_e) \tag{8}$$

Excitation only from level 1 has still been assumed. By comparing equations (2) and (8) it can be seen that the ratio of the fluxes in two lines given by these relations yields a measurement of the electron density, provided $C_{31} N_e \geqslant A_{31}$. In practice, since it is difficult to measure ratios which differ by more than an order of magnitude, the best cases for measuring N_e are where $C_{31} N_e \sim A_{31}$.

In the solar spectrum there are many sets of lines that can be used. (5,6) In IUE spectra only a few lines of interest can be observed. At low resolution ($\Delta\lambda \sim 6$ Å), and for material at $T_e \geqslant 10^4$ K, only the intersystem lines of Si III, at 1892 Å and C III, at 1909 Å are available. The atomic rate coefficients for these ions have received considerable attention in recent years and accurate data are now available (7, 8, 9). In main-sequence stars, the density is relatively high ($N_e \geqslant 10^{10}$ cm^{-3} at 5×10^4 K) and the C III] line is too weak to use, but the emission measure derived from the Si III] line as a function of N_e can be compared with the mean around $T_e \sim 3-5 \times 10^4$ K to give an estimate of the density. Even the Si III] line is difficult to observe above the continuum at low resolution and high resolution spectra are required (see Figure 2). However, as can be seen from Figures 3a and 3b, the density derived for ξ Boo A is higher than that derived for α Cen A.

In G-type giant and bright giant stars (see Figure 1) which can be

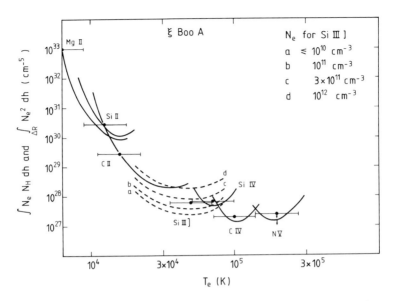

Figure 3a. Emission measure distribution for ξ Boo A (G8 V) (4).

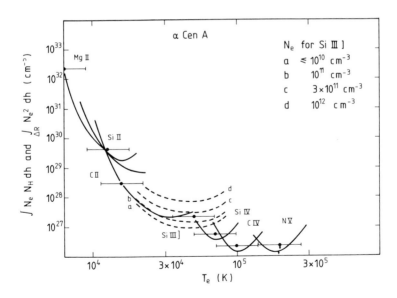

Figure 3b. Emission measure distribution for α Cen A (G2 V) (4).

quite strong X-ray sources, the density is lower and both Si III] and C III] are observed. Also intersystem lines of O III] (~1666 Å) and N IV] (~1486 Å) can sometimes be used. If high resolution IUE spectra can be obtained then the lines of O IV] (~1400 Å) can be resolved from the nearly resonance lines of Si IV and the ratio of O IV] to Si IV then provides another density diagnostic (10).

Another important indicator of high densities is apparent in the spectra of active Me dwarfs. Here the density may be sufficiently high ($N_e > 10^{12}$ cm^{-3}) to cause a reduction in the relative strength of the Si II resonance lines around 1810 Å. These permitted lines have unusually small A-values and collisional de-excitation can therefore become important. The atomic data have recently been improved (11).

The relative fluxes do not depend on the emitting area assumed but whilst regions of any N_e will contribute to permitted lines, high density regions contribute relatively less to the intersystem lines. When total fluxes are compared the ratio will still give an 'average' density, but regions at higher and lower than 'average' may be present. However, Dupree et. al. (12) found that the electron density varies only a little between solar supergranulation cell interiors and boundaries in spite of the large flux contrast.

2.3 Other diagnostics.

The Bα transition of He II at 1640 Å can be excited by several mechanisms, e.g. by recombination from He III, by collisional excitation from the ground state of He II and by line-leakage from an optically thick He II Ly β line. Because a coronal X-ray flux can photo-ionize He II the 1640 Å line could indicate the presence of an overlying hotter corona (13). Solar spectra have sufficient resolution to distinguish a broad, hot, collisional component and narrower recombination lines (14). The IUE resolution is not sufficient to separate components in stellar spectra, but this might be possible with the Hubble Space Telescope. When collisions dominate the 1640 Å line can also be sensitive to excitation by electrons corresponding to higher coronal temperatures (15).

One way of detecting the presence of inhomogeneous atmospheres is to use spherically symmetric models to compute the optical depth in strong lines. High resolution observations can then be used to examine the ratios of peak fluxes in multiplet components and the shape and widths of the lines. It is sometimes possible to at least conclude that $\tau > 1$ or $\tau < 1$, even if τ cannot be measured. (See (16) for details).

The optical depth is proportional to $\int N_H$ dh and the emission measure to $\int N_e N_H$ dh. Thus the ratio τ_0/F is proportional to P_e^{-1}, and knowing limits on τ_0 can at least limit P_e (16). However, if there are inhomogeneities in the atmosphere, they will affect τ_0 and F in different ways. If A is $< 2\pi R_*^2$ then the true surface flux will be corresponding larger. But τ_0 depends only on the <u>thickness</u> of the region. Then if the emission <u>is</u> restricted in area, the 'observed' value of τ_0/F will be larger and P_e will be smaller than the true value. But P_e measured from line ratios will not be affected by the area factor. If the latter are larger than the former this could indicate inhomogeneous structure. Again, the method is marginal with the resolution available from IUE but could be used with HST spectra.

3. THE USE OF LINE WIDTHS

The observed widths of optically thin ultraviolet emission lines exceed the expected thermal Doppler widths. The additional width is usually attributed to a 'turbulent' broadening, which may be associated with the passage of waves through the atmosphere. The observed FWHM, $\Delta\lambda$, can be expressed as

$$\Delta\lambda/\lambda = 7.1 \times 10^{-7} (T_e/M_i + (\xi_o^2 \; m_p/ \; 2 \; k))^{1/2} \quad (9)$$

where ξ_o is the non-thermal most probable velocity and $<V_T^2> = 3/2 \; \xi_o^2$ is the line of sight r.m.s velocity. The energy density can be written as

$$E = \text{const.} \; \rho \; <V_T^2> \quad (10)$$

where a constant of 3/2 is appropriate to isotropic mass motions. Observations of line widths in the solar spectrum, at a variety of positions on the disc and at the limb, indicate that the broadening in the quiet sun is isotropic.(17) The energy flux carried is then

$$F_m = 3\rho \; <V_T^2> \; V_p \quad (11)$$

where V_p is the wave propagation velocity. Thus the ultraviolet line profiles contain valuable information regarding the amount of energy available to the overlying corona in the form of a wave flux.

4. MODELS AND ENERGY BALANCE

4.1 Models.

The emission measure distribution Em(0.3) versus T_e, is the starting point of further modelling. At present the distribution between ~10^5 K and coronal temperatures is known only for the sun. Given a space mission to study the region between ~300 Å and 1000 Å it would be possible to fill in this gap for a small but important sample of near by stars.

For the higher gravity stars that have emission up to at least 10^5 K, and excluding stars which show detectable evidence of winds, the methods developed for treating the solar atmosphere can be applied.

The emission measure given by equation (7) can be re-written as

$$\overline{dT/dh} = P_e^2 / \sqrt{2} \; \overline{Em(0.3)} \; T_e \quad (12)$$

where $P_e = N_e T_e$, and P_e and dT/dh are taken as constant over the region of line formation.

It can be seen that no further progress can be made without knowing the electron density at some temperature. Thus the density diagnostics are crucial. For the stars considered, hydrostatic equilibrium is a good approximation and the hydrostatic pressure gradient can be combined with the temperature gradient from equation (12) to give

$$P_e^2 = P_{ref}^2 \pm 2 \times 10^{-8} g_* \int_{T_e}^{T_{ref}} Em(0.3) dT_e \qquad (13)$$

where P_{ref} is the electron pressure at a chosen reference point. Thus the pressure distribution between $\sim 10^4$ K and $\sim 2 \times 10^5$ K can be found from IUE spectra.

To include X-ray data assumptions must be made. The X-ray surface flux, F_x, can be attributed to a coronal emission measure, $Em(T_c)$ at a coronal temperature, T_c, by analogy with the <u>average</u> solar corona. Replacing Δh by the isothermal density squared scale height, where most of the line flux is formed, gives

$$Em(T_c) = P_c^2 \; 7 \times 10^7 / T_c \; g_* \qquad (14)$$

and P_c^2 can be found and used as a coronal reference pressure in equation (13). For main sequence stars the pressure difference between the corona and transition region is not large and does not depend critically on the assumed form of the emission measure between $T_0 = 2 \times 10^5$ K and T_c (4). For example an emission measure with a power law dependence on T_e such that $d \log Em / d \log T_e = b$, leads to

$$P_0^2 / P_C^2 \approx 1 + [(b + 1) (\sqrt{2})^b]^{-1}$$

which varies slowly for $0 < b < 3$. With $b \sim 3/2$ (as in the Sun), $P_0 \sim 1.1 \; P_C$. Below 2×10^5 K the pressure does not rise significantly until the emission measure increases rapidly at temperatures below $\sim 2 \times 10^4$ K. Once one value of P_e is known models of P_e and T_e can be made on a relative height scale (which must be matched on to detailed chromospheric models), but obviously dT/dh <u>above</u> 2×10^5 K depends entirely on any assumed form for $Em(0.3)$.

If variations of X-ray fluxes and temperatures could be observed then it would be possible to separate 'average' and 'active' components and then model the latter in terms of closed loop structures as carried out for solar observations (18,19). At present this approach requires the assumption of an area factor $f = A(loop)/4\pi R_*^2$ before the pressure can be determined. For the five dwarfs (4) the spherically symmetric approach (f=1) gives pressures consistent with uv spectral diagnostics, but for ϵ Eri and α Cen B (20) or α Cen A (21) the loop models do not give consistent results.

4.2 Radiation and Conduction.

Below $\sim 10^5$ K the temperature gradient derived from equation (12) can be used with the equation for the conductive flux, equation (16), below, to show that the energy deposited by conduction is negligible compared with local radiation losses. Then the radiation losses must be matched by deposition of non-thermal energy from below. Fortunately the radiation losses below ~10^5 K can be determined in a way that is not dependent on the detailed modelling. Since

$$dF_R/dh = 0.8 \, N_e^2 \, P_{rad}(T_e) \text{ and } \Delta F_R(0.3) \approx 0.8 Em(0.3) \, P_{rad}(T_e) \quad (15)$$

where $P_{rad}(T_e)$ is the power loss ($cm^3 s^{-1}$) of the plasma, (e.g. 22, 23) the energy lost over a range of $\Delta logT = 0.3$, typical of line formation can be found as a function of T_e and $\Delta F_R(0.3)$ can then be summed to give the total radiation losses.

The total radiation loss from the corona can be found from the broad-band X-ray observations if T_c can be estimated and the emissivity as a function of wavelength is known. In 'active' main sequence stars (e.g. ξ Boo A) the coronal radiation losses can exceed those from the region between 2×10^4 K - 2×10^5 K, but the reverse is true in less active stars eg. α Cen A. This change is also reflected in flux-flux correlations.

The conductive flux back from the corona at ~ 2×10^5 K can be found by combining equation (12) and

$$F_c(T_e) = -K_0 \, T_e^{5/2} \, dT/dh \quad (16)$$

i.e. the <u>local</u> values of Em(0.3) and P_e determine dT/dh and $F_c(T_e)$ at 2×10^5 K, not assumptions about the region <u>above</u> 2×10^5 K. In main sequence stars the fluxes implied are larger than the coronal radiation losses. This conclusion does depend on the assumption of a uniform, spherically symmetric atmosphere. Reducing the emitting area can increase Em(T_c) and the coronal radiation losses and if even smaller areas are involved at 2×10^5 K the value of $F_c(T_e)$ could be decreased.

4.3 Energy Input.

One can examine what type of wave mode could carry enough energy up to T_e ~ 2×10^5 K to account for at least the coronal radiation losses and match the energy flux implied by the non-thermal line broadening below ~10^5 K. Waves propagating at the sound velocity could carry a flux which in the five dwarfs (4) <u>is</u> sufficient to account for their radiation losses above ~2×10^4 K, but not to account also for the conductive flux back from their coronae.

Waves propagating at the Alfvén velocity involve the magnetic field B, which is not determined and can be chosen to be high enough to account for the total fluxes, implying quite modest fields. (B ≤ 25 gauss at ~10^4 K). In the region where $P_e \approx$ const., and in a narrow transition region where B ≈ const., the variation of the Alfvén wave flux with T_e can be found from

$$F_A = 8.5 \times 10^{-13} \, P_e^{1/2} \, B \langle V_T^2 \rangle \, T_e^{-1/2} \qquad \text{erg cm}^{-2} \text{ s}^{-1} \quad (17)$$

provided the variation of $\langle V_T^2 \rangle$ with T_e is known. In the solar transition region (17) and in main sequence stars (4) it has been found that $\langle V_T^2 \rangle$ is $\simeq \propto T_e^{\frac{1}{2}}$, close to the condition for constant flux. If $dF_A/dT = -dF_R/dT$, then dF_A/dT can also be expressed in terms of the variation of the emission measure with T_e, such that

$$Em(0.3) = \frac{F_A(T_o)}{0.8\sqrt{2}} \frac{(\beta-x-\frac{1}{2})}{P_{rad}(T_e)} \left[\frac{T_e}{T_o}\right]^{\beta-x-\frac{1}{2}} \quad (18)$$

where $B = B_o(T_o/T_e)^x$ and $V_T^2 = V_{To}^2 (T_e/T_o)^\beta$. The energy flux required at $\sim 2 \times 10^5$ K in x^1 Ori suggests $(\beta-x-\frac{1}{2}) \sim -0.05$, consistent with x small and $\beta \sim 0.5$. Then the product $Em(0.3) P_{rad}(T_e)$ is approximately constant, as satisfied by the observed distributions between $\sim 2 \times 10^4$ K and 10^5 K. i.e. only a small amount of energy deposition from an MHD wave travelling through to heat the corona is required to account for the radiation losses from this region. Moreover, this result explains why between $\sim 2 \times 10^4$ K and 10^5 K all the G-K main sequence stars examined in detail have essentially the same <u>shape</u> of emission measure distribution, a situation which is also found within the different regions of the solar atmosphere. From equation (18) it can be seen that the absolute level of the emission measure is directly proportional to the base mechanical flux.

5. SCALING LAWS

The chromospheric fluxes and transition region fluxes scales roughly as (24)

$$F_{Tr} \propto F_{Mg\ II}^{3/2} \quad . \quad (19)$$

The direct correlation between C II and C IV is consistent with the similar shapes of emission measures between 2×10^4 K and 10^5 K. Although area factors are not taken into account above, in the sun it is known that the emission is concentrated to the supergranulation boundaries. The area factor appears to be remain constant through these relatively thin layers of the transition region (25) and it is therefore understandable why these correlations hold quite widely. But the supergranulation is not apparent in the inner 'average' solar corona, suggesting that the magnetic field diverges, increasing the area factor. This must be contained within apparent scaling between transition region and x-ray fluxes (26). Using an empirical relation between $F_{Mg\ II}$ and the transition region pressure, P_o and stellar gravity (See (4) for details), such that

$$F_{Mg\ II} \propto P_o^{3/5}/g_* \quad (20)$$

and combining relations (19) and (20) gives

$$F_{Tr} \propto P_o^{9/10} g_*^{-3/2} \quad (21)$$

a relation which fits solar transition region observations (27).

Hearn (28) has proposed that, on the basis of a minimum energy loss hypothesis,

$$P_C \propto T_C^2 \, g_* \quad \text{and} \quad Em(T_C) \propto T_C^3 g_* \tag{22}$$

relations which fit the five dwarfs (4) and the sun remarkably well. Moreover, relation (22) gives a reasonable fit to other samples of X-ray data (29) provided the stellar gravity is taken into account. Combining relations (19), (20) and (22) leads to a <u>predicted</u> scaling between X-ray and transition region and chromospheric fluxes such that

$$F_X \propto F_{Tr}^{5/3} \, g_*^2 \quad \text{and} \quad F_X \propto F_{Mg\ II}^{5/2} \, g_*^2 \tag{23}$$

which lie within the range of values discussed in the literature. Again these fluxes do not allow for area factors.

If a scaling between flux and rotation rate, Ω, is known for one of the fluxes (30)

$$F_X \propto \Omega^3 \, T_{eff}^n \tag{24}$$

then relations (23) and (24) predict

$$F_{Tr} \propto \Omega^{1.8} \, g_*^{-6/5} \, T_{eff}^{3n/5}$$

and

$$F_{Mg\ II} \propto \Omega^{1.2} \, g_*^{-4/5} \, T_{eff}^{2n/5} . \tag{25}$$

These agree well with data in the literature (31, 32, 33).

Finally, one can introduce the coronal magnetic field B_C by <u>speculating</u> that the coronal thermal density energy is provided by the deposition of a wave flux propagating at V_A. (See (4) for details), ie that

$$F_m \propto P_C^{\frac{1}{2}} \, B_C \, T_C^{\frac{1}{2}} . \tag{26}$$

The scaling laws that follow are

$$P_C \propto P_0 \propto B_C^2 \quad \text{and} \quad T_C \propto B_C \, g_*^{-\frac{1}{2}} \tag{27}$$

$$B_C \propto \Omega \, g_*^{1/6} \, T_{eff}^{n/3} \tag{28}$$

Although other scalings may be preferred the above approach serves to illustrate how the combination of uv fluxes, X-ray fluxes and rotation rates may eventually allow specific heating mechanisms to be tested against available observations – the ultimate purpose of this field of research.

REFERENCES

1. Jordan, C. & Linsky, L.: 1987, in *'Exploring the Universe with the IUE Satellite'*, D. Reidel Publ. Co. p.259.
2. Brown, A., Jordan, C., Stencel, R.E., Linsky, J.L. & Ayres, T.R.: 1984, *Ap.J.* **283**, 731.
3. Ayres, T.R., Linsky, J.L., Simon, T, Jordan, C. & Brown, A.: 1983, *Ap J.* **273**, 784.
4. Jordan, C., Ayres, T.R., Brown, A., Linsky, J.L. & Simon, T.: 1987, *M.N.R.A.S.* **225**, 903.
5. Dere, K.P. & Mason, H.E.: 1981, in *'Solar Active Regions'* (Ed. Orral, F.Q.) Colorado Assoc. Univ. Press, p.129.
6. Feldmann, U.; 1981, *Phys. Scripta* **24**, 681.
7. Dufton, P.L., Hibbert, A., Kingston, A.E. & Doschek, G.A.: 1983, *Ap. J.* **274**, 420.
8. Baluja, K.L., Burke, P.G. & Kingston, A.E.: 1980, *J.Phys. B.* **13**, L543.
9. Baluja, K.L., Burke, P.G. & Kingston, A.E.: 1981, *J.Phys. B.* **14**, 1333.
10. Nussbaumer, H. & Storey, P.J.: 1982, *A. & Ap.* **115**, 205.
11. Dufton, P.L. & Kingston, A.E.: 1985, *Ap. J.* **289**, 844.
12. Dupree, A. K., Foukal, P & Jordan C.: 1976, *Ap. J.* **209**, 621.
13. Hartmann, L., Davis, R., Dupree, A.K., Raymond, J., Schmidtke, P.C. & Wing, R.F.: 1979, *Ap. J. (Letts.)* **223**, L59.
14. Kohl, J.L.: 1977, *Ap. J.* **211**, 958.
15. Jordan, C.: 1975, *M.N.R.A.S* **170**, 429.
16. Brown, A. & Jordan, C.: 1981, *M.N.R.A.S.* **196**, 757.
17. Mariska, J.T., Feldman, U. & Doschek, G.A.: 1978, *Ap. J.* **226**, 698.
18. Jordan, C.: 1975, in *'Solar Gamma, X-and EUV Radiation'* (Ed. S.R. Kane) Reidel, p.109.
19. Rosner, R., Tucker, W.H. & Vaiana, G.S.: 1978, *Ap. J.* **220**, 643.
20. Giampapa, M.S., Golub, L., Peres, G., Serio, S. & Vaiana, G.S.: 1985, *Ap. J.* **289**, 203.
21. Landini, M., Monsignori-Fossi, B.C., Paresce, F. & Stern, R.A.: 1985, *Ap. J.* **289**, 709.
22. Raymond, J.C., Cox, D.P. & Smith, B.W.: 1976, *Ap. J.* **204**, 209.
23. Summers, H.P. & McWhirter, R.W.P.: 1979, *J. Phys. B.* **12**, 2387.
24. Oranje, B.C.: 1986, *A.& Ap.* **154**, 185.
25. Reeves, E.M., Vernazza, J.E. & Withbroe, G.L.: 1976, *Phil. Trans. Roy Soc. Lond.* **A281**, 319.
26. Ayres, T.R., Marstad, N.C. & Linsky, J.L.: 1981, *Ap. J.* **247**, 545.
27. Kjeldseth Moe, O., Andreassen, O., Maltby, P., Bartoe, J.-D.F., Brueckner, G.E. & Nicolas, K.R.: 1984, *Adv. Space Res.* **Vol. 4**, 63.
28. Hearn, A.G.: 1977, *Sol. Phys.* **51**, 159.
29. Schrijver, C.J., Mewe, R & Walter, F.M.: 1984, *A. & Ap.* **138**, 258.
30. Mangeney, A. & Praderie, F.: 1984, *A. & Ap* **130**, 143.
31. Marilli, E. & Catalano, S.: 1984, *A. & Ap.* **133**, 57.
49. Simon, T., Herbig, G. & Boesgaard, A.M.: 1985, *Ap. J.* **293**, 551.
50. Hartmann, L., Baliunas, S.L., Duncan, D.K. & Noyes, R.W.: 1984, *Ap. J.* **229**, 778.

X-ray Emission from Normal Stars

J.H.M.M. Schmitt
Max-Planck-Institut für Extraterrestrische Physik
8046 Garching bei München
Federal Republic of Germany

Abstract. The occurrence of X-ray emission from normal stars, i.e., single stars located on the main-sequence or giant branch, is reviewed. The results of low-resolution X-ray spectroscopy of a large number of late-type stars are presented and the problems and possible interpretations of the X-ray spectra are discussed. Finally, theoretical models to explain the X-ray emission from early type stars are discussed and confronted with observations; future observations to discriminate between competing models are proposed.

1. Introduction

Space-based observations with high-sensitivity X-ray telescopes such as flown on board of the *Einstein* and EXOSAT satellites have demonstrated the ubiquity of X-ray emission and hence coronae among "normal" stars (Vaiana et al. 1981); in the context of this article, a "normal" star is a single star located on the main-sequence or giant branch, and consequently, some of the brighter coronal X-ray sources such as RS CVn, W UMa or Algol binaries will not be considered here. The majority of coronal X-ray sources is thought to be solar-like in the sense that the same physical processes as in the Sun are likely to be relevant for coronal formation and heating (Linsky 1985). However, it is important to realise that most of the newly discovered coronal X-ray sources emit X-rays at a level much higher than the Sun, and it is precisely this fact that made possible the detection of thousands of coronal X-ray sources with the *Einstein* satellite. The routine detection of stellar coronae represents a major observational break-through (as compared to the pre-*Einstein* era; cf. Mewe 1979) and has established stellar X-ray astronomy as a new and vital branch of X-ray astronomy (see Rosner, Golub and Vaiana 1985).

The purpose of this article is not to review the subject of stellar coronae in general; since a number of such reviews have recently appeared (cf., Rosner, Golub and Vaiana 1985; Linsky 1985; Vaiana and Sciortino 1987 and references therein), this is neither necessary nor possible within the present space limitations. Rather, I have selected three topics within the general context of stellar coronae for a more detailed investigation: I will first discuss the occurrence of X-ray emission in the HR-diagram with particular emphasis on what can be considered well established fact and where further observations are urgently needed. Second I will extensively discuss low resolution X-ray spectroscopy (i.e., color photometry) for a large number of coronal sources and the problems and possible interpretation of such spectra, and third I will discuss the status of X-ray observations of early-type stars as well as the theoretical attempts to explain and model the observed X-ray emission.

2. Occurrence of X-ray Emission in the HR diagram

The basic result of the stellar surveys carried out with the *Einstein Observatory* can be summarised by saying that almost all normal stars with the exception of late-type giants and supergiants as well as main-sequence A-dwarfs possess envelopes with hot X-ray emitting gas, i.e., stellar coronae (Rosner, Golub and Vaiana 1985). The majority of stellar X-ray luminosities has been derived from *Einstein Observatory* IPC data, i.e., from data with very low spectral resolution. For the computation of X-ray luminosities from X-ray count rates a conversion factor had to be used that depends on the spectral characteristics of the incident X-ray radiation. Despite our rather limited understanding of the coronal temperature stratification of other stars (see section 3), the uncertainties in this conversion are of the same order or less as those in the distances for typical objects, unless coronal temperatures are very low, i.e., below $\sim 2\ 10^6\ K$; in this case the conversion becomes quite uncertain and most of the coronal energy loss will occur outside the IPC band pass.

2.1 M-dwarfs

M-dwarfs are the most numerous class of stars in the immediate solar neighborhood. The high space density of these objects implies that a substantial number of near-by objects can be studied at very low detection threshold; in fact, the X-ray faintest M-dwarf known so far, Gl 699 with $log\ L_X = 26.1$, emits at least ten times below solar levels, and is the weakest known X-ray source outside the solar system. Bookbinder (1985) and Schmitt and Rosso (1988) present the X-ray luminosities of a volume-limited sample of nearby M-dwarfs. This M-dwarf sample is about 85 percent complete; the data is consistent with all stars being X-ray emitters, i.e., the lowest X-ray detection is below the smallest upper limit derived. The X-ray luminosity distribution function varies over three orders of magnitude, the bulk of the sources having $L_X \sim 10^{27}\ erg\ s^{-1}$, whereas some stars have X-ray luminosities in excess of $10^{29}\ erg\ s^{-1}$. These high X-ray luminosities are typically found for flare stars and show up as a high-luminosity tail in the X-ray luminosity distribution function. Two points are worth mentioning in this context: First, large X-ray luminosities are found in quiescence, i.e., no time variability seems to indicate large flares; during flares the X-ray luminosity increases even further. Second, if flare stars exist throughout the Galaxy, they will show up in copious amounts in X-ray surveys at fainter flux levels such as the ROSAT all-sky survey.

2.2 Dwarf stars of spectral type F, G and K

Einstein Observatory and EXOSAT observations have resulted in a large number of X-ray detections of late-type main sequence stars of spectral type F, G and K. The observed spread in X-ray luminosity among late-type stars is enormous and is displayed in the X-ray luminosity distribution functions for the various spectral classes (Schmitt *et al.* 1985 for F-stars; Maggio *et al.* 1987 for G stars; Bookbinder 1985 for K and M stars). Although the mean and median X-ray luminosity of F stars is higher than that of the other late-type stars, the observed spread in X-ray luminosity for given spectral type is still substantial, i.e., an X-ray bright M-star can be far more X-ray luminous than an X-ray faint F-star. This turns out to be a rather unfortunate fact for all studies of X-ray emission of binary systems; in such systems it is therefore not possible to unambigously assign the X-ray flux to the individual stellar components and studies requiring such an unambigous assignment of X-ray flux necessitate the use of carefully selected samples for which additional optical information is available. At present, the data is consistent with all stars being X-ray emitters, and examples of upper limits below the lowest detection are extremely rare.

2.3 A dwarfs

X-ray emission from A-type stars is particularly important for a characterisation of activity in the HR diagram. The commonly used indicators for chromospheres and transition regions in the IUE spectra of late-type stars cannot be applied to stars earlier than about F2. The ultraviolet continuum increases rapidly with effective temperature and makes the detection of weak chromospheric or transition region lines exceedingly difficult; these difficulties are aggravated by the presence of substantial rotational line broadening since many of these stars rotate rather rapidly.

Observations at X-ray wavelengths do not suffer from such observational biases and provide direct evidence for the existence of coronae. Unfortunately, the interpretation of the X-ray observations is fraught with some difficulty as well. First, it is extremely important to consider only single stars for reasons explained above; second, the strong photospheric UV radiation from A-stars causes problems in the interpretation of the X-ray data. Schmitt et al. (1985) showed that the detections of the nearby A-stars Sirius (A1Vm) and Vega (A0Va) in the *Einstein Observatory* HRI detector can be attributed to photospheric UV radiation and that a hot corona need not be introduced to explain the observations. On the other hand, the UV transmission of the HRI is quite uncertain; small changes in filter thickness correspond to many optical depths at UV wavelengths and may thus lead to large changes in the filter transmission. Therefore the data does also not exclude a hot corona around Sirius or Vega, i.e., we have a case of absence of evidence rather than evidence of absence, which can only be solved by further X-ray observations.

Walter (1983) and Schmitt et al. (1985) have studied X-ray emission from late A and early F stars; this color range is particularly interesting because the onset of convection, i.e., the rapid increase in the thickness of subphotospheric convection zones, is thought to occur exactly in this spectral range. The data of Schmitt et al. (1985), who carefully distinguished between single stars and binaries, is consistent with the idea that stellar activity as measured through X-ray emission increases with increasing convection zone size; the earliest late-type star in the solar neighborhood to show unambigous evidence for a hot corona is Altair (A7V). On the other hand, Schmitt et al. (1985) identified two field stars and Micela et al. (1985) a Pleiades A-star as X-ray sources, all not known as binaries by optical studies; it would be important to know whether these - usually serendipitously found - X-ray sources are really single stars or whether they turn out to be Ap stars with strong magnetic fields for which X-ray emission is also found (Cash and Snow 1982).

2.4 O/B-stars

O stars were among the first coronal X-ray sources to be detected with the *Einstein Observatory* (Harnden et al. 1979); further observations (Cassinelli et al. 1981) showed that essentially all O stars are X-ray sources with X-ray luminosity being proportional to bolometric luminosity, i.e., $L_X \sim 10^{-7} L_{bol}$; this relation holds over four orders of magnitude in bolometric luminosity with about an order of magnitude scatter (see Cassinelli 1984). Note that for late-type stars both the scatter as well as the absolute values of the L_X/L_{bol} ratios are far greater. It would be of great interest to know how far in spectral type the $L_X \sim 10^{-7} L_{bol}$ relation holds. If one formally extends the relationship to A dwarfs an X-ray luminosity of $\sim 3 \; 10^{28} \; erg \; s^{-1}$ would be expected at spectral type A0; such X-ray luminosities are typical for late-type stars, and hence a meaningful study can be undertaken only on carefully selected single stars. The star HD 93695 is sometimes quoted as the latest X-ray emitting early type star (Cassinelli 1984), however it should be kept in mind that some examples of (peculiar ?) X-ray emission among magnetic A and B stars are known (Cash and Snow 1982). In section 4 I will turn to the question of theoretical models for the X-ray emission of O/B stars.

2.5 Giants

X-ray emission from giants is not only observed from the well known RS CVn binaries (Majer et al. 1985), but also from apparently single giants or supergiants of spectral type A, F and G (Ayres et al. 1981; Golub et al. 1983). The activity observed in yellow giants is generally interpreted as solar-like, whereas the X-ray emission from earlier giants such as $\alpha \; Car$ (F0Ib) or $\alpha \; Oph$ (A5III) is more difficult to interpret (Linsky 1985). A remarkable feature about the X-ray emission from giants is the existence of a Coronal Dividing Line (Ayres et al. 1981; Haisch 1986) which separates the X-ray emitting yellow gaints from the apparently X-ray dark red giants and supergiants; a similar Transition Region Dividing Line had previously been proposed by Linsky and Haisch (1978) on the basis of IUE spectra. Roughly speaking, yellow giants possess transition regions and coronae, whereas red gaints have extended warm winds, but no transition region or corona. However, the discovery of hybrid stars which show both the signatures of warm winds and transition regions by Hartmann, Dupree and Raymond (1980) and Reimers (1982) demonstrated that the topology of the "dividing line" is more complicated than originally thought. At X-ray wavelengths there are no unambiguous examples of X-ray emission on the "forbidden" side of the dividing line; the best case so far appears to be an EXOSAT detection of $\alpha \; Tra$ (K2IIb-IIIa), which however has been claimed as binary (Ayres 1985). At any rate, the hybrid stars are naturally the most obvious candidates for searches of X-ray emission on the "forbidden" side of the dividing line.

3. Coronal Temperatures of Late-type Stars

So far I have utilised only the total fluxes in the *Einstein* pass band $0.2 - 4.0 keV$; however, the IPC also provided rather coarse X-ray spectra with $\Delta E/E \sim 1$ at $1 \; keV$; this energy resolution is insufficient to resolve lines in the incident spectra.

Therefore plasma diagnostics in a strict sense is not possible with proportional counters operating at soft X-ray energies. Only a very small number of crystal and grating observations are available on stellar coronae; a somewhat larger number of SSS spectra has been taken (Swank 1984), however with some bias towards RS CVn systems and no coverage in the carbon window. Therefore for a general characterisation of the X-ray spectra of stellar coronae only the IPC data is available at present.

For a physical interpretation of the observed low resolution X-ray spectra one has to use models of the theoretically calculated X-ray emission from optically thin plasmas. Raymond (1988) discusses in detail the many assumptions and uncertainties entering the theoretical computation of such X-ray spectra. The most important point to remember is that at temperatures below $\sim 10^7$ K the emission is dominated by line radiation, and thus the predicted X-ray emission in a broad band instrument such as the IPC depends on how exactly the strongest lines are treated with respect to ionisation equilibrium and the atomic physics parameters. In the simplest case an isothermal optically thin source with temperature T and emission measure EM is assumed. Given T and EM, one can then fold in one's knowledge about the instrument and compute the expected number of counts from this hypothetical source. This process is actually a rather non-trivial exercise as demonstrated by Zombeck (1988); for the case of line spectra the problem is aggravated by the fact that the computed response depends only on the calibration of a few selected points, i.e., the positions of strong lines.

Finally, the observed number of counts in all the instrument channels $n_{i,ob}$ is compared to the expected number of counts $n_{i,ex}(T, EM)$ through some test statistic (usually χ^2). If the value of χ^2_{min} found through a two-dimensional minimisation procedure is sufficently small, the solution T_{min} and EM_{min} is accepted as a (statistically) valid description of the observed X-spectrum; if not, the solution is rejected and different model spectra involving possibly two or more spectral components are used until a sufficiently low value of χ^2_{min} is obtained.

This is the general procedure used throughout X-ray astronomy for the interpretation of low and modest resolution X-ray spectra. It leads - in the case of optically thin spectra - to a parametrised description of the observations in terms of parameters T_i, EM_i, $i = 1, n_{comp}$. However, it is worthwhile keeping in mind that such a parameterised description does first depend on the signal-to-noise ratio (SNR), does second not necessarily have to be unique and does third not necessarily have to correspond to a physically adequate or appropriate description of the source.

With these caveats in mind I present the results of spectral fits to IPC observations of stellar X-ray emission. All stars are shown for which more than 200 counts were observed and which could be identified with an object in either the Bright Star Catalog (Hoffleit 1982) or Wooley Catalog (1970). A careful distinction was made between single stars, binaries and RS CVn systems on the one hand and between main sequence stars, subgiants and giants on the other hand; here only the results for single stars are shown. In figure 1 I plot the resulting temperature for successful single component fits as a function of color. Whereas stars redder than ~ 0.9 seem to have a large temperature dispersion, main sequence and subgiant stars in the color range $0.2 < B - V < 0.9$ appear to be confined (with one exception) to a relatively narrow temperature range around $\log T \sim 6.4$; thus they cluster at values not too far away from the quiet Sun. Giants on the other hand are - with two exceptions - much hotter and show temperatures at $\log T \sim 7.0$. In figure 2 I show the results for those stars requiring two-component fits again in the form T_{low} and T_{high} vs. $B - V$; as can be seen from figure 2, the majority of two-component fits occurs for main sequence stars later than about G5. The low temperature component is always $\log T_{low} \sim 6.4$, where as the high temperature component shows $\log T_{high} \sim 7.15$, with no intermediate temperatures found; similar results have been reported by Schmitt (1984), Majer et al. (1985) and Schmitt et al. (1987), here the results are shown for the first time for a sample complete in the sense defined above.

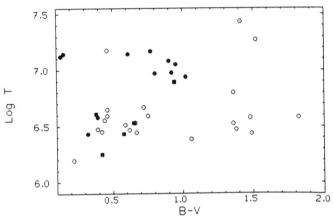

Fig.1: Coronal temperatures as a function of B-V color from single component fits for single stars; circles denote main sequence stars, filled squares subgiants and filled circles giants.

When interpreting the results shown in figure 1 and 2, two questions immediately come to mind: Is the division into one- and two-component fits real or simply caused by the data quality ? Do the two components represent physical components or can simpler descriptions of the incident spectrum be found ?

The first question can be answered by inspecting the number of counts in the X-ray spectra for which single and double component fits were obtained. It turns out that the division is largely caused by the data quality, but at least in the case of yellow giants with their rather high single temperature fits there are good reasons to believe that their X-ray spectra are intrinsically different from those of red dwarfs.

As far as the second question is concerned it has been argued by Schmitt et al. (1987) that a description in terms of a differential emission measure distribution provides a more natural and physically more reasonable description of the observed X-ray spectra. Defining the differential emission measure through

$$Q(T) \sim \frac{n^2}{T} \frac{ds}{d\ln T} = A \left(\frac{T}{T_{max}}\right)^\alpha, \qquad (1)$$

i.e., a power law distribution with index α up to some temperature T_{max} and zero otherwise, one can try to estimate the parameters α, T_{max} and the normalisation constant A from the data. The advantage of using equation (1) over the usual two-compenent description lies in the fact that only three instead of four parameters are required. More importantly, power law differential emission measure distribution are commonly observed on the Sun (Raymond and Doyle 1981) and can be naturally explained by plasma whose temperature structure is determined by balancing radiative and conductive energy losses (Antiochos, Haisch and Stern 1986); the parameter α is then related to the radiative cooling function and one expects $\alpha \sim 1$, the value found for the Sun.

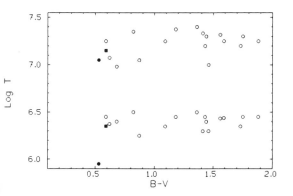

Fig.2: Coronal temperatures as a function of B-V color from two component fits for single stars; circles denote main sequence stars, filled squares subgiants and filled circles giants.

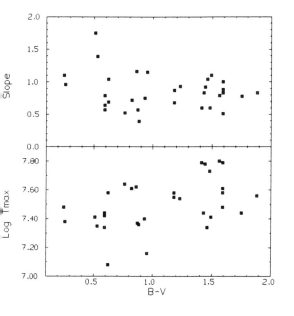

Fig.3: Maximum temperature T_{max} (lower panel) and power law slope α (upper panel) for continuous emission measure distribution fits; only stars with unacceptable one-component fits were used in the analysis.

In figure 3 I plot for those sample stars that require a two-component description the resulting fit parameters α and T_{max} if a continuous emission measure distribution as in equation (1) is adopted. The values of T_{max} lie consistently at temperatures $Log\ T_{max} \sim 7.3$, much higher than the solar values. This is obviously due to a selection effect, since only those stars requiring a two-component description, i.e., stars for which there is already evidence for plasma hotter than $10^7 K$ (see figure 2), were subjected to the differential emission measure distribution analysis. The values of the power law slope α cluster - with two exceptions - at values $\alpha \sim 0.8$, close to the observed solar values and consistent with expectations from the radiative cooling function.

4. Models of X-ray Emission from O/B stars

As already discussed in section 2, essentially all O-stars appear to be X-ray sources. The observed correlation of X-ray and bolometric luminosity together with the fact that X-ray emission appears to be absent for stars of spectral type $\sim B5 - A7$ as well as theoretical arguments about the possible role of magnetic fields in early type stars (see Linsky 1985) have led to the assumption that the X-ray emission from early type stars is intrinsically different from that observed in late type stars. The outer envelopes of early type stars do not resemble those of late-type main sequence stars; the ultraviolet spectra reveal the presence of very massive winds with mass loss rates of $\sim 10^{-6} M_\odot/yr$ and supersonic flow speeds of up to $\sim 3000 km/sec$. Plasma at a variety of temperatures is observed, ranging from material radiating at CIV, OV and NVI temperatures to material with temperature in excess of $10^7 K$, and a theory is required to explain the observed winds and their ionisation structure in a consistent and comprehensive fashion. Such a theory does not exist at present, rather theory has been developed along three different lines of thought to explain parts of the observation, i.e., the radiatively driven wind theory, the recombination stellar wind model and the distributed shock model. The theory of radiatively driven winds does not address the problem of X-ray production and will therefore not be considered here; however, radiatively driven wind theory does now offer an understanding of the dynamics and ionisation of O star winds (Pauldrach 1987).

Figure 4 demonstrates the basic problem that any explanation of the X-ray emission for O/B stars is facing: it shows the loci of unity optical depth to an observer looking from the outside through the wind of a typical O star ($\zeta\ Pup$); the range of energies chosen spans the band pass of the IPC, and for guidance the radius R_{50} at 50 stellar radii is also shown. For the calculation of the photoabsorption cross section a gas with hydrogen and helium completely ionised and carbon, nitrogen and oxygen in ionisation stages CIV, NV and OVI has been somewhat whimsically assumed. From figure 4 it is apparent that at high energies the wind is almost completely transparent, whereas at lower energies one can hardly see within 50 stellar radii. If therefore the X-ray emission is associated with the star and is located on or near the stellar surface, a tremendous attenuation problem arises and the observed X-ray flux will in general only be a small fraction of the actual X-ray flux present at the stellar surface. If on the other hand the X-ray emission is primarily associated with the wind the attenuation problem is greatly reduced. However, one then has to solve a radiative transfer problem; in particular, at high energies most of the wind is visible in X-rays and thus a global model is required for the interpretation of broad-band X-ray data.

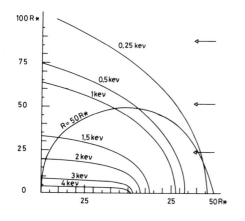

Fig.4: Loci of unit optical depth in a spherically symmetric wind. The stellar parameters are those of ζPup; each curve is labelled by energy, the circle at 50 stellar radii R_{50} is also shown.

4.1 Base Coronae

A base corona, i.e., a geometrically thin hot region between the stellar photosphere and the wind proper, was introduced by Hearn (1975) and Cassinelli and Olson (1979) in order to help accelerating the wind to the observed supersonic speeds as well as explaining the observed superionisation. The recent success of radiatively driven wind theory makes the original introduction of base coronae obsolete since they are not required for explaining the wind acceleration or superionisation. However, the introduction of base coronae led Cassinelli and Olson (1979) to predict O stars as X-ray sources prior to the launch of the *Einstein Observatory*, a prediction that was qualitatively although not quantitatively verified by *Einstein* observations. The difficulty lay in the observed softness of the low resolution IPC spectra which led Cassinelli *et al.* (1981) to conclude that the originally proposed concept of a slab corona together with a cool wind could not account for the observed IPC spectra. The theory was then refined by Waldron (1984) with his recombination stellar wind (RSW) models. The soft X-ray opacity of hot gas is well known to depend sensitively on the assumed ionisation state of matter (Krolik and Kallman 1984); any additional source of ionisation over the degree of ionisation due to collisional equilibrium can substantially decrease the soft X-ray opacity and hence increase soft X-ray transmission. By taking into account the ionisation of the wind by the proposed base corona Waldron (1984) was able to fit the IPC spectra of four O stars by a simple model involving an isothermal base corona with temperature T_c and emission measure EM_c.

Despite this success a number of problems still remain with the RSW model, some of which I mention here. First, the high temperatures determined by Cassinelli and Swank (1983) are difficult to reconcile with the RSW assumptions. Second, the SSS spectra show no signatures of an oxygen absorption edge predicted by the RSW model; on the other hand, a firm upper limit to such an edge has not been established from the SSS observations. Third, Baade and Lucy (1987) show that the derived RSW parameters for ζ *Pup* are inconsistent with their upper limit of the FeXIV 5303 Angstrom line; on the other hand, due to its spectral variability (Collura et al. 1988), ζ *Pup* may have been in a high-temperature state during the optical observations. Fourth, Baade and Lucy (1987) point out a dynamical inconsistency (unaccounted momentum flux) in the RSW model; however, since the RSW model leaves the mechanism of coronal energy input unspecified anyway, the same mechanism might just as well be asked to also provide the required momentum flux. On the other hand, the RSW model appears to be sucessful in accounting for IRAS observations of O/B-stars (Wolfire, Waldron and Cassinelli 1985), so the question of base coronae in O/B-stars has to be left open.

4.2 Distributed Shock Models

Lucy and White (1980) and Lucy (1982) developed an alternative explanation for the X-ray emission from OB-stars. Their basic conjecture is that motions supersonic with respect to the mean flow can be generated and maintained; the observed X-ray emission is assumed to be produced in the wake of these shocks. They justify this conjecture by pointing out the instability of line-driven winds (Owocki and Rybicki 1984), the large differences in radiative acceleration in non-monotonic flows as well as the rapid cooling of the shocked gas. The formation of shocks has so far not been demonstrated; only linear instability analyses have been performed and the non-linear development of the instability is unknown.

In his current model Lucy (1982) assumes that at any point in the wind there is a characteristic time scale between the passage of two consecutive shocks; further he assumes the radiative cooling zones to be infinitesimally thin and cooling proportional to the cooling function. The latter assumptions have been relaxed by Krolik and Raymond (1985) who calculated the detailed structure of a shock travelling in the wind of an O/B-star; however, they considered only isolated shocks and as pointed out above for a comparison between theory and observations one needs a global wind model. The most important quantity in Lucy's (1982) stationary model is the mean volume emissivity, i.e., the generated radiative energy density at energy E and radius r in the wind which depends - in his model - on sound speed, wind speed and gradient in wind speed respectively as well as a dimensionless parameter ν measuring the shock strength.

With the emissivity being specified one only needs to know the opacity κ_E to solve the equation of transfer for the emergent flux. In contrast to Waldron (1984), Lucy and White (1980) and Lucy (1982) do not consistently compute the opacity, but rather assume a plausible reduction of the cold gas photoabsorption cross sections. With this approach Lucy (1982) was able to account for the energetics of the observed X-ray emission from O/B-stars in a natural way. However, a number of problems also remain with the distributed shock models, most of them being noted already by Lucy (1982) himself. With the canonical parameters for the wind of ζ *Pup* one expects X-ray temperatures $< 10^6$ K, contrary to the observed X-ray spectra; in particular, the X-ray emission of ζ *Pup* is variable in the soft energy band (Collura et al. 1987), the changes - when interpreted in temperature - corresponding to temperatures of more than 10^7 K. The same point was made by

Cassinelli and Swank (1983) who found temperatures in excess of 10^7 K necessary to account for the observed SSS spectrum of ϵ Ori. If one were to interpret these high temperatures within the framework of Lucy's (1982) shock model, a distribution of shocks would have to be assumed ad hoc, with the X-ray emission being due to rather infrequent but strong shocks. In conclusion, the distributed shock model in its simplest form can account for the energetics but not for the observed X-ray spectra of O/B-stars and is therefore also unacceptable.

4.3 Future Work

As demonstrated in sections 4.2 and 4.3, none of the proposed models for X-ray emission for early type stars can account for all the observations. More theoretical work is definitely needed on base coronal models (cf., the dynamical inconsistency pointed out by Baade and Lucy 1987) as well as shock models (production of larger shock speeds). From an observational point of view a number of experiments could be performed to decide which of the proposed models (if any) applies: The presently proposed shock theories predict most of the energy output at rather low energies; a distribution of shocks might then show up as soft excess in the X-ray and XUV spectra. Eclipse measurements on suitable objects can tell whether the X-ray emission originates from near the stellar surface (in which case the X-ray light curve should follow the optical light curve) or from far out in the wind (in which case the optical light curve would be irrelevant). Finally, once X-ray spectroscopy with a resolution of $\lambda/\Delta\lambda \sim 100$ becomes available for O/B-stars, Doppler broadening of the emission lines in the X-ray spectrum could be detected, if they are indeed generated far out in the wind.

Acknowledgments: It is a pleasure to acknowledge the collaboration with the group headed by Dr. Vaiana at the Osservatorio Astronomico di Palermo, Dr. Harnden at SAO and Dr. A. Collura at MPE in the IPC spectral analysis; a detailed account of this work will be given elsewhere. Furthermore, I enjoyed enlightening discussions on O stars with Drs. R. Kudritzki, A. Pauldrach and J. Raymond.

References

Antiochos, S.K., Haisch, B.M. and Stern, R.A., 1986, *Ap. J. Lett.*, **307**, L55.
Ayres, T.R., Linsky, J.L., Vaiana, G.S., Golub, L and Rosner, R., 1981, *Ap. J.*, **250**, 293.
Ayres, T.R., 1985, *Ap. J. Lett.*, **291**, L7.
Baade, D. and Lucy, L.B., 1987, *Astron. Ap.*, **178**, 213.
Bookbinder, J.A. 1985, Ph. D. Thesis, Harvard University.
Cash, W. and Snow, T. 1982, *Ap. J. Lett.*, **263**, L59.
Cassinelli, J.P. 1984, in **The Origin of Nonradiative Heating/Momentum in Hot Stars**, ed. by A.B. Underhill and A.G. Michalitsianos, NASA Conference Publication 2358.
Cassinelli, J.P. and Olson, G.L., 1979, *Ap. J.*, **229**, 304.
Cassinelli, J.P. *et al.* 1981, *Ap. J.*, **250**, 677.

Cassinelli, J.P. and Swank, J.H., 1983, *Ap. J.*, **271**, 681.
Collura, A. , 1988, *Ap. J.*, submitted.
Golub, L., Harnden, F.R., Jr., Rosner, R., Vaiana, G.S. and Cash, W. 1983, *Ap. J.*, **271**, 264.
Hearn, A.G., 1975, *Astron. Ap.*, **40**, 277.
Haisch, B. 1986, *Irish Astron. J.*, **17**, 200.
Harnden, F.R. Jr. et al. 1979, *Ap. J. Lett.*, **234** L51.
Hartmann, L., Dupree, A.K. and Raymond, J.C., 1980, *Ap. J. Lett.*, **236**, L143.
Hoffleit, D. 1982, **The Bright Star Catalog**, Yale University Observatory.
Krolik, J.H. and Kallman, T.R. 1984 *Ap. J.*, **286**, 366.
Krolik, J.H. and Raymond, J.C., 1985, *Ap. J.*, **298**, 660.
Linsky, J.L. 1985, *Solar Physics* , **100**, 333.
Linsky, J.L. and Haisch, B.M., 1978, *Ap. J. Lett.*, **229**, L27.
Lucy, L.B., 1982, *Ap. J.*, **255**, 286.
Lucy, L.B. and White, R.L. , 1980, *Ap. J.*, **241**, 300.
Maggio, A . et al. 1987, *Ap. J.*, **315**, 687.
Majer, P. et al. 1985, *Ap. J.*, **300**, 360.
Mewe, R., 1979, *Sp. Sc. Rev.* , **24**, 101.
Micela, G. et al. 1985, *Ap. J.*, **292**, 172.
Owocki, S.P. and Rybicki, G.B. 1984, *Ap. J.*, **284**, 337.
Pauldrach, A., 1987 *Astron. Ap.*, in press.
Raymond, J.C. 1988 these proceedings.
Raymond, J.C. and Doyle, J.G. 1981, *Ap. J.*, **245**, 1141.
Reimers, D., 1982, *Astron. Ap.*, **107**, 292.
Rosner, R., Golub, L. and Vaiana, G.S. 1985, *Ann. Rev. Astron. Astrophys.*, **23**, 413.
Schmitt, J.H.M.M. 1984, in **X-ray Astronomy 84**, ed. by M.Oda and R. Giacconi, ISAS, 17.
Schmitt, J.H.M.M., Golub, L., Harnden, F.R., Jr., Maxson, C.W., Rosner, R., and Vaiana, G.S., 1985, *Ap. J.*, **290**, 307.
Schmitt, J.H.M.M., Pallavicini, R., Monsignori-Fossi, B.C. and Harnden, F.R., Jr., 1987, *Astron. Ap.*, **179**, 193.
Schmitt, J.H.M.M. and Rosso, C., 1988, *Astron. Ap.*, in press.
Swank, J.H. 1984, in **The Origin of Nonradiative Heating/Momentum in Hot Stars**, ed. by A.B. Underhill and A.G. Michalitsianos, NASA Conference Publication 2358.
Vaiana, G.S. et al. , 1981, *Ap. J.*, **245** , 163.
Vaiana, G.S. and Sciortino, S. , 1987, in IAU Symposium 122, **Circumstellar Matter**, ed. by I. Appenzeller and C. Jordan, 333.
Waldron, W.L., 1984, *Ap. J.*, **282**, 256.
Walter, F.M. 1983, *Ap. J.*, **274**, 794.
Wolfire, M.G., Waldron, W.L. and Cassinelli, J.P. 1985, *Astron. Ap.*, **142** , L25.
Woolley, R. et al. 1970, *Royal Obs. Ann.* , 5.
Zombeck, M.Z. 1988 these proceedings .

SOME EXOSAT RESULTS ON STELLAR CORONAE

R. Pallavicini
Osservatorio Astrofisico di Arcetri
Largo E. Fermi 5
50125 Firenze, Italy

ABSTRACT.   The EXOSAT Observatory has contributed substantially to our understanding of X-ray emission from stellar coronae for both single stars and close binaries. We present the highlights of EXOSAT results on stellar coronae with emphasis on the following topics: a) the temperature structure of coronae, as derived from low- and medium-resolution spectral data; b) the spatial structure of coronae, as derived from observations of eclipsing binary systems; 3) the time variability of coronal emission, as detected with the long, uninterruped observations allowed by the highly eccentric orbit of EXOSAT.

1. INTRODUCTION

In the previous paper, Dr. Jurgen Schmitt has given us an excellent overview of our present knowledge of stellar coronae. Quite understandably, the picture he has presented was based almost exclusively on the results obtained with the EINSTEIN Observatory. There are many good reasons for that. The EINSTEIN Observatory was the first one to give us a comprehensive picture of stellar coronae throughout the HR diagram, and was the only one to provide a sufficiently large sample of data as to allow statistical studies. With the observations from EINSTEIN we could study for the first time the luminosity function of X-ray emitting stars, and we could address fundamental questions such as the heating mechanism of coronae for stars of different spectral types and luminosity classes. However, in the post-EINSTEIN era, there was another satellite, EXOSAT, that, although much less sensitive than EINSTEIN, also provided many new and interesting results on stellar coronae. The purpose of this contribution is to present the highlights of these EXOSAT observations.
   The effective area of the Low Energy (LE) experiment on EXOSAT was about one order of magnitude smaller than that of the imaging experiments on EINSTEIN, so one could not expect to obtain with EXOSAT high sensitivity surveys and/or to detect new classes of objects. On the other hand, for fairly bright sources, EXOSAT could perform types of observations that were not possible with EINSTEIN, thus complementing in a very effective way the more sensitive EINSTEIN observations. First of all, EXOSAT had three different instruments on board, that were operated simultaneously: two of them, the Low Energy (LE) and Medium Energy (ME) experiments were particularly relevant for stellar observations (see Taylor 1985 for a description of EXOSAT instrumentation). With the combination of these two instruments, it was possible to cover simultaneously a much larger spectral range than with EINSTEIN (more specifically, from ≈ 0.04 KeV to more than ≈ 10 KeV, i. e. over the entire range ≈ 1 to 300 Å). Secondly, EXOSAT had a very highly eccentric orbit which allowed continuous observations for periods of up to four days, without the usual data gaps -due to Earth eclipses and passages through regions of high background- that were associated with previous low-orbit satellites. The importance of this fact in order to be able to study stellar variability and flares is self-evident.

Finally, for stars of early-spectral types, EXOSAT suffered some UV contamination in the low-energy detectors, but was completely free of such a problem for stars later than ≈ F0. Thus, while EXOSAT could not add much to our understanding of early-type stars (a problem that had to be postponed to future X-ray missions), it contributed significantly to our understanding of cool stars.

In this paper, I will briefly discuss three main topics: a) the temperature structure of stellar coronae, as derived from EXOSAT low- and medium-resolution spectral observations; b) the spatial structure of coronae, as derived from observations of eclipsing binary systems; c) the time variability of coronal X-ray emission.

2. TEMPERATURE STRATIFICATION

In order to infer the temperature structure of stellar coronae it would be desirable to obtain medium to high-resolution spectral observations: unfortunately, only a limited number of such observations were performed by the EINSTEIN Observatory, using the Solid State Spectrometer (SSS) and the Objective Grating Spectrometer (OGS) on board. In the vast majority of cases, only some coarse information on temperature could be derived from low-resolution observations obtained with the Imaging Proportional Counter (Schmitt et al. 1987b). The same is largely true also for EXOSAT, which obtained only a few moderate-resolution coronal observations with the Transmission Grating Spectrometer (TGS), in addition to a much larger sample of spectral data obtained with the ME experiment. In all other cases, broad-band observations were performed using filters. In this section, I will discuss the spectral data obtained by EXOSAT, and their relevance for the understanding of the temperature stratification in stellar coronae. I will discuss first broad-band observations obtained with the LE telescope, and then the medium-resolution spectral data obtained with the TGS and the ME experiment.

After the failure of the Position Sensitive Detectors (PSD) early in the mission, the LE telescopes on EXOSAT were used most of the time in combination with Channel Multiplier Array (CMA) detectors, which have no spectral resolution. In order to get some crude spectral resolution, the CMA can be used in conjunction with different filters and the measured count rates can be compared with those predicted on the basis of model spectra (broad-band photometry). The filters that were used most commonly for stellar observations were the Thin-Lexan (3-Lex), Thick-Lexan (4-Lex), Parylene-N + Aluminium (Al-Pa) and Boron. They have different spectral responses, and the ratio of the count rates obtained in two different filters is in principle a measure of coronal temperature, if the source can be assumed to be *isothermal*.

Unfortunately, the derivation of coronal temperatures from broad-band EXOSAT observations is by no means a trivial task. The basic difficulty is well illustrated by Fig. 1, which shows the computed ratio of count rates for various pairs of EXOSAT filters as a function of temperature and for hydrogen column densities $N_H = 1 \times 10^{18}$ cm^{-2} and $N_H = 5 \times 10^{18}$ cm^{-2}, respectively. For the typical filter ratios found in nearby stellar coronal sources (Al-Pa/3-Lex ≈ 0.3-0.5), one usually gets two and even three solutions over the temperature range log T ≈ 6.0 - 8.0 K pertinent to coronal sources. Further solutions could also formally be obtained at lower temperatures (≈ 5×10^5 K). This is an intrinsic limitation of filter spectroscopy with EXOSAT, which would persist even for a truly isothermal source. Moreover, different filter pairs usually give different temperatures. What is the physical meaning of the various temperatures derived in this way ? As will be discussed later, probably *none* of the temperatures derived from broad-band EXOSAT observations is the correct one if the source is characterized by a continuous emission measure distribution vs. temperature. The most likely interpretation of the available data is that the source is not isothermal, and, therefore, different filters sample different plasma regions (Pallavicini et al. 1987a, Schmitt and Rosso 1987).

The conclusions above is further supported by a comparison of EXOSAT LE and EINSTEIN IPC observations of the same sources (Schmitt et al. 1987b, Pallavicini et al. 1987a). One such comparison is shown in Fig. 2. The EXOSAT fluxes appear to be systematically higher than the IPC fluxes by a factor 2 to 3. This is far in excess of what would be expected on the basis of the different spectral bands observed by the two satellites (shaded area). Since it is

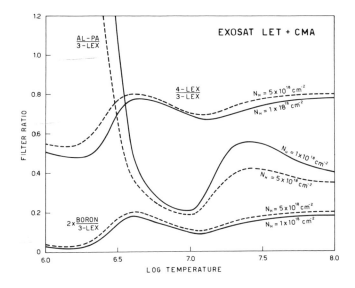

Fig. 1 : Computed ratio of count rates for various pairs of EXOSAT filters as a function of temperature and for two values of interstellar absorption (from Pallavicini et al. 1987a).

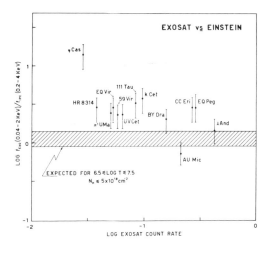

Fig. 2 : Observed ratio of 3-Lex to IPC fluxes for stellar sources observed by both EXOSAT and EINSTEIN. The dashed area is what would be expected from the different passbands of the two satellites (from Pallavicini et al. 1987a).

unlikely that almost all stars in the sample were found in a high-state by EXOSAT, a more plausible interpretation is that the EXOSAT 3-Lex and the IPC were not looking at the same plasma volume and that the effective temperatures seen by the two satellites were substantially different. The excess flux is likely provided by plasma at temperatures $T \leq 3 \times 10^6$ K which contributes mostly to the 3-Lex flux, and little to the IPC spectral band.

Better constraints on source temperature have been provided by the medium-resolution observations obtained by the Transmission Grating Spectrometer (TGS) and the ME energy experiment onboard EXOSAT. These observations, however, were possible only for very bright and sufficiently hard sources, mostly binaries of the RS CVn type. As an example, Fig. 3 a,b shows the spectra of two bright RS CVn binaries (Capella and σ CrB) obtained with the TGS (Mewe et al. 1986; see also Schrijver 1985). Although individual lines were still not well resolved, the broad line complexes discernible in these spectra are already sufficient to constrain quite well the range of temperatures present in the source. In both cases, a best fit of the data requires two components, one at $\approx 5 \times 10^6$ K and the other at $\approx 20 \times 10^6$ K.

The results shown in Fig. 3a,b are typical of what found by EXOSAT and EINSTEIN for bright stellar coronal sources. In all cases in which the spectral resolution and the S/N ratio were sufficiently high, it was invariably found that spectra of stellar coronae require at least two-temperatures, with one component at a temperature of a few to several million degrees, and the other one at a temperature about one order of magnitude higher. This is true for the EXOSAT TGS observations of Capella and σ CrB discussed above, as well as for the larger sample of stars (mostly RS CVn binaries and Algol-type systems) observed with the SSS experiment on EINSTEIN (Swank et al. 1981) and with the ME experiment on EXOSAT (Singh et al. 1987, Pasquini et al. 1987). This is also true for the lower resolution observations of both RS CVn and single active stars obtained with the IPC on EINSTEIN (Majer et al. 1986, Schmitt et al. 1987b).

Taken at face value, these two-temperature models suggest the co-existence on stars of separate regions in two distinct temperature regimes. More physically, these two regions could be constituted by two different families of loops, characterized by different physical conditions. Alternatively, the two temperature solutions could be a consequence of limited spectral resolution and energy dependence of detector response, if the source were characterized, as it is on the Sun and probably also in other single stars, by a continuous emission measure distribution with temperature. Observations of the Sun show that coronal emission comes predominantly from magnetically confined loop-like structures with a continuous emission measure distribution, rather than from isothermal regions. Furthermore, even considering the Sun -as we should do- as a mixture of active and quiet regions, the average temperatures of the various coronal structures do not differ by more than a factor of 2, and there is no indication on the Sun of the presence of two families of loops different in temperature by nearly one order of magnitude (Vaiana and Rosner 1978). We note in passing that the steady high-temperature component (at $T \approx 10^7$ K), seen outside flares by the HXIS experiment on the Solar Maximum Mission (Schaade et al. 1983), contributes negligibly ($\approx 10^{-3}$) to the integrated X-ray flux of the Sun.

Different authors have argued convincingly for either one of the above alternatives. For instance, Schmitt (1984) and Majer et al. (1986) have pointed out the existence of systematic differences in the temperatures derived by observing the same sources with different instruments: this suggests a dependence of the results of spectral fits on detector response. Moreover, Schmitt et al. (1987b) and Pasquini et al. (1987) have shown with the help of MonteCarlo simulations that two temperature fits can be obtained also for a continuous emission measure distribution and, therefore, two temperature fits do not necessarily imply physically distinct regions. More specifically, Schmitt et al. (1987b) and Pasquini et al. (1987) have made MonteCarlo simulations of EINSTEIN IPC data and of EXOSAT LE and ME data, using a continuous emission measure distribution of the form $\sim (T/T_M)^\alpha$, which closely mimics the emission measure distribution in magnetically confined loops (T_M is the maximum temperature in the loop). When the simulated data are analyzed spectrally in the same way as real data, it is found that the derived temperatures cluster around two well separated values.

On the other hand, Mewe et al. (1986) and Schrijver (1987) have found it difficult to explain with a single family of loops EXOSAT TGS observations of RS CVn binaries. The

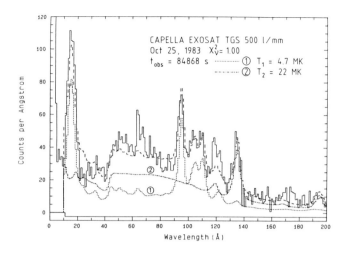

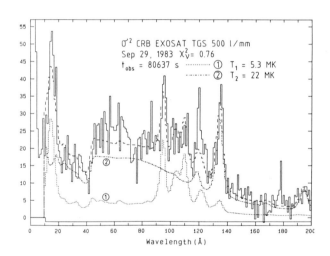

Fig. 3a,b: EXOSAT TGS observations of the RS CVn stars Capella and σ^2 CrB (from Mewe et al. 1986). Two-temperature fits of the data are also shown.

differential emission measure distribution they derive from the data has two well separated peaks, centered at the two temperatures found by fitting the data with the usual two-temperature isothermal models. White et al. (1986, 1987) have provided evidence, on the basis of eclipse observations, that two distinct families of loops, different in both temperature and size, may coexist in binaries. These eclipse observations, to which we will come back to in the next section, indicate that the higher temperature component most likely originates from very large structures which extend up to more than a stellar radius. There is no evidence on the Sun that large structures of this type -if they exist- contribute significantly to the total coronal X-ray emission.

At present, the situation appears by and large unclear. Pasquini et al. (1987), while analyzing a sample of EXOSAT ME observations of RS CVn binaries, found it possible to fit some, but not all of them, with a continuous emission measure distribution. Whereas they were able to fit with a continuous distribution the LE and ME observations of UX Ari and other sources, they were unable to do so for ME observations of Capella obtained simultaneously with the TGS data discussed by Mewe et al. (1986) and Schrijver (1987). It is possible that different situations (one or two families of loops) may hold for single and binaries systems, or even for different stars within the same class: certainly, the results of eclipse observations to be discussed below seem to suggest the presence of two families of loops in close binary systems. If confirmed, these observations may indicate the existence of fundamental differences between the coronae of single stars and the coronae of close binaries. At any rate, the above uncertainties clearly demonstrate the need we have of further medium to high-resolution spectral observations to really resolve the temperature structure of stellar coronae: such observations will hopefully be obtained by the next generations of X-ray Observatories, including NASA's AXAF and the ESA's XMM.

3. SPATIAL STRUCTURES

Except in a few very special cases (e.g. eclipsing binary systems), it is usually not possible to obtain direct information on the spatial structure of stellar coronae. An X-ray observation of a stellar source will provide a disk-integrated flux, from which we can infer indirectly some limited, and model-dependent, information about spatial structures. What is usually done is to assume a certain model configuration -as suggested by observations of magnetic structures on the Sun- and try to determine the physical parameters of this configuration by a χ^2 fit of the disk-integrated data. This approach is not without pitfalls and, what is even more important, it does not guarantee that an acceptable solution (in a purely statistical sense) is also a physically appropriate description of the coronal source (see, for instance, Schmitt et al. 1985b, Giampapa et al. 1985, Landini et al. 1985a,b, Stern et al. 1986).

Fortunately, there is at least one class of objects (the eclipsing binary systems) for which direct information on coronal structures can in principle be obtained. This is an area in which the EXOSAT contribution has been particularly important. If one has an eclipsing binary, one can use the periodic eclipses of one star by the other to roughly infer the location and size of coronal structures. For the technique to be effective, one needs high S/N ratio data which should extend continuously over long periods of time: in addition, several consecutive cycles should be observed, in order to separate the modulation due to the eclipse from possible intrinsic changes in the emission of the component stars. Even with the best available data, there is always the problem of the uniqueness of the solution, as well known to those who try to infer the properties of photospheric starspots from the observed light curves of BY Dra and RS CVn stars.

The technique of using eclipse observations to infer the structure of stellar coronae was first applied by Swank and White (1980) and by Walter et al. (1983), using the EINSTEIN Observatory. They observed the RS CVn binary AR Lac, which is formed by a G2 IV primary (with R=1.52 R_o) and a K0 IV secondary (with R=2.77 R_o), separated by 9.1 R_o. In spite of the many gaps present in the data owing to the low orbit of the satellite, Walter et al. (1983) were able to observe a deep primary eclipse (when the G star is occulted by the K star) and a shallow secondary eclipse (when the K star is behind the G star). From this they concluded that compact

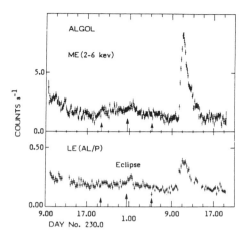

Fig. 4 : EXOSAT LE and ME observation of Algol centered on the secondary eclipse. No evidence of an X-ray eclipse is apparent (from White et al. 1986).

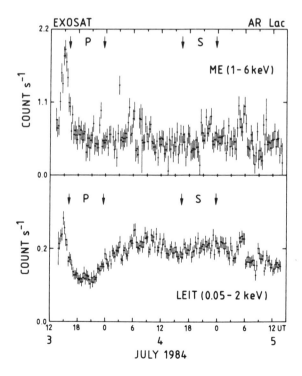

Fig. 5 : EXOSAT LE and ME observation of the eclipsing binary AR Lac throughout a full orbital cycle. The primary eclipse is seen in the LE data, but not in the ME data (from White et al. 1987).

coronal structures existed on both stars and, in addition, that an extended corona, highly inhomogeneous in longitude, was around the K0 IV component. They also tentatively identified the extended component with the high temperature solution found by Swank et al. (1981) from spectral fits of SSS data: this identification, however, could only be considered hypothetical at that time.

As a matter of fact, it has only been with the long, uninteruped observations provided by the EXOSAT Observatory that it has become possible to fully exploit the technique of eclipse observations (White et al. 1986, 1987). EXOSAT has obtained observations of several systems, including a 35 hour continuous observation of Algol, centered on the secondary eclipse, and complete coverage of a full orbital period for AR Lac (P=1.98 days) and TY Pyx (P=3.20 days). Unfortunately, no source has been observed so far for more than one orbital cycle. The observation of Algol (White et al. 1986) failed to reveal any eclipse when the K0 IV X-ray bright component was behind the X-ray dark B8 V primary (see Fig. 4; note that the two components have about the same radius). From this it was inferred that an extended high temperature corona with a scale-height of at least $\approx 1\ R_*$ was around the K star. The observation of AR Lac is even more interesting (White et al. 1987; see Fig. 5). The primary eclipse was observed in the Low Energy detector (as it was with the IPC on EINSTEIN), but there was no obvious eclipse in the Medium Energy data: this indicated that the extended corona was at a higher temperature than the more compact structures close to the surface of the stars.

White et al. (1987) have carried out extensive model simulations of the EXOSAT light curve of AR Lac. The best fit to the primary eclipse requires an X-ray emitting structure on the G star on the side facing the K0 IV component. This structure has an angular half-width of $\approx 27°$. The shallow secondary eclipse requires a much larger structure (with a hight of $\approx 1\ R_*$) above the K star which extends for $\approx 90°$. Extending the fit to include the whole light curve shows that there must be also a third X-ray structure on the rear face of the G star to account for the large uneclipsed excess flux around secondary eclipse. The height of both structures on the G star is small ($< 0.1\ R_*$).

The picture which emerges from these eclipse observations indicates that structures of different sizes, temperatures and pressures are likely to coexist in close binaries. The more compact structures close to the star surface are apparently at a lower temperature and higher density than the more extended structures whose size are comparable and even larger than the stellar radius. We do not know what these extended components are. They might be associated with loops connecting the two stars, as has been suggested by extrapolations of photospheric magnetic fields (Uchida and Sakurai 1983). Direct evidence for the existence of extended magnetospheres embracing the entire binary system has been recently provided by VLBI radio observations (Mutel et al. 1985, Massi et al. 1987). Alternatively, the extended components may represent the upper end of a range of solar-like loops whose larger dimensions are allowed by the lower gravity and higher temperature (and hence larger pressure scale-height) of RS CVn binaries. If so, the two-temperature components found in spectral fits of SSS and EXOSAT TGS data (Swank et al. 1981, Mewe et al. 1986) may really refer to spatially distinct regions. At any rate, the ensemble of the available data suggests that the coronae of RS CVn and other close binary systems may be quite different from the solar corona, whose structures are typically ≈ 0.1 of the solar radius. This possibility has to be taken seriously into account when trying to extend to binary systems the same type of concepts usually adopted for the solar case.

4. TIME VARIABILITY

Until quite recently very little was known on time variability of coronal X-ray sources. The observations from the EINSTEIN Observatory were usually quite short ($\approx$ a few thousands seconds) and longer observations -when available- were interrupted by Earth eclipses and passages through high background regions. Furthermore, systematic effects and the poor knowledge of source temperature make it difficult to compare observations obtained with different satellites in search of long-term variations (as could be expected from stellar activity cycles). For instance, fluxes measured with the EINSTEIN IPC and the EXOSAT LE may have

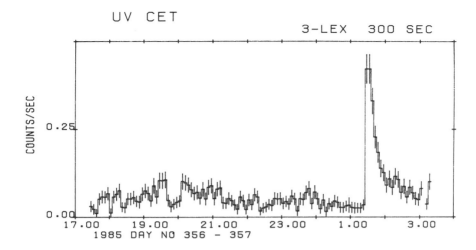

Fig. 6: EXOSAT LE observation of UV Cet showing the occurrence of a large flare and more gradual variations during the preflare phase (from Pallavicini et al. 1987c). The observation covers continuously a period of more than 9 hours.

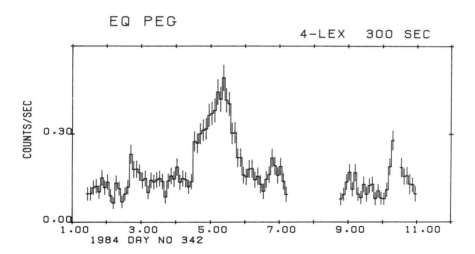

Fig. 7 : EXOSAT LE observation of a flare on EQ Peg characterized by a slow rise and a more rapid decay (from Pallavicini et al. 1987c; see also Haisch et al. 1987).

systematic differences of up to a factor of ≈ 2, owing to the different spectral bands and to the temperature dependence of the detector response (Pallavicini et al. 1987a). The most comprehensive study of stellar X-ray variability before the advent of EXOSAT is probably that of Ambruster et al. (1987). They have analyzed a sample of active late-type stars (mostly flare stars), using a new and substantially improved version of the classical χ^2 test (Collura et al. 1987b). They found that variability is ubiquitous in their sample of K and M stars, with a typical amplitude of ≈ 30% and time scales ranging from a few hundred seconds to more than 1000 sec. However, the presence of many data gaps, typical of EINSTEIN observations, makes the physical nature of the observed variability rather unclear. In addition, long time scales (including the entire evolution of long-lived transient events) could not be adequately studied with EINSTEIN (except in a few very special cases, cf. Haisch et al. 1983).

The EXOSAT satellite has dramatically increased our knowledge of time variability of stellar coronal sources. With the new EXOSAT observations we are starting to become aware of the rich variety of transient phenomena observed on X-ray stars, and we are now in a better position to make comparisons with similar phenomena observed on the Sun. The quality of the data is such as to allow for the first time realistic modelling of the observed emission and the derivation of meaningful physical parameters. In this section, I will discuss both flares and the more subtle variations which may occur during relatively quiescent periods.

Flares have been observed by EXOSAT on a variety of stars (Brinkman et al. 1984, de Jager et al. 1986, Landini et al. 1986, Pallavicini et al. 1986, White et al. 1986, Nelson et al. 1987, Haisch et al. 1987, Butler et al. 1987, Kundu et al. 1987, Pallavicini 1987a and b, Doyle et al. 1987, Tagliaferri et al. 1987, van den Oord et al. 1987). They have been observed from classical dMe flare stars (UV Cet, AT Mic, YZ CMi, EQ Peg, YY Gem, Wolf 630 etc., cf. Figs. 6, 7 and 9), on RS CVn and Algol-type systems (σ CrB, Algol, II Peg, TY Pyx etc., cf. Fig. 4) and even on a single solar-type G star (π^1 UMa, cf. Landini et al. 1986). The latter observations may not appear as particularly surprising, since variability and flares are commonly observed in integrated X-ray observations of the Sun. What is surprising, however, is that the flare on π^1 UMa was seen against a background quiescent emission that was two orders of magnitude higher than the quiescent X-ray luminosity of the Sun. This implies that the π^1 UMa flare released, in the X-ray band alone, at least a factor of ≈ 10 more energy than the total energy released by the largest solar flares over the entire electromagnetic spectrum (the total energy released in the X-ray band was ≈ 10^{33} erg). This indicates that activity on young rapidly-rotating solar-type stars (such as π^1 UMa) does not result simply from larger areas of the star covered by magnetic regions: it must also be intrinsically more powerful. An even more interesting observation is that shown in Fig. 8. It is a 3-Lex observation of the bright star Castor (α Gem), which is constituted by two (unresolved in X-rays) A-type stars (A1V + A2Vm, both spectroscopic binaries, with components of similar spectral types). Since, as shown by the EINSTEIN Observatory, A-type dwarfs are not strong X-ray emitters (if at all, cf. Schmitt et al. 1985a), the observed EXOSAT emission during the quiescent phase is mostly due to contamination of the EXOSAT LE detector by the ultraviolet radiation from the photosphere and the chromosphere of the stars. The data in Fig. 8 show the occurrence of a flare (Pallavicini 1987a), whose time behavior closely resembles that of flares observed on the Sun and dMe stars. It is difficult to separate a possible X-ray contribution during the flare from UV contamination: Schmitt (1987, private communication) has found evidence from summed signal histograms that the flare may be a genuine X-ray event, with a total energy of at least 10^{33} ergs above the background quiescent emission. At any rate, even if the flare was mainly due to UV contamination, it indicates the occurrence of magnetic activity on an A star: it is thought that these stars do not possess a subphotospheric convection zone and, hence, are not expected to have the high level of turbulent surface motions that is required to stress surface magnetic fields. The observation of a flare on Castor indicates, therefore, that some of our current expectations are probably oversimplified. It would be interestig to search for similar transient events in the hundreds of early-type stars observed serendipitously by the EXOSAT Observatory: those sources were dominated by UV contamination and for this reason have not yet been adequately studied.

The largest number of flares observed by EXOSAT occurred on dMe stars. These flares cover a broad range of total X-ray energies (from ≈ 2×10^{30} erg to ≈ 1×10^{34} erg) and have a

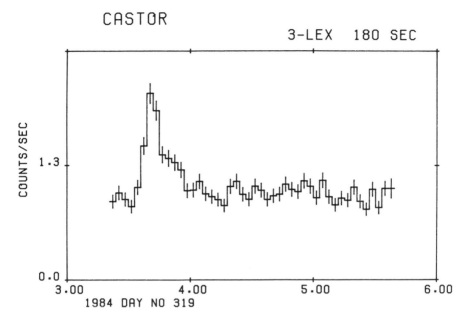

Fig. 8: EXOSAT LE observation of a flare on the A-type star Castor. The quiescent emission is almost entirely UV radiation from the photosphere and chromosphere of the star. Some UV contamination is likely to be present also during the flare (from Pallavicini 1987a).

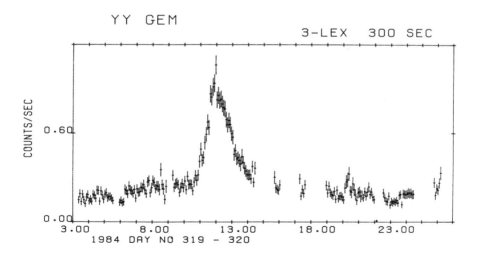

Fig. 9 : EXOSAT LE observation of the eclipsing binary flare star YY Gem throughout a full orbital cycle. Note the large long-decay flare (from Pallavicini 1987a).

variety of different time scales (from a few minutes to hours). There is clear evidence in the data of different types of flares on dMe stars, as also observed on the Sun. More specifically, the EXOSAT Observatory has shown the existence of at least three types of stellar events (Pallavicini 1987b):

a) <u>impulsive flares</u> (with rise times of a few minutes and decay times of tens of minutes), which are reminiscent of *compact* flares on the Sun. Examples are the events observed on AT Mic: 25 May 1985, Wolf 630: 25 Aug 1985, UV Cet: 23 Dec 1985, and others (cf. Fig. 6).

b) <u>long-decay flares</u> (with decay times of the order of ≈ 1 hour or longer), which are reminiscent of solar long-duration *2-ribbon* flares. Examples are the flares observed on YY Gem: 14 Nov 1984 and EQ Peg: 6 Aug 1985 (cf. Fig. 9).

c) <u>flares with a gradual rise and a more rapid decay</u>, which have apparently no solar analogy. The best example of this class is the flare observed on EQ Peg on 1984 Dec 7 (cf. Fig.7 ; see also Haisch et al. 1987)

As we shall discuss below, these morphological differences probably indicate real physical differences in the energy release process, as is also true for solar compact and 2-ribbon flares. It is interesting to note that different types of flares may occur on the same star. For instance, EQ Peg showed flares of all three types. We also note in passing that although stellar flares are strongly reminiscent of solar compact and 2-ribbon flares in their time behaviour, the total energies involved are often orders of magnitude higher than in typical solar flares.

The simplest way of modelling flaring events is to assume a single magnetically-confined loop, which remains unchanged throughout the flare evolution. We assume that energy is released impulsively close to the top of the loop, and that there is no energy input during the decay phase. These assumptions are the same as those believed to be valid for *compact* flares on the Sun (Moore et al. 1980). Under these assumptions, the flare will decay through radiative and conductive losses whose characteristic times depend on temperature, density and loop length (or, more correctly, on a characteristic length-scale for the temperature gradient at the footpoints of the coronal loop, cf. Landini et al. 1986). In order to estimate the physical parameters of the flaring region, the observed decay time is usually equated to the radiative time, and the further assumption is made that conductive and radiative times are approximately equal. This approach, as crude as it might be, gives reasonable numbers when applied to flares on dMe stars. Densities, temperatures and volumes derived in this way ($T \approx 2-3 \times 10^7$ K, $n \approx 10^{11}$-10^{12} cm^{-3}, $V \approx 10^{27}$-10^{28} cm^3) are not very different from those of flares on the Sun. However, as mentioned above, the total energy released may be much larger than in compact flares on the Sun (most typically, on the order of 10^{31}-10^{33} erg, as compared to $\approx 10^{29}$-10^{31} erg in compact solar flares). Note that the flare on Algol shown in Fig. 4 released a much larger energy ($\approx 10^{35}$ erg , White et al. 1986) and involved a larger volume ($\approx 10^{31}$ cm^3) than typical flares on dMe stars. It also had a higher temperature ($T \approx 6 \times 10^7$ K) and a longer decay time ($\approx 7 \times 10^3$ sec) than most flares on the Sun and on dMe flare stars. Apparently, this flare on Algol had more in common with large long-decay events than with compact impulsive events. However, it remains unclear at present whether flares on RS CVn and Algol-type binaries bear a real physical relationship with flares on dMe stars.

Are the above order of magnitude estimates realistic? Recently, Schmitt et al. (1987a) have tested this question by applying the above formalism to *solar flares* observed with the EINSTEIN Observatory by looking at the X-ray radiation scattered by the Sun-lit Earth. The parameters derived for these flares were in good agreement with those derived directly from spatially resolved solar observations, thus supporting the basic correctness of the method, at least to a first approximation. Of course, the physics involved in real flares is much more complex than the simple order of magnitude estimates discussed above. A flaring loop is not an isolated system: it is rooted in the dense chromospheric and photospheric layers, and the flare evolution will depend on the complex hydrodynamic phenomena which result from this coupling. More specifically, if energy is deposited at the loop top, it will be transferred to lower levels either by accelerated particles or by heat conduction. When the chromosphere receives more energy than can be radiated away, it expands upwards (chromospheric evaporation) filling the loop with high density plasma. This in turn will profoundly affect the subsequent evolution of the flare. Full hydrodynamic calculations of this type have been carried out for the Sun and have been successfully applied to flares observed from the Solar Maximum Mission (Pallavicini et al. 1983, Cheng et al. 1983, Peres et al. 1987). In principle the same type of modelling can be applied to stellar flares as well: this should allow us to get a better insight into the physics of the flare phenomenon. A first attempt in this direction has already been made by Reale et al. (1987) for a flare observed on Prox Cen by the EINSTEIN Observatory: the many high quality observations of stellar flares obtained with EXOSAT provide an ideal sample for further pursuing this modelling effort.

All the above considerations apply to flares which occur in magnetically closed structures which remain unchanged throughout the event. This, however, is completely different from what occurs in long-duration solar two-ribbon flares (Priest 1981). These flares are believed to occur as a consequence of a disruptive phenomenon which suddenly opens a magnetic field structure (Kopp and Pneuman 1976, Kopp and Poletto 1984). The open field lines, under the action of an unbalanced Lorentz force, relax back to a closed lower-energy configuration and energy is released gradually by magnetic reconnection as the field lines reconnect at progressively higher altitudes during the flare decay. An interesting question is whether the same physical processes are also responsible for the large long-decay flares observed by EXOSAT from some dMe stars.

In order to address this question, Poletto et al. (1986, 1987) and Pallavicini et al. (1987b) have applied the reconnection model developed by Kopp and Poletto (1984) for solar two-ribbon flares to two long-duration flares observed on the stars EQ Peg and Prox Cen. The first of these events was observed by EXOSAT ; the other one was observed by Haisch et al. (1983) with the EINSTEIN Observatory, and is one of the very few relatively long ($\approx$ 5 hours) uninterrupted observations that was possible to obtain with EINSTEIN (owing to the favorable position of the source in the sky). The comparison of model predictions and observations show that the reconnection model is indeed capable of reproducing the main characteristics of the decay phase of the observed flares, thus supporting the identification of these long-duration events as stellar analogs of solar two-ribbon flares. An important by-product of this modelling exercise is the possibility of deriving constraints on the physical parameters of the emitting region: it was derived that the flare on EQ Peg probably occurred in a small localized region, covering $\approx$ 1% of the stellar surface, where the photospheric magnetic field was on the order of $\approx$ 3600 G. The flaring loops reached a hight of $\approx$ 40,000 Km, with an initial upward velocity of $\approx$ 5 Km sec^{-1}. The average density at the beginning of the decay phase was $\approx$ 6x10^{12} cm^{-3}. On the contrary, the flare on Prox Cen may have involved lower magnetic fields ($\approx$ 1000 G) and a lower density ($\approx$ 1x10^{12} cm^{-3}). The smaller magnetic field strength derived for the Prox Cen flare is consistent with the fact that this flare had a peak X-ray luminosity a factor $\approx$ 150 lower than the flare on EQ Peg, and the quiescent X-ray luminosity of Prox Cen is more than 2 orders of magnitude smaller than the quiescent X-ray luminosity of EQ Peg.

In addition to flares, the EXOSAT satellite has given us the opportunity to study lower amplitude fluctuations which may occur during relatively quiescent periods. Such fluctuations are expected from the emergence of magnetic flux at the stellar surface, as observed in the case of the Sun. Moreover, long term variations of the integrated X-ray emission from a star could result from the existence of activity cycles similar to the 11-year sunspot cycle. Recently, it has also been suggested that the heating of solar and stellar coronae may result from continuous low-amplitude "microflaring" activity, produced by a large number of discrete events when magnetic fields lines, shuffled around by random fluid motions at the footpoints, reconnect and dissipate their energy. Some authors (Butler and Rodonò 1985, 1986, Butler et al. 1986) have claimed to have detected a signature of this microflaring activity in observations of flare stars obtained with EXOSAT.

More specifically, Butler et al. (1986) have compared EXOSAT observations of flare stars with spectroscopic observations obtained simultaneously in the Hγ line. They noticed that some of the peaks observed in the X-ray light curve -when binned at very short time intervals ($\approx$ 30 to 60 sec)- were correlated with simultaneous Hγ peaks, which led them to suggest that most fluctuations seen in X-rays might be statistically significant. This was particularly true for an observations of UV Cet obtained on 1984 Dec 6. A comparison of the EXOSAT LE data with simultaneous Hγ data obtained at ESO (Butler et al. 1986) showed that some of the main peaks were indeed associated at the two wavelengths, although this was not obvious for all of them. From this observation the authors concluded that the quiescent corona of dMe stars likely originates from a continuous succession of "microflares" lasting from tens of seconds to several minutes and with characteristic energies of $\approx 2 \times 10^{30}$ erg. In a subsequent observation of YZ CMi, however, there was little, if any, correlation between X-ray and optical fluctuations (Doyle et al. 1987). At any rate, the energy involved in these *soft X-ray* "microflares" is orders of magnitude larger than typically observed in solar hard-ray and UV microflares (the latter is on the order of $\approx 10^{24}$ erg sec^{-1}, cf. Lin et al. 1984, Porter et al. 1987).

In order to test the above suggestion for a much larger sample of data, and with the aim of getting a better idea of time variability in coronal X-ray sources, I have undertaken an extensive analysis of all flare star observations obtained with EXOSAT (Pallavicini 1987b, Pallavicini et al. 1987c). In total, there are 36 separate observations in the EXOSAT archives, pertaining to 21 different stars. About 20 out of 36 are long continuous observations lasting from $\approx$ 5 hours to $\approx$ 1 day. A number of objects (UV Cet, YY Gem, YZ CMi, AD Leo, CM Dra, Wolf 630, BY Dra, EV Lac, EQ Peg) were observed on more than one occasion, often at time intervals several months apart. This allows us to study variability on time scales longer than those typically observable in the course of one observation. One source (YY Gem) is an eclipsing binary that was monitored throughout a full rotational period. In addition, four sources (UV Cet, EQ Peg, YZ CMi and AD Leo) were observed simultaneously at the Very Large Array for continuous periods of 8 to 10 hours (Kundu et al. 1987).

In order to study the short-term variability expected from "microflaring" activity, I have applied to the data a rms variability analysis particularly appropriate for the study of continuous EXOSAT data (Stella 1985). This technique is substantially different from that used previously by Ambruster et al. (1987) for the time-analysis of EINSTEIN data and which is basically a modified χ^2 test (Collura et al. 1987b). It is important, therefore, to briefly outline the main features of the method.

We have used binned data, with no phase average, and we have run all variability tests for a number of different bin sizes (Δt = 60, 120, 180, 240, 300 and 600 sec). For the flare stars observed by EXOSAT, the source count rates during quiescent conditions were in the range $\approx$ 0.05 - 0.2 counts/sec, so bin sizes shorter than $\approx$ 60 sec could not reliably be used. We have used background subtracted data, with the background taken by averaging over a much larger area than the source cell. When scaled to the source cell, the background amounted at most to $\approx$ 30%-40% of the source count rate. We have initially excluded the most obvious flares detected at a significance level greater than 5σ; however, in order to test the method, we have also run the variability analysis including all flares. In every case, the variability analysis was done separately for the background subtracted source count rate and for the background alone.

We define a rms variability (expressed as percentage variability of the average source count rate) as rms = $(\sigma_{obs}^2 - \sigma_{exp}^2)^{1/2}$ / <c>, where σ_{obs}^2 is the observed variance, σ_{exp}^2 is the expected variance for a constant source and <c> is the background subtracted average source count rate. We have computed the rms and the associated error for all our data and we have established the significance of the measured variability in terms of the number of standard deviations (we have taken the variability as detected when the significance level exceeded 3σ). For the cases when significant variability was detected, we have computed the autocorrelation function (ACF) for both the background subtracted source count rate and for the background alone, and we have further computed the cross-correlation function (CCF) between the two. If, based on the CCF, no clear correlation was found between the source and background variations, the relevant time scale of the detected variability was determined from the exponential decay time of the ACF, assuming that the variability was due to random shot noise (see Stella et al. 1984 for details on the method).

The results of this study for the flare stars observed with EXOSAT can be summarized as follows :

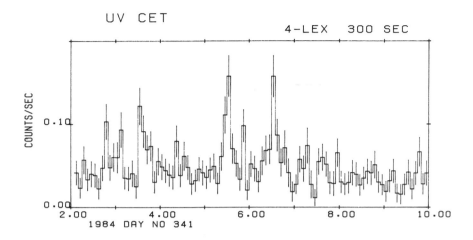

Fig.10a : EXOSAT LE observation of UV Cet on December 6, 1984 (from Pallavicini 1987a; see also Butler et al. 1986). Note the absence of significant variability during the last three hours of the observation.

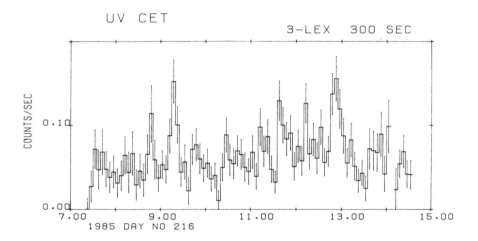

Fig. 10b: EXOSAT LE observation of UV Cet on August 4, 1985 (from Pallavicini 1987a; see also Kundu et al. 1987). Note the similarity of this observation and the one shown in Fig. 10a with regard to rapid time variability. By contrast, the same source observed on December 22-23, 1985 showed more gradual variations, in addition to a major flare (cf. Fig. 6).

a) We find substantial variability for most flare stars over a variety of time scales (from a few minutes to hours).

b) The observed variability is in the form of both individual *usually sporadic* flares and of more gradual variations (on times scales of tens of minutes to hours), and appears to be stochastic. Flares show a wide range of amplitudes with respect to the quiescent emission level (up to a factor of ≈ 10), while the more gradual variations on times scales of hours do not exceed amplitudes of 50% and are usually substantially less than that.

c) For those stars observed by EXOSAT on more than one occasion, we did not see any large variation of the quiescent X-ray flux. The observed variations were always less than a factor of ≈ 2 over periods of several months, and more typically did not exceed amplitudes of $\approx 20\%$-30%.

d) The small data sample available from EINSTEIN and EXOSAT, and the uncertainty inherent in the comparison of fluxes from different instruments, do not allow the determination of possible long-term variations on time scales of years. The available data suggest, however, that these variations, if they occur, are probably small.

e) In contrast with previous claims by others, we do not find evidence in the EXOSAT data for continuous low-amplitude short-time scale variability as might be expected from "microflaring" activity. More specifically, we do not confirm the continuous low-level rapid variability reported by Butler and Rodonò (1985, 1986) and Butler et al. (1986).

f) For all stars in our sample, continuous periods of several hours were observed with no significant variability, in contrast to what could be expected from an interpretation of the quiescent emission of dMe flare stars as the superposition of many discrete events. If these events occur, they must be washed out in disk-integrated soft X-ray observations.

g) For those stars which have been observed simultaneously at X-ray and radio wavelengths (Kundu et al. 1987), there was very little correlation between the variability observed in the two wavelength domains.

As an example of the high variability observed occasionally with EXOSAT, we show in Fig. 10a,b two observations of UV Cet, including the one previously analyzed by Butler et al. (1986). In both cases, the source appeared quite variable (at significance levels exceeding 7σ and 5σ, respectively). However, the observed variability could be interpreted in both cases as due to individual flares with durations of ≈ 10 to 20 min each and total energies on the order of $\approx 5 \times 10^{30}$ to $\approx 1 \times 10^{31}$ erg. The time scales of these events and the energy involved were both substantially larger than those reported by Butler et al. (1986) for their "microflares". In addition, there was no evidence of *continuous* lower amplitude variations outside these relatively major events. For instance, the last three hours of the observation of UV Cet on Dec 6, 1984 were completely free of significant variability.

The variability observed in all other cases was substantially less than that shown in Fig. 10a,b. More typically, individual events were separated by relatively long periods (several hours) with no obvious variability or with only gradual fluctuations on time scales of tens of minutes to hours (see, for instance, Fig. 6). We also note in passing that in order to observe fluctuations in disk integrated soft X-ray observations, the rate of energy release in these fluctuations must be comparable to, or greater than, the total X-ray luminosity of the star, i.e. much larger than the typical energies associated with solar "microflares". It is not clear, therefore, whether there should be any relationship between the impulsive, localized phenomena observed on the Sun and any possible, short-term variability that might be detected in disk-integrated stellar observations.

To summarize, I conclude that, on the basis of our analysis, there is no evidence in the EXOSAT data that quiescent emission of dMe (and possibly other) stars may originate from continuous low-amplitude flaring activity. Identical conclusions have been reached independently by Collura et al. (1987a) who have recently analyzed a subset of the same EXOSAT observations using the optimized χ^2-test previously employed by Ambruster et al. (1987) in the analysis of EINSTEIN data. The results of both studies do not necessarily imply that "microflaring" activity of the type observed on the Sun may not be relevant for heating stellar coronae: they simply indicate that soft X-ray disk-integrated observations have not yet been able to give us a clear observational proof that this is actually occurring. We also note that if continuous rapid microflaring activity were a general property of flare stars, and if an observable signature of it were present in disk-integrated soft X-ray observations, it could have been detected much more easily with the EINSTEIN Observatory, whose higher sensitivity would have allowed variability to be studied on time scales as short as ≈ 10 sec for most of our sources. There is no indication of the existence of this type of variability in the data reported by Ambruster et al. (1987).

It is a pleasure to acknowlege enlightening discussions on stellar coronae with Dr. J.H.M.M. Schmitt and Dr. N.E. White. The present contribution is largely based on a review talk given by the author at the 9th Sacramento Peak Summer Workshop on "Solar and Stellar Coronal Structure and Dynamics" held in Sunspot, New Mexico on 17-21 August 1987. This work has been supported by CNR-Piano Spaziale Nazionale.

REFERENCES

Ambruster, C.W., Sciortino, S. and Golub, L.: 1987, Ap. J. Suppl. 65, 273.
Brinkman, A.C., Gronenschild, E.H.B.M., Mewe, R., McHardy, I. and Pye, J.P.: 1985, Adv. Space Res. 5, No. 3, 65.
Butler, C.J. and Rodonò, M.: 1985, Irish Astron. J. 17, 131.
Butler, C.J. and Rodonò, M.: 1986, Lecture Notes in Phys. 254, 329.
Butler, C.J., Doyle, J.G., Foing, B.H. and Rodonò, M.: 1987, in Activity in Cool Star Envelopes (O. Havnes ed.), in press.
Butler, C.J., Rodonò, M., Foing, B.H. and Haisch, B.M.: 1986, Nature 321, 679.
Cheng, CC., Oran, E.S., Doschek, G.A., Boris, J.P. and Mariska, J.T.: 1983, Ap. J. 265, 1090.
Collura, A., Pasquini, L. and Schmitt, J.H.M.M.: 1987a, in Activity in Cool Star Envelopes (O. Havnes ed.), in press.
Collura, A., Sciortino, S., Maggio, A., Serio, S., Vaiana, G.S. and Rosner, R.: 1987b, Ap. J. 315, 340.
de Jager, C., Heise, J., Avgoloupis, S., Cutispoto, G., Kieboom, K., Herr, R.B., Landini, M., Langerwerff, A.F., Mavridis, L.N., Melkonian, A.S., Molenaar, R., Monsignori-Fossi, B.C., Nations, H.L., Pallavicini, R., Piirola, V., Rodonò, M., Seeds, M.A., van den Oord, G.H.J., Vilhu, O. and Waelkens, C.: 1986, Astron. Ap. 156, 95.
Doyle, J.G., Butler, C.J., Byrne, P.B. and van den Oord, G.H.J.: 1987, Astron. Ap., in press.
Giampapa, M.S., Golub, L., Peres, G., Serio, S. and Vaiana, G.S.: 1985, Ap. J. 289, 203.
Haisch, B.M., Butler, C.J., Doyle, J.G. and Rodonò, M.: 1987, Astron. Ap. 181, 96.
Haisch, B.M., Linsky, J.L., Bornmann, P.L., Stencel, R.E., Antiochos, S.K, Golub, L. and Vaiana, G.S.: 1983, Ap. J. 267, 280.
Kopp, R.A. and Pneuman, G.W.: 1976, Solar Phys. 50, 85.
Kopp, R.A. and Poletto, G.: 1984, Solar Phys. 93, 351.
Kundu, M.R., Pallavicini, R., Jackson, P. ans White, S.M.: 1987, Astron. Ap., in press
Landini, M., Monsignori-Fossi, B.C. and Pallavicini, R.: 1985a, Space Science Rev. 40, 43.
Landini, M., Monsignori-Fossi, B.C., Pallavicini, R. and Piro, L.: 1986, Astron. Ap. 157, 217.
Landini, M. , Monsignori-Fossi, B.C., Paresce, F. and Stern, R.A.: 1985b, Ap. J. 289, 709.
Lin, R.P., Schwartz, R.A., Kane, S.R., Pelling, R.M. and Hurley, K.C.: 1984, Ap. J. 283, 421.
Majer, P., Schmitt, J.H.M.M., Golub,L., Harnden, F.R.Jr. and Rosner, R.: 1986, Ap. J. 300, 360.
Massi, M., Felli, M., Pallavicini, R., Tofani, G., Palagi, F. and Catarzi, M.: 1987, Astron. Ap., in press.
Mewe, R., Schrijver, C.J., Lemen, J.R. and Bentley, R.D.: 1986, Adv. Space Res. 8, No. 6.
Moore, R. et al.: 1980, in Solar Flares (P.A. Sturrock ed.), p. 341.

Mutel, R.L., Lestrade, J.F., Preston, R.A. and Phillips, R.B.: 1985, Ap. J. 254, 641.
Nelson, G.J., Page, A.A., Slee, O.B. and Denby, B.: 1987, submitted to M.N.R.A.S.
Pallavicini, R.: 1987a, in Solar and Stellar Activity (E.H. Schroter and M. Schussler eds.), p. 98.
Pallavicini, R.: 1987b, in Activity in Cool Star Envelopes (O. Havnes ed.), in press.
Pallavicini, R., Kundu, M.R. and Jackson, P.D.: 1986, Lecture Notes in Phys. 254, 225.
Pallavicini, R., Monsignori-Fossi, B.C., Landini, M. and Schmitt, J.H.M.M.: 1987a, Astron. Ap., in press.
Pallavicini, R., Peres, G., Serio, S., Vaiana, G.S., Acton, L., Leibacher, J. and Rosner, R.: 1983, Ap. J. 270, 270.
Pallavicini, R., Poletto, G. and Kopp, R.A.: 1987b, in Activity in Cool Star Envelopes (O. Havnes ed.), in press.
Pallavicini, R., Stella, L and Tagliaferri, G.: 1987c, in preparation.
Pasquini, L., Schmitt, J.H.M.M. and Pallavicini, R.: 1987, in Activity in Cool Star Envelopes (O. Havnes ed.), in press.
Peres, G., Reale, F., Serio, S. and Pallavicini, R.: 1987, Ap. J. 312, 895.
Poletto, G., Pallavicini, R. and Kopp, R.A.: 1986, Adv. Space Res. 12, 895.
Poletto, G., Pallavicini, R. and Kopp, R.A.: 1987, submitted to Astron. Ap.
Porter, J.G., Moore, R.L., Reichmann, E.J. and Harvey, K.L.: 1987, Ap. J., in press.
Reale, F., Peres, G., Serio, S., Rosner, R. and Schmitt, J.H.M.M.: 1987, Ap. J., in press.
Rosner, R., Golub, L. and Vaiana, G.S.: 1985, Ann. Rev. Astron. Ap. 23, 413.
Schadee, A., de Jager, C. and Svestka, Z.: 1983, Solar Phys. 89, 287.
Schmitt, J.H.M.M.: 1984, X-Ray Astronomy '84 (M.Oda and R.Giacconi eds.), p. 17.
Schmitt, J.H.M.M., Harnden, F.R. Jr. and Fink, H.: 1987a, Ap. J. 322, 1023..
Schmitt, J.H.M.M., Golub, L., Harnden, F.R.Jr., Maxson, C.W., Rosner, R. and Vaiana, G.S.: 1985a, Ap.J. 290, 307.
Schmitt, J.H.M.M., Harnden, F.R.Jr., Peres, G., Rosner, R. and Serio, S.: 1985b, Ap. J. 288, 751.
Schmitt, J.H.M.M., Pallavicini, R., Monsignori-Fossi, B.C. and Harnden, F.R. Jr. 1987b, Astron. Ap. 179, 193.
Schmitt, J.H.M.M. and Rosso, C.: 1987, Astron. Ap., in press.
Schrijver, C.J.: 1985, Space Science Rev. 40, 3.
Schrijver, C.J.: 1987, in Solar and Stellar Coronal Structure and Dynamics (R.C. Altrock ed.), in press.
Singh, K.P., Slijkhuis, S., Westergaard, N.J., Schnopper, H.W., Elgaroy, O., Engvold, O. and Joras, P.: 1987, M.N.R.A.S. 224, 481.
Stella, L.: 1985, EXOSAT Express 14, 31.
Stella, L., Kahn, S.M. and Grindlay, J.E.: 1984, Ap. J. 282, 713.
Stern, R.A., Antiochos, S.K., and Harnden, F.R.Jr. : 1986, Ap. J. 305, 417.
Swank, J.H. and White, N.E.: 1980, in Cool Stars. Stellar Systems and the Sun (A.Dupree ed.), SAO Special Rep. 389, p. 47.
Swank, J.H., White, N.E., Holt, S.S. ans Becker, R.H.: 1981, Ap. J. 246, 208.
Tagliaferri, G., White, N.E., Giommi, P. and Doyle, J.G.: 1987, in Fifth Cambridge Workshop on Cool Stars. Stellar Systems and the Sun (J. Linsky and R. Stencel ed.), in press.
Taylor, B.G.: 1985, Adv. Space Res. 5, No. 3, 35.
Uchida, Y. and Sakurai, T.: 1983, in Activity in Red-Dwarf Stars (P.B.Byrne and M. Rodonò eds.), p. 629.
Vaiana, G.S. and Rosner, R.: 1978, Ann. Rev. Astron. Ap. 16, 393.
van den Oord, G.H.J., Mewe, R. and Brinkman, A.C.: 1987, submitted to Astron. Ap.
Walter, F.M., Gibson, D.M. and Basri, G.S.: 1983, Ap. J. 267, 665.
White, N.E., Culhane, J.L., Parmar, A.N., Kellett, B.J., Kahn, S., van den Oord, G.H.J. and Kuipers, J.: 1986, Ap.J. 301, 262.
White, N.E., Shafer, R., Parmar, A.N. and Culhane, J.L.: 1987, in Fifth Cambridge Workshop on Cool Stars. Stellar Systems and the Sun (J. Linsky and R. Stencel ed.), in press.

THE MAGNETIC FIELDS ON COOL STARS AND THEIR CORRELATION WITH CHROMOSPHERIC AND CORONAL EMISSION

Steven H. Saar
Smithsonian Astrophysical Observatory
60 Garden St., Cambridge, MA 02138, USA

ABSTRACT. I discuss the results of recent measurements of magnetic fields on cool stars and how these measurements relate to stellar "activity". Special emphasis is given to the correlations between the magnetic field strength, surface filling factor, and the non-thermal emission from the hot outer atmospheres of these stars.

1 INTRODUCTION

One of the major lessons we have learned from solar research is the critical role of magnetic fields in governing the structure and energy balance of the solar atmosphere. Evidence for the importance of magnetic fields is widespread. A wide variety of phenomena on the Sun, including spots, plages, granules, flares, and loops, are closely associated with magnetic fields. These inhomogeneous structures grow and decay in an 11 year cycle, consistent with the continual regeneration of magnetic fields through the operation of a magnetic dynamo. And, directly pertinent to this conference, magnetic fields are crucial for the existence and stability of the million degree solar corona through their role in the heating and confinement of the hot plasma. The dominant heating mechanism for the hot plasmas of the chromosphere and transition region (TR) is also likely magnetic in nature.

By analogy with the Sun, magnetic fields should also be important for other cool stars. Indeed, less than 2% of the solar surface is covered by magnetic fields; many stars show evidence for considerably more extensive magnetic "activity". The fraction of the surfaces of RS CVn variables covered by spots can frequently be 100 times larger than the corresponding filling factor of solar umbrae (Vogt 1983). The 11 year cycle of magnetic structures on the Sun is also observed in stars: at present some 60% of cool dwarfs have well determined cycles, based on the modulation of chromospheric Ca II emission (Baliunas 1986). The levels of emission from hot plasma in stellar chromospheres and coronae, however, is often orders of magnitude larger than the Sun (e.g., Linsky 1985). A thorough study of stellar magnetic properties would seem to be in order.

Unfortunately, the measurement of magnetic fields on solar-like stars is not trivial. The most direct means of detecting magnetic fields is through observation of their effect on spectral lines. The presence of a magnetic field will, in general, lift the degeneracy of certain quantum levels in an atom, separating them slightly in energy (the Zeeman effect). The exact number and spacing of the newly accessible levels (and corresponding allowed transitions) depend on the detailed quantum mechanics of the configuration. In the simplest case, triplet splitting, the shift in wavelength of the two Zeeman split lines (the circularly polarized σ components) from unshifted one (the linearly polarized π component) is given by $\Delta\lambda_B(\text{Å}) = \pm 4.667 \times 10^{-13} g_{eff} \lambda^2 B$, where B is the magnetic field strength (in gauss) and g_{eff} is the effective Landé value of the transition. For typical fields of a few kilogauss or less, the Zeeman shift is smaller than the Doppler width of the line and the Zeeman transitions overlap, making them difficult to detect. Use of the contrasting polarization properties to aid the detection of the Zeeman components, unfortunately, is thwarted by the bipolar nature of solar and stellar active regions. If mixed polarities exist within the observed region, a proportionate part of the circularly polarized magnetic signal is cancelled, lowering the inferred magnetic flux. In solar work, this is not always an important effect, since the regions of opposite polarity are often spatially resolvable. For the completely unresolved surfaces of stars, on the other hand, the effect of mixed polarities is ruinous. Almost complete cancellation of the circular polarization signal results when the complex, mixed polarities on the Sun are is observed as a point source, for example. It is not suprising then, that very few cool stars show *any* net circular polarization (e.g., Borra, Landstreet, and Mayor 1984).

2 TECHNIQUES AND PRELIMINARY RESULTS

So how *does* the intrepid astronomer directly measure stellar magnetic fields? Linear polarization does not cancel completely in integrated light, but is a second order effect and difficult to interpret (Landi Degl'Innocenti 1982). We are left with the unpolarized line profile, with its complex blending of not only σ and π components from magnetic regions, but contributions from non-magnetic portions of the stellar surface as well. The analysis of these various contributions is a difficult problem. Robinson (1980) devised a scheme to deconvolve magnetically sensitive (large Landé g) lines from low g lines. The ratio of the Fourier transforms of the line profiles is modeled; the period of the ratio is proportional to B and the amplitude of the ratio indicates the fraction (f) of the stellar surface occupied by magnetic regions. Several improvements on this basic technique have since been developed (Marcy 1982; Gray 1984; Saar 1988; Basri and Marcy 1988).

Preliminary results of stellar magnetic measurements are beginning to emerge. Marcy (1984) found that the magnetic parameters change as one moves down the main sequence from G to K dwarfs. Gray (1984) found similar magnetic parameters, but did not detect fields on any F dwarfs. The most rapidly rotating star in his sample, π^1 UMa, showed substantially larger surface averaged field (i.e., the product fB) than the others. Marcy and Bruning (1984) could not detect fields on active giants, a fact they ascribe to lower field strengths in these stars with lower surface gravities (see also Gondoin, Giampapa, and Bookbinder 1985). Consistent with this idea, Saar and Linsky (1986) and Saar, Linsky, and Giampapa (1987) reported that among dwarf stars, B increased towards later spectral types and higher surface gravities. They also found an increase in magnetic flux with rotation rate (see also Linsky and Saar 1987) as first suggested by Gray (1985). Finally, f appears to decrease with stellar age (Linsky and Saar 1987), implying that changes in the filling factor dominate the time evolution of activity.

3 MAGNETIC FIELDS AND UPPER ATMOSPHERIC EMISSION

The apparent connection between magnetic fields and rotation is not surprising, for many authors have demonstrated evidence for a connection between rotation and (presumably magnetic-related) emission from hot plasmas in stellar chromospheres and coronae (e.g., Pallavicini et al. 1981; Mangeney and Praderie 1983; Vilhu 1984; Simon and Fekel 1987). The age dependence of chromospheric activity has also been extensively studied (Skumanich 1972; Soderblom 1983; Hartmann and Noyes 1987), and thus the observed decrease in the magnetic parameters with age is not unexpected. It is important, however, to determine the direct connections between the magnetic fields and the outer atmospheric activity in order to understand the underlying physics of the heating.

Marcy (1984) reported correlations between magnetic parameters, X-ray, and Ca II H+K flux. Saar and Schrijver (1987) have combined preliminary magnetic field measurements (derived using the methods of Saar 1988) with X-ray and Ca II H+K fluxes. They find an almost linear dependence between F_x and fB (Fig. 1), independent of color: $F_x \propto (fB)^{0.93}$. Taken in isolation, this result might only indicate that more active regions (and hence a larger surface-averaged field, fB) beget more the X-ray flux, without giving any real information on the heating mechanism. However, individual active regions on the Sun show a similar relation between X-ray and magnetic fluxes (Schrijver 1987b), suggesting that the coronal heating mechanism is roughly linear in magnetic flux density (see, however, Golub 1983). Note that since the magnetic analysis methods work best for stars with considerable magnetic flux, the Sun is an important calibration point at low activity levels. The strong relationship between B and gravity (Saar and Linsky 1986) implies that f is the primary parameter affecting the level of stellar activity on the main sequence.

The case for the chromospheric Ca II H+K emission lines is more complex, due to presence of a substantial photospheric component to the measured flux. Schrijver (1987a) obtains the tightest correlations between chromospheric and TR or coronal radiative losses when a color-dependent "basal" flux (probably acoustic in origin) is subtracted from the chromospheric fluxes, yielding the excess Ca II flux, ΔF_{CaII}. With the relation $F_x \propto \Delta F_{CaII}^{1.5}$ (Schrijver 1987a), the F_x - fB relation can be rewritten as $\Delta F_{CaII} \propto (fB)^{0.62}$. This power law fits the solar and stellar data provided fB $\leq$ 300 Gauss (Figure 2). The Ca II H+K excess flux density appears to saturate for values of fB above this level (i.e., at f $\sim$20% for a G-type dwarf); a similar saturation may be seen in the $F_x - \Delta F_{CaII}$ relation for log F_x > 6.2 (see Schrijver 1987a). The large body of existing Ca II observations indicate an upper limit of $\Delta F_{CaII} \sim 3$ for G dwarfs, which also coincides with the maximum observed in a solar active region (Schrijver and Coté 1987, corresponding to an average magnetic flux density of $\sim$ 300 Gauss). The similarity of the solar and stellar saturation levels suggests that the saturation is real (not a selection effect), perhaps the result of a change in the heating mechanism. X-rays show no such saturation, however, suggesting a different heating mechanism operates in stellar coronae.

I have also studied correlations between the preliminary magnetic parameters and

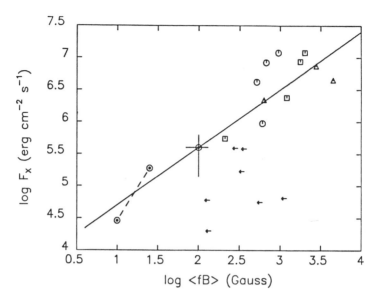

Figure 1: Soft X-ray flux density F_x vs. the mean magnetic flux density fB. The solid line is $F_x \propto (fB)^{0.93}$. Magnetic and X-ray fluxes at solar activity minimum and maximum are connected with a dotted line. The mean value and range for solar active regions is marked with the symbol at fB = 100 G (Schrijver 1987b). Symbols: circles, squares and triangles mark G-, K- and M-type dwarfs respectively; arrows are stars with upper limits for fB.

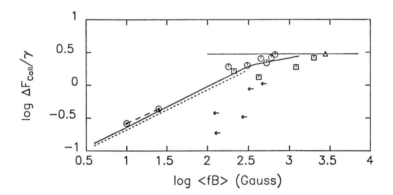

Figure 2: Ca II H+K excess flux density ΔF_{CaII} vs. fB. Symbols as in Fig. 1. The left segment of the solid line is $\Delta F_{CaII} \propto (fB)^{0.62}$, while the right segment is an approximation to the $F_x - \Delta F_{CaII}$ saturation seen in active stars (Schrijver 1987a). The dotted line is the relation seen for solar active regions (Schrijver and Coté 1987), with the horizontal line segment showing the maximum value observed in solar active regions (Schrijver and Coté 1987) and for G dwarfs. The decrease in chromospheric emission from M dwarfs has been corrected for by the deficiency factor γ (Schrijver and Rutten 1987).

other diagnostics of chromospheric and TR plasma. Stellar surface fluxes in O I (1304 Å), C II (1335 Å), C IV (1550 Å), and Si II (1808 Å) measured with the IUE satellite and analogous solar measurements were gathered from Oranje (1983) and a number of other sources. The preliminary results (based on a small sample of stars) are: $F_{CIV} \propto (fB)^{0.55}$; $F_{CII} \propto (fB)^{0.44}$; $F_{SiII} \propto (fB)^{0.40}$; $F_{OI} \propto (fB)^{0.36}$ (see Figure 3). The cooler chromospheric lines (e.g., O I, with $T_{form} \approx 6 \times 10^3$ °K) show a reduced dependence on fB relative to lines formed higher in the atmosphere (C II, $T_{form} \approx 2 \times 10^4$ °K, and C IV, $T_{form} \approx 10^5$ °K), consistent with the F_x and F_{CaII} results. Note, however, that no "basal" correction has been applied to the O I and Si II data.

Once a larger set of magnetic measurements has been made, it should eventually be possible to begin to test various atmospheric MHD wave heating models for chromospheres, TR and coronae (e.g., Ulmschneider and Stein 1982), test empirical activity theories (Vilhu 1984) and predictions of the range of dynamo time behavior (e.g., Noyes, Weiss and Vaughan 1984) and guide the development of more detailed theories.

4 ROTATIONAL MODULATION OF MAGNETIC AND RADIATIVE FLUXES

Observations of magnetic fields and a wide range of other activity-related diagnostics (Ca II, IUE line emission, He I D3, broadband linear polarization) on the active G8 dwarf, ξ Boo A, were made in June of 1986. Once again, as no compensation for weak blends was made, the resulting magnetic fluxes should be regarded as preliminary.

Figure 4 shows the normalized magnetic ($\propto$ fB) and ultraviolet fluxes for ξ Boo A plotted versus rotational phase (P = 6.2 days, Noyes et al. 1984; $\phi \equiv 0$ defined as the phase of maximum Ca II emission). Rotational modulation of the emission line and magnetic fluxes, in phase with one another, is clearly visible. Lines formed at higher temperatures (C IV, Si IV, He II, C II) appear to show greater modulation than the cooler lines formed in the chromosphere (Ca II, Mg II, Si II, C I, O I). The greater sensitivity of hotter lines to magnetic flux in rotationally modulated *plages* is consistent with the results in §3 for *entire stars*, suggesting that perhaps the main difference between active and quiet stars is the *number* or *size* of plages, and not a major difference in their physical properties. The reduced sensitivity of the chromospheric lines to changes in magnetic flux suggests once more that for stars as active as ξ Boo A (fB ≈ 600 Gauss), the heating mechanism for the chromosphere "saturates", while the heating mechanisms operating higher in the atmosphere still are strongly affected by the variable magnetic flux.

Further information on the nature of the magnetic regions on ξ Boo A can be derived from the contemporaneous measurements of broadband linear polarization (P_L). While the polarization amplitude varied by a factor of 3, the angle which the linear polarization vector made with respect to the N-S axis never varied from the 0 − 180 degree line by more than about 10 degrees. The simplest configuration that can reproduce this result (for fields normal to the surface) is a series of active regions which cross the stellar disk near disk center. Since sin $i \approx 0.6$ for ξ Boo A, the active regions would then be located at intermediate latitudes on the average. The line perpendicular to the polarization angle (i.e., at ≈ 90°) then corresponds to the stellar rotation axis, entirely consistent with the known orbital axis (at 78°) of the ξ Boo A+B system.

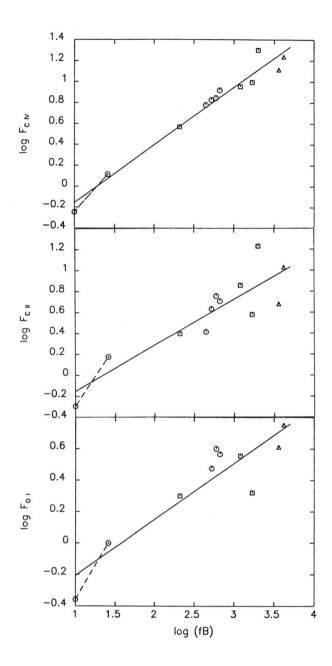

Figure 3: IUE emission line surface flux densities vs. magnetic flux density. The solid lines are the power law fits $F_{CIV} \propto (fB)^{0.55}$, $F_{CII} \propto (fB)^{0.44}$, and $F_{OI} \propto (fB)^{0.36}$. Data for the quiet and active Sun are connected by dashed lines. Symbols as in Figure 1.

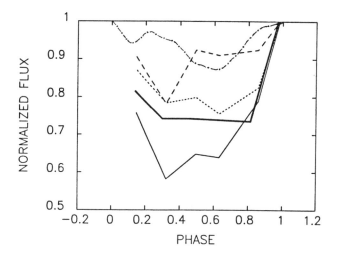

Figure 4: The magnetic flux ($\propto$ fB, heavy solid), various ultraviolet emission line fluxes (C IV, solid; C II, dashed; O I, long dashed) and the smoothed, Ca II excess flux (dash-dotted) of ξ Bootis A, normalized and plotted versus rotational phase.

The amplitude of P_L measures the *net* tangential component of the magnetic field and is maximized when significant magnetic flux is distributed asymmetrically, near the stellar limb (e.g., Landi Degl'Innocenti 1982). The combination of the variation of fB (weighted towards *disk-center* regions because of projection and limb-darkening effects) with the amplitude of P_L therefore contains positional information, which Saar *et al.* (1987) used to construct a crude "magnetic image" of ξ Boo A. Using more sophisticated modeling, we hope to map the active region geometry of ξ Boo, which, when combined with the radiative and magnetic fluxes as a function of phase, will allow us for the first time to derive realistic, two-component (quiet + active) models for an outer atmosphere of a solar-like star.

This research was funded in part by the Smithsonian Institution Center Fellowship program and by NASA grant NAGW-112. Much of the research described was done while the author was a Postdoctoral Research Associate at the Joint Institute for Laboratory Astrophysics in Boulder, Colorado. Work there was supported by NASA grant NGL-06-003-057 to the University of Colorado. It is a pleasure to thank the National Optical Astronomy Observatories for ample telescope time and travel support.

REFERENCES

Baliunas, S. L. 1986, in *The Fourth Cambridge Workshop on Cool Stars, Stellar Systems, and the Sun*, eds. M. Zeilik and D. M. Gibson (New York: Springer-Verlag), p. 3.
Basri, G. S., and Marcy, G. W. 1988, *Ap. J.*, submitted.
Borra, E. F., Edwards, G., and Mayor, M. 1984, *Ap. J.*, **284**, 211.
Durney, B. R., and Robinson, R. D. 1982, *Ap. J.*, **253**, 290.

Golub, L. 1983, in *Activity in Red Dwarf Stars*, eds. P. Byrne and M. Rodonó (Dordrecht: Reidel), p. 83
Gondoin, Ph., Giampapa, M. S., and Bookbinder, J. 1985, *Ap. J.*, **297**, 710.
Gray, D. F. 1984, *Ap. J.*, **277**, 640.
Gray, D. F. 1985, *Pub. A. S. P.*, **97**, 719.
Hartmann, L. W., and Noyes, R. W. 1987, *Ann. Rev. Astr. Ap.*, **25**, 271.
Landi Degl' Innocenti, E. 1982, *Astr. Ap.*, **110**, 25.
Linsky, J. L. 1985, *Solar Phys.*, **100**, 333.
Linsky, J. L., and Saar, S. H. 1987, in *The Fifth Cambridge Workshop on Cool Stars, Stellar Systems, and the Sun*, in press.
Mangeney, A., and Praderie, F. 1984, *Astr. Ap.*, **130**, 143.
Marcy, G. W. 1982, *Pub. A. S. P.*, **94**, 989.
Marcy, G. W. 1984, *Ap. J.*, **276**, 286.
Marcy, G. W., and Bruning, D. H. 1984, *Ap. J.*, **281**, 286.
Noyes, R. W., Hartmann, L., Baliunas, S. L., Duncan, D. K., and Vaughan, A. H. 1984, *Ap. J.*, **279**, 793.
Noyes, R. W., Weiss, N. O., and Vaughan, A. H. 1984, *Ap. J.*, **287**, 769.
Oranje, B. J. 1983, *Astr. Ap.*, **124**, 43.
Pallavicini, R., Golub, L., Rosner, R., Vaiana, G. S., Ayres, T., and Linsky, J. L. 1981, *Ap. J.*, **248**, 279.
Robinson, R. D. 1980, *Ap. J.*, **239**, 961.
Saar, S. H. 1988, *Ap. J.*, Jan. 1.
Saar, S. H., Huovelin, J., Giampapa, M. S., Linsky, J. L., and Jordan, C., 1987, in *Activity in Cool Star Envelopes*, in press.
Saar, S. H., and Linsky, J. L. 1986, *Advances in Space Physics*, **6**, No. 8, 235.
Saar, S. H., Linsky, J. L., and Giampapa, M. S. 1987, in Proc. of *27th Liegé International Astrophysics Colloquium on Observational Astrophysics with High Precision Data*, in press.
Saar, S. H., and Schrijver, C. J. 1987, in *The Fifth Cambridge Workshop on Cool Stars, Stellar Systems, and the Sun*, in press.
Schrijver, C. J. 1987a, *Astr. Ap.*, **172**, 111.
Schrijver, C. J. 1987b, in press.
Schrijver, C. J., and Coté, J. 1987, in *The Fifth Cambridge Workshop on Cool Stars, Stellar Systems, and the Sun*, in press.
Schrijver, C. J., and Rutten, R. G. M. 1987 *Astr. Ap.*, **177**, 143.
Simon, T., and Fekel, F. 1987, *Ap. J.*, **316**, 434.
Skumanich, A. 1972, *Ap. J.*, **171**, 565.
Soderblom, D. R. 1983, *Ap. J. Suppl.*, **53**, 1.
Ulmschneider, P., and Stein, R. F. 1982, *Astr. Ap.*, **106**, 9.
Vilhu, O. 1984, *Astr. Ap.*, **133**, 117.
Vogt, S. S. 1983, in *Activity in Red Dwarf Stars*, eds. P. B. Byrne and M. Rodonó (Dordrecht: Reidel), p. 137.

RE-ANALYSIS OF THE CORONAL EMISSION FROM RS CVn TYPE BINARIES

Osman DEMİRCAN*
Middle East Technical University
Physics Department
06531 Ankara
Turkey

ABSTRACT. The coronal emission from nineteen bright double lined eclipsing RS CVn systems with reliable absolute elements have been re-examined with the aim to explore the activity parameters in these systems. It is well established that the x-ray luminosity L_x is correlated with the Roche lobe filling percentage RL and the radius R where I belive, RL and R represent the depth l of the subphotospheric convective layer. It is known that l is the basic parameter of stellar dynamo efficiency and its value increases together with RL and R through the evolution off the MS. Further qualitative results obtained such as $L_x \propto L_{bol}$, $F_x \propto P$ and $F_x \propto R^2$ need confirmation by using extended more accurate data.

1. INTRODUCTION

The RS CVn systems are known to be binaries formed of two late type stars; The cooler component is slightly more massive and evolved to the base of the giant branch while the other component is evolved to near central hydrogen exhaustion (Morgan and Eggleton, 1979). The UV and x-ray satellite observations have shown that RS CVn's have among the most active upper atmospheres known in late type stars. The soft x-ray (0.1 - 3 keV) coronal luminosities in observed bright RS CVn systems were found to be ranging from $\sim 10^{30}$ to a few times 10^{31} ergs per second (Walter and Bowyer, 1981). The atmospheric activity of the late type stars in general, were found to be rotation rate dependent phenemonon (e.g. Wilson 1966, Kraft 1967, Skumenich 1972, Middelkoop and Zwaan 1981, Pallavicini et al. 1981) and thus widely believed that atmospheric activity is magnetic in origin and power house lies in the magnetic dynamo inside the star's subphotospheric convective shell.

The expected correlations between rotation periods P and different activity indicators for the RS CVn systems have been found by many authors (e.g. Ayres and Linsky, 1980; Walter and Bowyer, 1981; Barden, 1985; Basri et al 1985; Fernandez,Figueroa et al., 1986). In all these studies the normalized apparent activity

---

* Now at the University of Ankara, Department of Astronomy, Beşevler, 06100 Ankara, Turkey.

flux by apparent bolometric flux f_{bol} have been used as activity indicator and taken to be equivalent to surface activity flux (i.e. $f/f_{bol} = L/L_{bol} = F/F_{bol}$). It is known that bolometric luminosities L_{bol} and rotational periods P of Roche lobe filling stars in binaries are correlated (Gratton, 1950; Young and Koniges, 1977; see also Rengarajan and Verma, 1983) as L_{bol} increases with increasing P. It was shown by Rengarajan and Verma that Walter and Bowyer's correlation ($L_x/L_{bol} \propto P^{-1.2}$) for RS CVn systems is essentially due to the correlation between L_{bol} and P- a conclusion that has been further strengthened by the findings of Majer et al. (1986). Thus the rotation period dependence of the activity in RS CVn systems became, in general, questionable.

To further explore the activity parameters in these systems the interdepence between coronal emission and a number of relevant stellar parameters of nineteen bright double lined eclipsing RS CVn systems have been studied after decoupling the emission into the component stars.

2. DATA, ANALYSIS AND RESULTS

The interdependences between coronal emission and other stellar parameters in RS CVn systems may well be camouflaged by the complications due to (i) time variabilty of the activity, even flaring occasionally, (ii) emission from two close component stars at different levels (decoupling problem), and (iii) large uncertainties in the measurements and in the stellar parameters. Although the first complication will be resolved by continuous futur observations it is unavoidable in the HEAO1-A2 and Einstein observations. In order to suppress the complicatio due to second and third causes the bright double lined and eclipsing RS CVn systems with relatively more reliable absolute elements have been selected from Hall's catalogue. Thus, only nineteen RS CVn systems with existing x-ray data were found to have reliable absolute elements. These binaries are SS Boo, SS Cam, RZ Cnc, AD Cap, RS CVn, WW Dra, Z Her, AW Her, MM Her, GK Hyd, RT Lac, AR Lac, VV Mon, AR Mon, LX Per, SZ Psc, RW UMa, $\mathcal{E}$ UMi and ER Vul. In addition, two other binaries β Per (Algol) and V471 Tau from two different classes of binary stars and active Sun have been included in the analysis for comparison. The absolute elements were mostly taken from three sources; Brancewicz and Dworak 1980; Popper 1980; and Hall et al. 1985. The x-ray data are from the HEAO1-A2 and Einstein surveys and borrowed from Walter and Bowyer 1981.

A plot of the total x-ray luminosities $L_{x_1} + L_{x_2} = 4\pi d^2 f_x$ from the systems against the sum of the Roche lobe filling percentages $RL_1 + RL_2$ of the component stars revealed a close correlation between these two quantities of respective RS CVn systems (Figure 1). It is interestin that the scatter in the plot of the $L_{x_1} + L_{x_2}$ against RL_1 of the more evolved components is much more larger. Thus, it becomes clear that not

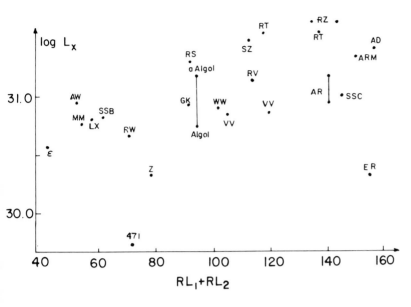

Figure 1. Total soft x-ray luminosity L_x versus the sum of the Roche lobe filling percentages RL_1+RL_2 for the RS CVn systems. The binary systems in this and subsequent figures are identified by parts of their variable star names. The components of ER Vul and Z Her are relatively lower mass earlier spectral type MS stars for which L_x value is lower. The other lower values of L_x's belong to the mass transfering systems.

only cooler more evolved but both components of the RS CVn systems are active and the level of activity is proportional to the Roche lobe filling percentage. Such property is used to decouple the total x-ray emission into the component stars. Then, the L_x and F_x values were plotted against the other stellar parameters such as RL, radius R, L_{bol}, mass M, equatorial rotation speed v and rotational period P ($\pm$ orbital period for RS CVn's). See Figures 2-7.
   The results from the plots shown in the figures can be given as follows:
1- There is no significant L_x-P correlation (Figure .). Only with the inclusion of contact binaries, which fill the gap around ER Vul in Figure 2, an increasing trend of L_x values with increasing P becomes significant. This correlation is induced by radius effect.

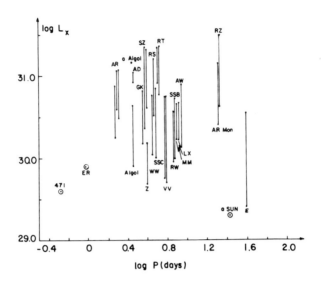

Figure 2. Total L_x versus the rotation period P (assumed to be equal to the orbital period). The component stars are connected by straight lines. The symbols are the same as in Fig. 1.

2- The L_x values are correlated tightly with RL (Figure 3) and R (Figure 4). In both correlations the MS stars, giants and the components of mass transfering systems have relatively lower values of L_x.

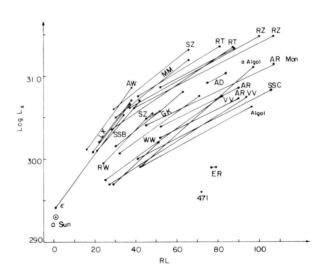

Figure 3. L_x versus the Roche lobe filling percentage RL for the same binaries as in Fig.1.

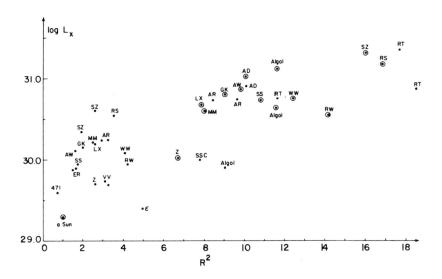

Figure 4. L_x versus the radius R for the components of same binaries as in Fig.1.

3- A plausible loose correlation was found between L_x and L_{bol} (Figure 5). The existence of such correlation is also implied by L_x-R^2 correlation. Simply, $L_x \propto R^2$ and $L_{bol} \propto R^2$ reveal that $L_x \propto L_{bol}$. Thus, it seems the activity in late type stars is also powered -as in the case of early type stars- by photospheric radiation probably through the stellar wind and microflares.

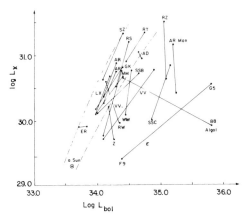

Figure 5. L_x versus L_{bol} for the components of same binaries as in Fig.1. L_x value is lower for the MS stars, earlier type stars, giants and the components of mass transfering systems.

4- A loose correlation between L_x and the equatorial rotation speed v is not real but induced by radius effect. Since $v = 2\pi R/P$ and the correlation disappears complately when L_x is replaced by F_x.

5- A qualitative period-activity correlation was obtained when surface x-ray fluxes were used as activity indicator (Figure 6). Similar correlations were found by many authors (e.g., Basri et al. 1985; Rutten and Schrijver 1985; Vilhu and Rucinski 1983) by using transition region chromospheric and/or coronal emissions from different samples of late type stars (single or components of synchronised binaries). The characteristic feature of the correlation is the saturation of activity at about $P \simeq 3$ days, a conclusion were pointed out first by Vilhu and Rucinski although they used relative fluxes as activity indicator in their analysis. For $P \gtrsim 3$ days, F_x decreases with decreasing rotation rate. The three-day limit was found to be some what smaller (~ 1 day) for transition region line emission fluxes by Rutten and Schrijver.

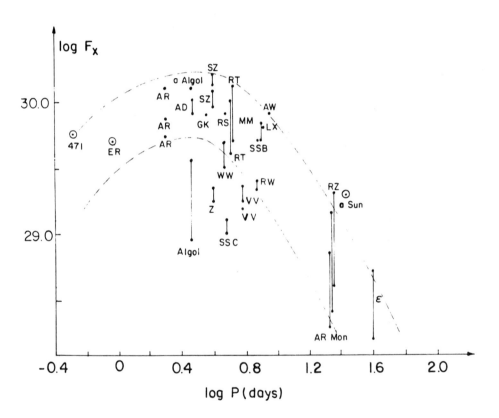

Figure 6. Surface x-ray flux F_x (ergs per sec per solar area) versus rotation period for the component stars of same binaries as in Fig. 1. The position of active Sun is also shown among the giants. The dashed lines are the boundary estimates of the correlation (see text).

6- The radius effect is not eliminated from L_x by considering the surface fluxes as $F_x = L_x/4\pi R^2$. The F_x values obtained in this way are still R dependent quantities (Figure 7). Thus, it seems $L_x \propto R^\alpha$ and the power $\alpha \neq 2$ and probably has different values for MS subgiant and giant stars. The $F_x - R^2$ correlation shown in Figure 7 represents, in fact, the evolutionary effect on the x-ray emission. For constant mass stars ($M = 1.4 \pm 0.7\ M_\odot$ for our sample) R and consequently depth of the convective layer increases by evolution off the MS, and F_x increases through the evolution until the star reach the giant branch, then the F_x drops rapidly. The F_x also drops significantly whenever the component star fills the Roche lobe during the evolution. The probable deformation of the magnetic structure by the Roche lobe mass transfer diminishes the energy transfer to the upper atmospheres and causes a drop in the x-ray emission flux.

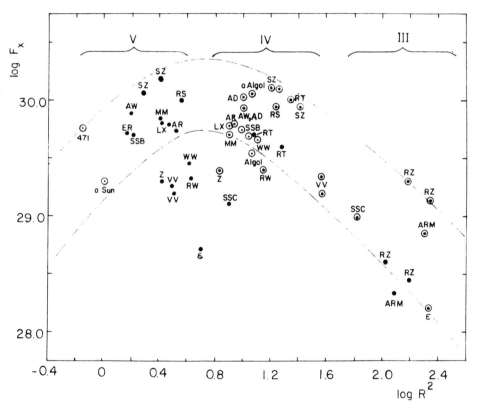

Figure 7. F_x versus R for the component stars of same binaries as in Fig. 1. Similarly, the dashed lines are boundary estimates of the correlation. The position of the active Sun is among the MS stars. The correlation represents the evolutionary effect on F_x (see text).

3. CONCLUSION

The RL and R dependence of L_x for the components of RS CVn systems has been well established. It is quite convincing that both RL and R here represent the total depth l of the subphotospheric convective layer which increases together with RL and R through the evolution off the MS. It is known, on the other hand, that the efficiency of dynamo processes and concequently the stellar activity is proportional to $\Omega^\alpha l^\beta$ (see e.g. Gilman 1980) with powers α and β to be settled by futur studies.

An evidence is found that activity in late type stars is also powered -as in the case of early type stars- by photospheric radiation. This result- however, is not conclusive and needs confirmation.

A loose rotational dependence of coronal activity is obtained by using x-ray surface fluxes F_x as activity indicator. The F_x values are also found to be R dependent quantities. These results are all agree qualitatively with the stellar dynamo models and more accurate data are needed for any quantitative result.

REFERENCES

1. Ayres, T. and Linsky, J. 1980, Ap. J., 241, 279.
2. Basri, G., Laurent, R. and Walter, F. M. 1985, Ap. J., 298, 761.
3. Barden, S. C. 1985, Ap. J., 295, 162.
4. Brancewicz, H. K., and Dworak, T. Z. 1980, A.A., 30, 501.
5. Fernandez-Figuroa, M. J., Sedano, J. L., and Castro, E. 1986, A.&Ap., 169, 237.
6. Gilman, P. 1980, in D. Gray and J. Linsky (eds), Stellar Turbulance, Springer, p19.
7. Gratton, L. 1950, Ap. J. 111,31.
8. Hall, D.S., Zeilik, M., Nelson, E.R. and Eker, Z.1985, Hall Catalogue of RS CVn Star Systems.
9. Kraft, R.P. 1967, Ap. J 150,551
10. Majer, P., Schmitt, J.H., Golub, L., Harnden, F.R. and Rosner, R.1986 Ap J. 300,360
11. Middelkoop, F. and Zwaan, C. 1981, A.&Ap. 101,26.
12. Morgan, J.G. and Eggleton, P.P. 1979, MN,187,661
13. Pallavicini, P. Golub, L., Rosner, R.and Viana, G.S.1981, Ap. J. 248,279
14. Popper,D.M. 1980 , Ann. Rev. A.& Ap. 18,115.
15. Rengarajan T.N., and Verma R.P. 1983, MN, 203, 1035
16. Rutten, R.G.M., and Schrijver, C.J. 1985, in M.Zeilik and D.M. Gibson (eds.), Cool Stars Stellar Systems and the Sun, Springer, p120.
17. Skumenich, A. 1972, Ap.J. 171,565.
18. Vilhu, O. and Rucinski, S.M. 1983, A.&Ap. 127,5.
19. Walter, F.M. and Bowyer, S. 1981, Ap.J. 245,671
20. Wilson, O.C. 1966, Ap.J. 144, 695
21. Young, A. and Koniges, A. 1977, Ap. J. 211,836

4. RELATED (OPTICALLY-THICK) STELLAR SOURCES

HIGH RESOLUTION SOFT X-RAY SPECTROSCOPY OF HOT WHITE DWARFS

Frits Paerels and John Heise
Laboratory for Space Research Utrecht
Beneluxlaan 21, 3527 HS Utrecht, the Netherlands

ABSTRACT. We present the results of the high resolution X-ray spectroscopy of hot white dwarfs, obtained with the 500 lines mm^{-1} Transmission Grating Spectrometer on EXOSAT. These data, which are unique in wavelength coverage, spectral resolution, and statistical quality, have offered us the first detailed look at the photospheric X-ray spectrum of these objects. The shape of the soft X-ray spectrum is highly sensitive to the spectral parameters; we use this sensitivity to derive accurate estimates of the effective temperature, surface gravity, and chemical composition of the atmosphere. Combining all available spectral data for these stars, from the optical to the X-ray band, a fully consistent description of the complete electromagnetic spectrum, in terms of a single model atmospheres spectrum, can be obtained.

INTRODUCTION

In 1975 soft X-rays were discovered from Sirius with the ANS Soft X-ray Experiment /2,3/. Originally it was thought that these X-rays were emitted by a corona. A more natural explanation for the X-ray emission was put forward by Shipman in the following year /4/. He pointed out that a pure hydrogen atmosphere at high effective temperature, such as the one of Sirius B, would be highly ionized, and consequently very transparent at X-ray wavelengths. We would observe the thermal emission of hot atmospheric layers far below the optical photosphere. Model atmospheres calculations showed that this model could quantitatively account for the soft X-ray flux observed with ANS. This explanation was vindicated shortly after by the discovery of intense soft X-ray and EUV emission from two other hot DA white dwarfs, HZ 43 and Feige 24, and a number of others since then.
    The shape of the soft X-ray spectrum, and the total soft X-ray flux of hot DA white dwarfs are very sensitive to the photospheric parameters. The total X-ray flux is a steep function of effective temperature, and trace amounts of elements other than hydrogen will produce strong absorption edges in the X-ray/EUV spectrum (at high temperatures, these elements will be highly ionized, and the strong ground state absorption

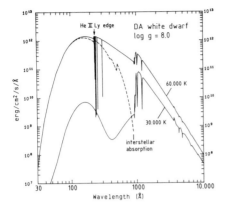

Figure 1. Spectrum emerging from a hot, pure hydrogen atmosphere at log g = 8, from optical to X-ray wavelengths, for T_e = 30,000 K, and T_e = 60,000 K; in this last model the atmosphere contains a trace of helium (He/H = 10^{-5}). The dashed line shows the attenuation of the stellar spectrum by absorption by neutral gas of column density n_H = 10^{18} cm^{-2}.

edges and the resonance lines will be in the extreme short-wavelength range). In contrast, the optical and UV spectrum becomes increasingly insensitive to the photospheric parameters with increasing effective temperature. These properties of the X-ray and optical spectra of hot DA white dwarfs are illustrated in Figure 1, which shows the flux (ergs cm^{-2} s^{-1} Å^{-1}) emerging from (1) a pure hydrogen atmosphere at T_e = 30,000 K, and (2) an atmosphere at T_e = 60,000 K, with a fractional helium-abundance He/H = 10^{-5}; the location of the He II Lyman edge at 227 Å has been indicated. At T_e = 60,000 K a DA white dwarf radiates more than half of its total (photon) luminosity shortward of the (hydrogen) Lyman edge.

Soft X-ray observations are obviously of great value to the observational investigation of a number of fundamental questions concerning the physics of hot white dwarfs. Improved estimates of T_e, stellar radius, and surface gravity from combined X-ray/UV/optical spectroscopy should provide the information to construct luminosity functions and evolutionary sequences. Mass estimates obtained from the measured gravity and radius can be used to experimentally verify theoretical mass-radius-relations.

Since the X-ray and EUV spectra of hot DA white dwarfs are very sensitive to the exact chemical composition of their photospheres, X-ray spectroscopy and photometry can reveal trace amounts of helium and metals in a hot, hydrogen dominated white dwarfatmosphere that are a factor 100 below the present optical spectroscopic detection limit.

But the very detection of traces of heavier elements already poses problems in itself. Such traces are not expected to be present at all, since the timescale for downward diffusion of 'metals' in a pure hydrogen atmosphere is extremely short. In the early 1980's, however, evidence accumulated from EUV and X-ray observations that such traces are indeed present in the photospheres of at least some hot DA white dwarfs.

Malina et al. /5/ first reported the detection of a He II Lyman absorption edge at 227 Å in a spectrum of the hot white dwarf HZ43. Kahn et al. /6/ compared measured X-ray fluxes for four hot DA's (observed with the EINSTEIN IPC) with X-ray fluxes predicted from the visual magnitude, and an independent estimate of effective temperature. The measured X-ray fluxes, surprisingly, fell significantly short of the predictions, and these authors interpreted this result as indicating the presence of some X-ray absorbing material in the photosphere. Assuming this material to be helium, they derived photospheric helium abundances in the range He/H = 10^{-5}–10^{-3}. This positive detection of helium in hot DA atmospheres requires the presence of some mechanism counteracting the rapid downward diffusion. Petre et al. /7/ supplied helium abundances, determined likewise from X-ray fluxes measured with the EINSTEIN HRI, for a further 5 DA's.

With the launch of the European X-ray Observatory EXOSAT, the Low Energy Imaging Telescopes and Transmission Grating Spectrometers of which had a higher sensitivity and a larger dynamic range in the soft X-ray/EUV band than previous experiments /8,9/, it became possible to verify and extend these observations. We could measure the X-ray spectrum with high sensitivity between 44 and 400 Å, with typical spectral resolution ~6 Å; comparison with Figure 1 shows that we observe the full X-ray spectrum of a hot white dwarf, between the intrinsic cutoff at short wavelengths, to the interstellar absorption cutoff in the EUV band. We wanted to verify explicitly the conformity of the shape of the X-ray spectrum of hot DA white dwarfs with model atmospheres calculations, and to use the sensitivity of the X-ray spectrum to the stellar parameters to obtain improved estimates of these. The fact that the hot DA's, HZ 43, Sirius B, are very bright X-ray/EUV sources, and that they have been studied extensively in all wavelength bands, made them the natural 'first choice' for observation with the EXOSAT Transmission Grating Spectrometers (TGS).

The remainder of this paper is devoted to a discussion of the X-ray spectroscopic observations. The following section contains a discussion of the analysis of the spectral data on HZ 43, Sirius B; in the last section we summarize these specific results, and place them in the general hot white dwarf astrophysical context.

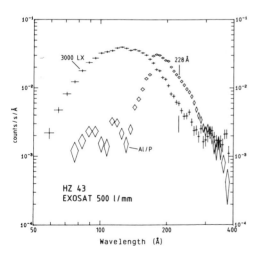

Figure 2. X-ray spectrum of HZ 43, measured with the 500 l mm^{-1} TGS, with the 3000LX and Al/P filters (background subtracted, positive and negative order spectra summed; error bars are 1σ photon counting errors). Location of the He II Ly edge has been indicated.

HIGH RESOLUTION X-RAY SPECTROSCOPY OF HZ 43

HZ 43 is a typical 'young' hot DA white dwarf. Previous estimates of T_e ranged between 45,000 and 60,000 K, of the bolometric luminosity between $L/L_\odot$ = 0.5-3.8 . Comparing this with theoretical white dwarf cooling curves /1/, we see that HZ 43 should still be cooling to a large extent by neutrino emission; its core is near becoming isothermal.

HZ 43 was observed with the 500 l mm^{-1} TGS on 28 June, 1983, for a total of 18,500 sec, with two different beam filters, the 3000LX and Al/P filters. The measured spectra are shown in Figure 2 as they appear in the spectrometer (counts s^{-1} A^{-1} vs. wavelength). As is evident from a comparison with the spectra in Figure 1, we have indeed observed the full X-ray/EUV spectrum of HZ 43 with the two filters.

The position of the He II Ly edge at 227 A has been indicated; no edge is detected in the Al/P spectrum (which has the highest statistical quality at the longer wavelengths). This absence can be converted into a sensitive upper limit on the fractional abundance of helium, He/H. The best fit is at He/H = 0, the 99% confidence upper limit is at He/H = 1.0×10^{-5} (by number), as derived from standard χ^2 analysis. Figure 3 shows the Al/P spectrum in the region around the edge, together with a model spectrum at He/H = 0, and at He/H = 2×10^{-5}, both convolved with the spectrometer response. This last model was chosen to demonstrate that such a value for He/H is already emphatically excluded by the measured shape of the spectrum of HZ 43.

The column density of neutral interstellar gas was determined likewise from the Al/P spectrum, and was found to be in the range n_H = 6-16 x 10^{17} cm^{-2}, dependent on effective temperature, with a typical uncertainty of 0.4 in log n_H (99%).

Allowing for the ranges in He/H and n_H given above, and taking values of log g between 8 and 9, we determined the constraints imposed on T_e and stellar radius $R/R_\odot$ by the integrated X-ray flux as measured in the two spectra, and in a photometric exposure with the Boron filter on the

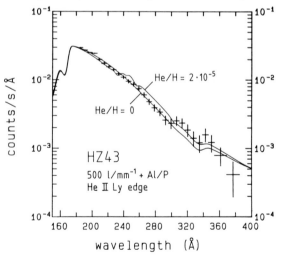

Figure 3. Al/P spectrum of HZ 43 between 170 and 380 A, together with model atmospheres spectra at T_e = 60,000 K, log g = 8.5, n_H = 1.4×10^{18} cm^{-2}, He/H = 0 and $2 \cdot 10^{-5}$, convolved with the spectrometer response. Best fit is for He/H = 0 ($\chi^2/26$ = 1.25), 99% confidence upper limit is at He/H = 1.0×10^{-5}.

Low Energy telescopes (through the ratio of measured X-ray flux to model flux at the stellar surface, using the distance, 63.3 pc /11/). Figure 4 shows these (99%) constraints, as well as the constraints imposed by the visual magnitude estimate V = 12.99 ± 0.03, taken from /12/. The intersection of the EXOSAT constraints and the 'optical' constraint gives the range of T_e and $R/R_\odot$ that yields models consistent with both the X-ray and optical absolute flux from HZ 43; we derive T_e = 45,000-54,000 K, $R/R_\odot$ = 0.0140-0.0165. At $R/R_\odot$ = 0.015, the spectra measured with the EINSTEIN OGS and the Voyager 2 EUV spectrometer /13/ yield effective temperatures that agree well with this solution (indicated at the top of Figure 4). Thus we determine the luminosity of HZ 43 to be $L/L_\odot$ = 1.0-1.5.

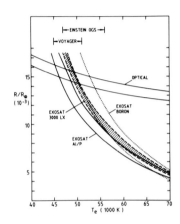

Figure 4. 99% Confidence constraints imposed on T_e and $R/R_\odot$ of HZ 43 by the measured total X-ray flux (in the 3000LX and Al/P spectra, and in photometric observation with the Boron filter), and by the visual magnitude. Intersection of X-ray and 'optical' constraint areas indicates parameter range for which consistent solution to all measured absolute fluxes is found, in terms of a single DA model spectrum. Horizontal bars at the top indicate 99% confidence estimates of T_e from EINSTEIN OGS and Voyager 2 spectra at $R/R_\odot$ = 0.015.

When we fit model atmospheres spectra with parameters in the above ranges, however, we find that the shape of the model spectra does not match the shape of the measured spectrum at the shortest wavelengths (λ < 90 A); χ^2 indicates that the models do not yield a statistically acceptable description of the shape of the 3000LX spectrum. At T_e = 50,000 K we have to assume a 15% systematic error in the model fluxes shortward of 90 A. This is shown in detail in Figure 5 (3000LX data, and a model at T_e = 50,000 K, log g = 8.5, He/H = 0, n_H = 2.10^{18} cm^{-2}, $R/R_\odot$ = 0.014; reduced χ^2 = 5.70; the lower panel shows the post-fit residuals). This systematic error cannot be traced to a systematic error in the calibration of the spectrometer, and must be ascribed to the model atmospheres calculations. This conclusion is confirmed by the fact that corresponding model atmospheres spectra, calculated with different codes based on the same input physics show 15% differences in monochromatic flux at the shortest wavelengths (these are the models published in /14,7/, and our own calculations).

At present the exact source of the differences remains unresolved, nor can any of the three codes be identified a priori as the one being free of the systematic uncertainties mentioned. However, the three codes agree in monochromatic flux longward of ~100 A, so that the Al/P spectrum can be used to measure He/H and n_H independent of the specific code used to generate the models. In addition, the codes agree to within a few

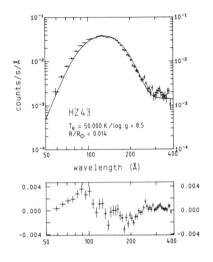

Figure 5. EXOSAT 500 1 mm^{-1} TGS (3000LX) spectrum of HZ 43, fitted with a model spectrum at T_e = 50,000 K, $R/R_\odot$ = 0.014, log g = 8.5, He/H = 0, n_H = 2.10^{18} cm^{-2}, that is consistent with measured optical/UV fluxes with respect to total X-ray flux. The spectral shape of the model is seen to exhibit a systematic difference with the measured shape of ~15% in monochromatic flux shortward of 90 Å, indicative of a systematic uncertainty in the model spectra (see text).

percent in integrated X-ray flux, so that our multicolor photometric determination of effective temperature and radius is valid, regardless of the systematic error diagnosed above.

A full discussion of the HZ 43 data and their analysis can be found in /15,16/.

HIGH RESOLUTION X-RAY SPECTROSCOPY OF SIRIUS-B

Sirius B is one of the very few white dwarfs for which an accurate astrometric mass determination exists: $M/M_\odot$ = 1.053 ± 0.028 /17/. Comparing this mass, and an independent estimate of the radius of Sirius B with theoretical mass-radius relations for electron-degenerate configurations is actually one of the oldest confrontations of theory with observations in the field of astrophysics of compact objects /18,19/. The difficulty has always been to obtain a good estimate of T_e, in order to derive $R/R_\odot$, the difficulty in observation being in the proximity and brightness of

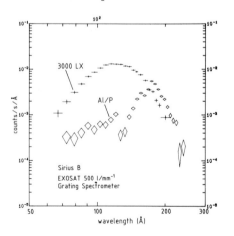

Figure 6. X-ray spectrum of Sirius B, measured with the 500 1 mm^{-1} TGS, with the 3000LX and Al/P filters (background subtracted, positive and negative order spectra summed; error bars are 1σ photon counting errors).

Sirius A /20/. With EXOSAT we have obtained the first resolved X-ray spectrum of Sirius B. Since Sirius A is ~100 times fainter than Sirius B at soft X-ray wavelengths, this spectrum can be said to be the first really 'clean' look at the photospheric spectrum of Sirius B. Analysis of this spectrum will provide an accurate estimate of T_e. Combination of all available data on Sirius B will produce an accurate estimate of its radius.

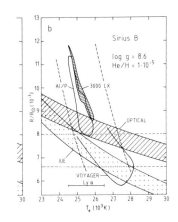

Figure 7. 99% Confidence constraints imposed on T_e and $R/R_\odot$ of Sirius B by the shape of the measured X-ray spectrum (3000LX and Al/P filters), by the visual magnitude, by the UV continuum ('IUE' and 'Voyager'); also shown is the T_e constraint derived from fitting the Ly α profile. Intersection of X-ray and 'optical' constraint areas indicates parameter range for which consistent solution to all measured absolute fluxes is found, in terms of a single DA model spectrum. Vertical dashed lines are 99% confidence constraints derived from two-color X-ray photometry, using total flux in 3000LX and Al/P spectra. Horizontal dashed lines are 99% confidence extremes on $R/R_\odot$ derived from astrometric mass and Hamada-Salpeter mass-radius relation for ^{12}C core.

Sirius B was observed with the 500 1 mm^{-1} TGS on EXOSAT on 19/20 October, 1983, for a total of 87,000 sec, with both the 3000LX and Al/P filters in the TGS. Figure 6 shows the two measured spectra as they appear in the spectrometer (counts s^{-1} Å^{-1} vs. wavelength). As with HZ 43, we have the full X-ray spectrum, including the important short-wavelength cutoff. We fitted model atmospheres spectra to these data, for varying T_e, log g, He/H, n_H, and $R/R_\odot$. χ^2 analysis of the shape of the X-ray spectrum, demanding consistency within the 99% confidence bounds of the estimates of the spectral parameters as derived from both measured spectra yields constraints on the photospheric helium abundance (He/H < $2 \cdot 10^{-5}$), and the gravity (log g > 8.0); T_e is in the range 24,500 - 26,000 K, the column density is restricted to n_H < $6 \cdot 10^{18}$ cm^{-2}. All these parameter estimates are coupled. The parameters can be further restricted in the following way. In Figure 7 we show the region of T_e and $R/R_\odot$ that yields models that are statistically consistent with the shape of both measured spectra (99% confidence). For this plot we set log g to 8.6, He/H to $1 \cdot 10^{-5}$, n_H ranges in its 99% confidence interval. The location of this 99% confidence area depends only weakly on He/H and log g. At the known mass of Sirius B we can therefore use it to calculate 99% confidence constraints on the gravity, from the constraints on $R/R_\odot$. For $R/R_\odot$ = 0.008-0.012 we find log g = 8.3-8.7. Also plotted in Figure 7 are the 1σ constraints on effective temperature and radius derived from optical and UV observations (see /21/ for a full

discussion of these data). Taking the intersection of all 99% confidence constraints, we identify the following set of parameters as a complete, consistent description of the complete electromagnetic spectrum of Sirius B: T_e = 25,000 - 26,000 K (mainly determined by the X-ray data), $R/R_\odot$ = 0.0079-0.0085 (mainly determined by the optical/UV data). With the known mass of Sirius B this yields log g = 8.5-8.7, and the photospheric helium abundance we found to be less than 2.10^{-5}. For these parameter values, the X-ray spectrum indicates n_H = 1.1-5.0 x 10^{18} cm^{-2}. These parameter estimates are insensitive to calibration uncertainties of the TGS, and to systematic uncertainties in the model atmospheres spectra (the vertical dashed lines in Figure 7 indicate the 99% limits on T_e derived from two-color X-ray photometry, using the integrated X-ray flux as measured in the two EXOSAT spectra - i.e. the parameter estimates resulting from not using the shape of the measured X-ray spectra, which is a very conservative assumption). Figure 8 shows two example fits to the 3000LX and Al/P spectra, with the model parameters as indicated in the Captions.
For a full description of the Sirius B X-ray data, we refer to /23/.

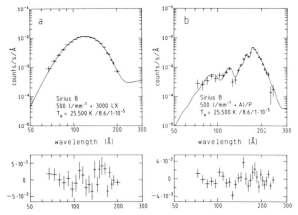

Figure 8. Example fits of a model atmosphere spectrum to the measured 3000LX and Al/P spectra, for T_e = 25,500 K, log g = 8.6, He/H = 1.10^{-5}. The measured spectra are shown as crosses, the model spectra (which have been convolved with the instrument response) as a solid line. (a) 3000LX, with n_H = 4.10^{18} cm^{-2}. $R/R_\odot$ = 0.0096. (b) Al/P, with n_H = 3.10^{18} cm^{-2}. $R/R_\odot$ = 0.0086. Below each spectrum we show the residuals (measured count rate minus model count rate, binned in the same wavelength intervals as the measured spectra).

CONCLUSIONS

The high sensitivity, high spectral resolution, and large wavelength coverage of the Transmission Grating Spectrometers on EXOSAT gave us the possibility, for the first time since the discovery of X-ray emission from hot DA white dwarfs, to make detailed observations of their complete photospheric X-ray/EUV spectrum. The conclusions that can be drawn from the analysis of the X-ray spectra of HZ 43 and Sirius B are summarized below.

(1) First and foremost, we have found that for ordinary hot DA stars the soft X-ray/EUV spectrum indeed conforms to theoretical models of this spectrum, calculated from homogeneous model atmospheres consisting of pure hydrogen, or hydrogen with trace helium, at high effective temperature and high gravity, both with respect to the shape of the spectrum and the absolute flux level. This is demonstrated powerfully by the measured X-ray spectra of HZ 43 and Sirius B.

(2) The presence of a He II Lyman absorption edge at 227 A in the spectrum of HZ 43, reported by Malina et al. /5/ on the basis of data obtained in a rocket EUV experiment, is not confirmed by the spectra obtained with EXOSAT. In view of the higher statistical quality of the EXOSAT spectra, and the larger wavelength coverage on both sides of 227 A we conclude that the original detection of the He II edge must be attributed to systematic errors.

(3) Using the high sensitivity of the X-ray spectrum to the presence of traces of helium in a hot DA atmosphere, through the size of the He II Ly edge, we put a tight upper limit on the photospheric helium abundance of HZ 43 (He/H $\leq 1.0 \times 10^{-5}$). Likewise, for Sirius B we find He/H < 2.10^{-5}.

(4) The most powerful procedure to obtain accurate estimates of the photospheric parameters of hot DA white dwarfs combines X-ray, EUV, UV, and optical spectroscopy, employing the different sensitivity of the spectral shape and absolute flux to these parameters in the different wavelength bands. This method has been applied to HZ 43 and Sirius B, and a fully consistent description of their complete electromagnetic spectrum, in terms of a single DA model spectrum, or range of model spectra, is the result. Prime result for HZ 43 is an improved estimate of its luminosity, $L/L_\odot$ = 1.0-1.5 . For Sirius B we made an accurate measurement of T_e; we find an improved estimate of its radius ($R/R_\odot$ = 0.0079-0.0085).

(5) In the case of HZ 43, however, in order to obtain a solution for the spectral shape that is consistent with all measured absolute fluxes, we have to assume the presence of a ~15% systematic error in the monochromatic fluxes of the model spectra at the shortest wavelengths (λ < 90 A). Since this error is larger than either the statistical uncertainty in the shape of the measured X-ray spectra, or the known systematic uncertainties in the calibration of the spectrometer, we identify this systematic error with a systematic error in the model atmospheres calculations at T_e ~ 50,000 -60,000 K.

REFERENCES

1. I. Iben and A. Tutukov, Ap.J., 282, 615 (1984).
2. R. Mewe et al., Nature, 256, 711 (1975).
3. R. Mewe et al., Ap.J.(Letters), 202, L67 (1975).
4. H. L. Shipman, Ap.J.(Letters), 206, L67 (1976).
5. R. F. Malina, S. Bowyer, and G. Basri, Ap.J., 262, 717 (1982).
6. S. M. Kahn et al., Ap.J., 278, 255 (1984).
7. R. Petre, H. L. Shipman, and C. Canizares, Ap.J., 304, 356 (1986).
8. B. Taylor, Adv. Space Res., 5, 35 (1985).
9. P. A. J. de Korte et al., Space Sci. Rev., 30, 495 (1981).
10. F. B. S. Paerels and J. Heise, Ap.J.,(submitted) (1987).

11. C. C. Dahn et al., Astron. J., **87**, 419 (1982).
12. J. B. Holberg, F. Wesemael, and J. Basile, Ap.J., **306**, 629 (1986).
13. J. B. Holberg et al., Ap.J.(Letters), **242**, L119 (1980).
14. F. Wesemael et al., Ap.J.Supplements, **43**, 159 (1980).
15. F. B. S. Paerels et al., Ap.J., **308**, 190 (1986).
16. J. Heise et al., Ap.J.,(submitted) (1987).
17. G. D. Gatewood and C. V. Gatewood, Ap.J., **225**, 191 (1978).
18. R. Marshak, in: Sixth Texas Symposium on Relativistic Astrophysics, New York Academy of Sciences, New York 1973 (p.5).
19. J. L. Greenstein and J. B. Oke, Q.J.R.A.S., **26**, 279 (1985).
20. J. L. Greenstein, J. B. Oke, and H. L. Shipman, Ap.J., **169**, 563 (1971).
21. J. B. Holberg, F. Wesemael, and I. Hubeny, Ap.J., **280**, 679 (1984).
22. F. B. S. Paerels et al., Ap.J.(in press) (1987).
23. F. B. S. Paerels et al., Ap.J.(Letters), **309**, L33 (1986).

PULSATING WHITE DWARFS

M.A.Barstow
X-ray Astronomy Group, Physics Department
University of Leicester
University Road
Leicester, LE1 7RH, UK

ABSTRACT. In recent years, this small, but important, class of highly evolved stars, have yielded exciting results from intensive ground and space based studies. Pulsational instabilities, which arise in three distinct regions along the white dwarf cooling sequence, have periods ranging from 200s to 2500s and offer unique opportunities for the study of the interior and thermal evolution of compact objects. These instabilities are interpreted as non-radial gravity modes arising in partial ionization zones of C&O (PG1159 objects near 100000K), He (DB white dwarfs near 30000K) and H (DA ZZ Ceti white dwarfs around 10000K). Many of these objects have been well studied optically. However, the hotter objects can be observed at X-ray wavelengths which adds an extra dimension to their study, particularly when such observations are combined with simultaneous optical data.

1. INTRODUCTION

In the late 1960s and early 1970s it was discovered that a small number of isolated DA white dwarfs (WDs) were multiperiodic variables, having amplitudes less than ≈ 0.1 magnitudes. It was established (Warner, 1975) that these ZZ Ceti objects lie near the cool end of the observed DA sequence and that the range of colours (T= 10500 − 13500K) is small, coinciding with an extension of the Cepheid instability strip across the WD evolutionary path. However, comparison of the observed periods, lying in the range 200s to 2000s, with theoretical models (van Horn, 1972) shows that they are two long to be explained by radial pulsation models (the classical Cepheid mechanism) but can be understood in terms of non-radial gravity (g) modes. Subsequent observations increased the number of known ZZ Ceti stars and led to the discovery of similar pulsations in hotter DB white dwarfs (Winget et al, 1982; T$\approx$ 28000K) and the He rich PG1159-035 objects (McGraw et al, 1979; T$\approx$ 100000K). A common practise is to refer to these objects as DAV, DBV and DOV respectively. It is currently accepted that the driving mechanism of pulsation is the presence of partial ionization zones in the non-degenerate mantles of the stars, comprising C&O, He and H in DOV, DBV and DAV stars respectively.

The existence of three distinct regions of pulsational instability on the WD cooling sequence presents us with a useful diagnostic tool, in addition to the usual spectroscopic techniques, for characterising WDs and studying their evolution. Observations of pulsations in WDs reveal atmospheric stratification and composition and allow determination of the stellar mass and global cooling rates. Inversely, the apparent stability against pulsation of other stars can be used to place constraints on their internal structure. The value of this knowledge is enhanced if the stars observed can then be placed in an evolutionary context by determination of effective temperature and surface abundances in the atmosphere, gathered by a combination of optical/UV spectroscopy and broad band X-ray photometry.

A brief outline of non-linear pulsation theory is presented, with the aim of helping the reader to understand how the light variations are interpreted in terms of the physical behaviour of the star. A summary of the characteristics of pulsating stars is followed by an outline of recent X-ray observations.

2. NON-RADIAL PULSATION THEORY

Excellent, detailed discussions of the theory of non-radial pulsation can be found in the literature (eg. Cox, 1980 and Unno et al, 1979) and only a summary of the basic principles is presented here. Radial pulsations are easily visualised as the cyclical inflation and deflation of a star but it is not as easy to understand the appearance of a stellar surface undergoing non-radial pulsations. In a spherically symmetric star the normal modes are characterised by the eigen functions that are proportional to the spherical harmonics

$$Y_l^m(\theta, \phi) = P_l^m(\cos\theta) e^{im\phi}$$

where: $P_l^m(x)$ is the associated Legendre polynominal, l $(= 0, 1, 2....)$ and m $(= -1, -1+1, ..0, ..l-1, 1)$ are integers and θ and ϕ are the normal spherical coordinates.

The radial modes of oscillation are thus the special cases of l=0. Modes are further divided by the number of nodes (k) in the radial component of displacement from the centre to the surface of the star. k=0 is the fundamental mode with k=1,2... being first and second overtones, etc. Consequently the normal modes are classified by radial quantum number k, angular quantum number l and, when 'splitting' is introduced by rotation or by a magnetic field, azimuthal quantum number m. Figure 1 illustrates qualitatively the appearance of a star undergoing l=1 and l=2 pulsations.

Prediction of the pulsation properties of real stars requires detailed theoretical modelling which must include assumptions about the composition and stratification of the stars in question. Examples of such analyses for DAV, DBV and DOV stars are publications by Cox et al (1987), Winget et al (1983) and Starrfield et al (1984) respectively. Analysis of the stellar pulsation periods does not necessarily require knowledge of the possible driving mechanisms. However, some discussion is necessary in order to determine what physical processes underlie the flux variations. For the most part, stellar pulsation is an envelope phenomenon and it is generally accepted that the driving mechanisms involve the partial ionization of an abundant element at some critical depth below the stellar surface.

A region in a star can be said to be driving if it is absorbing heat at maximum compression. Conversely, a region losing heat at maximum compression is damping. When, under proper conditions, the effects of driving outweigh those of damping the star is pulsationally unstable. Gains and losses of heat must be accomplished by modulation of the radiation flowing through these regions. A destabilising agent can be seen to be most effective if it lies in a transition region between the region of the star that is quasi-adiabatic and that where, as a result of low heat capacity, the flux variations are 'frozen-in' (Figure 2). In or near the transition region driving can take place but immediately above it damping may be eliminated by non-adiabatic effects.

Figure 1. Schematic illustration of l=1 and l=2 non-radial pulsations. Shaded and unshaded areas represent compressed and expanded regions respectively. The boundaries are the loci at equilibrium radius.

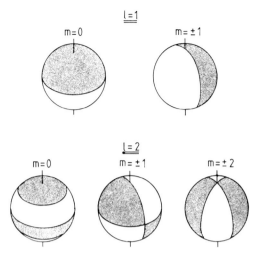

Figure 2. Schematic illustration of the transition region.

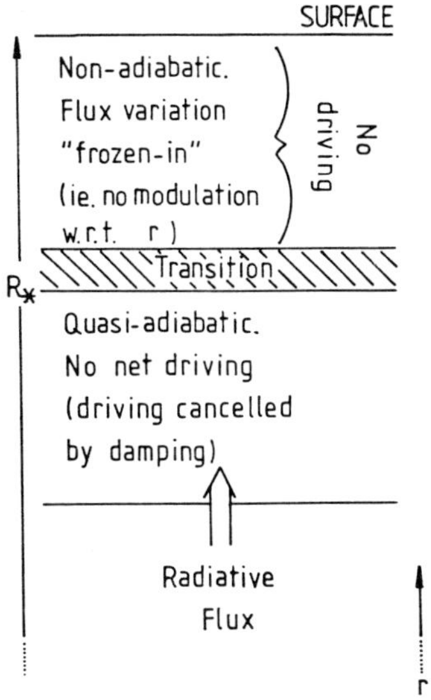

Figure 3. The pulsation cycle.

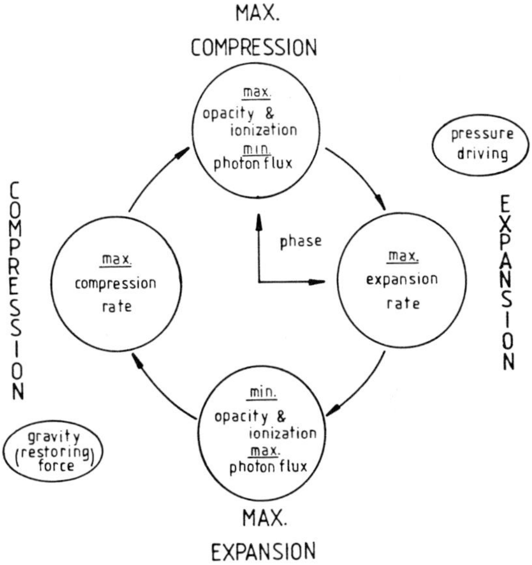

The presence of a sufficiently abundant, partially ionized element produces destabilisation by causing the local temperature to be lower than if ionization was not occurring. Any energy added to the gas is shared between further ionisation and kinetic (thermal) motions. Consequently, the local opacity κ increases upon compression (it normally decreases). This results in 'trapping' of the energy flux. The gas absorbs this energy and, on subsequent expansion, the pressure is larger than if the process had been adiabatic. Hence, pulsations are 'pumped up'. This is the κ mechanism. A similar overall effect is caused by the diminution of radiation upon compression, because of the relative coolness of this region - the γ mechanism. In addition the local radiating area of the star is reduced also trapping radiation inside the star - the r (radius) effect. The process is cyclical (figure 3) with the restoring force being gravity. In some models (Kawaler, 1987) destabilisation by nuclear shell burning (ϵ mechanism) is possible where the temperature oscillations on compression and expansion result in a cyclical change of the energy generation rate.

Changes in stellar luminosity must be caused by a combination of changes in the stellar surface area and temperature integrated across the stellar disc, as viewed along the line of sight. The phase of such variations relative to the driving cycle must be determined by the mechanical response and radiative timescale of the atmosphere, which are poorly understood. DV stars are sufficiently hot that optical pulsations occur on the relatively flat Raleigh-Jeans tail of the energy distribution. If pulsations were observed at wavelengths on the steep Planck portion of the spectrum, below the peak wavelength ($\approx$ 2900Å at 10000K, $\approx$ 1000Å at 3000K and $\approx$ 300Å at 100000K), the amplitude due to temperature fluctuations would be much larger, whereas area dependent flux changes are constant at all wavelengths. Observation and comparison of pulsation amplitudes at different wavelengths will serve as a means of determining the relative importance of each mechanism and, because photons of shorter wavelength arise from deeper regions within the atmosphere, allow the atmospheric structure to be studied. In DV stars T effects dominate.

3. OBSERVATIONAL ASPECTS OF PULSATING WHITE DWARFS

Most of the known DV stars are shown in Table I. It is clear that pulsation phenomena are comparatively rare in WDs, occurring in $\approx$ 20 out of $\approx$ 1500 known WDs. This must be due substantially to the narrowness of the temperature ranges in which pulsations may occur. However, stars of a temperature corresponding to a particular instability strip do not all pulsate. For example, only a third of stars in the DA instability region pulsate. This narrow temperature band allows tight constraints to be placed on the physical structure of the stars, contributing to the study of their evolution. Sion (1986) has recently published an excellent review of the current knowledge of these evolutionary processes. This reveals several important areas where our understanding is poor. A particular problem is the relationship between the He rich DO/DB and H rich DA WDs. There are no known DO or DB degenerates in the temperature range 30000-45000K and apparently the DO-DB path must somehow pass through the DA sequence. DOV and DBV stars lie on either side of this 'gap', thus pulsation studies are useful here.

TABLE I. Properties of pulsating white dwarfs

Star	T (K)	Dominant Periods (s)	Peak-Peak Amplitude (mag)	Reference
DAV				
R548	12550 ± 150	213,274	0.02	Robinson (1979)
HL TAU-76	13010 ± 350	494,626,661,746+others	0.34	Robinson (1979)
G38-29	11900 ± 1000	925,1020+others	0.21	Robinson (1979)
GD99	13350 ± 1000	260,480,590+others	0.13	Robinson (1979)
G117-B15A	13640 ± 350	216,272,308	0.06	Robinson (1979)
R808	11730 ± 250	813,830+others	0.15	Robinson (1979)
G29-38	12630 ± 150	694,820,930+others	0.28	Robinson (1979)
BPM 30551	10315 ± 400	607,745,823+others	0.18	Robinson (1979)
BPM 31594	12870 ± 400	311,404,617+others	0.21	Robinson (1979)
L19-2	12520 ± 400	114,192	0.04	Robinson (1979)

GD385		245	0.05	Robinson (1979)
G207-9		292,318,557,739	0.06	Robinson (1979)
GD154		790,1186+others	0.10	Robinson (1979)
DBV				
GD358	26000 ± 2000	≈ 28 142-952s	up to 0.30	Winget et al (1982), Koester et al (1983)
PG1654+160		578,851+others	0.05-0.18	Winget et al (1984)
PG1116+158	26000 ± 2000			Liebert et al (1986)
PG1351+489	25000 ± 2000			Liebert et al (1986)
DOV				
PG1159-035	> 80000	516,539+6 others	0.04-0.0184	Winget et al (1985)
K1-16	> 80000	1475-2500	0.02	Grauer and Bond (1984)
PG1707+427	≈ 100000	450+others	≈ 0.10	Bond et al (1984)
PG2131+066	≈ 100000	382,414	≈ 0.10	Bond et al (1984)

As can be seen in Table I, the DAV stars have quite well characterised temperatures forming a well defined population. This is not the case for the DBVs and DOVs where fewer are known and their higher temperatures cannot be easily measured. The temperatures of hot stars are best measured with observations in bands near the peak of their energy distribution, in the EUV or soft X-ray. At DB white dwarf temperatures high He opacity in the atmosphere prevents the emergence of significant soft X-ray flux and any EUV emission is highly attenuated by the ISM. Consequently, the easiest pulsating stars to study in high energy bands are the DOVs - the PG1159 objects. These are the very hottest helium dominated white dwarfs and they are characterised by the presence in their spectra of a broad, shallow absorption trough of HeII $\lambda 4686$ blended with the several transitions of CIV, CIII and NIII. A second spectral characteristic is the presence of OVI optical transitions, most prominently at $\lambda 3434$ (Winget et al, 1985). Frequently, sharp self-reversed emission components are present in many of the optical absorption features. In addition to emphasising the high ($> 10^5$K) effective temperature of these objects, the OVI features provide an evolutionary link between the 'OVI' central stars of planetary nebulae and the DO white dwarfs (Sion et al, 1985). The actual temperatures of PG1159 objects are not well determined. Composite Voyager, IUE and optical spectra of PG1159-035 (Basile and Holberg, 1985) can be fitted to spectral slopes with a temperature range of $10^5 - 2 \times 10^5$K. The atmospheric composition is not well known either. However, from the lack of HI Balmer lines and the presence of HeII $\lambda 4686$, He/H is found to be > 1. CNO metals are clearly present, although their abundances are not known, and these objects are certainly degenerate with log g > 7. Four of the eleven discovered objects (Table II) exhibit pulsations.

4. X-RAY OBSERVATIONS OF PG1159-035 STARS

4.1. Observation details

Five PG1159 objects were observed by EXOSAT. Two observations of both PG1159-035 and K1-16 were made, the second of each being of long duration to study X-ray pulsations. Simultaneous optical data were obtained for the long PG1159-035 observation with the Steward observatory 1.5m telescope by A.D. Grauer, as described by Barstow et al (1986). Table II summarises the mean EXOSAT CMA count rates in the 3000Å lexan (3Lx) and aluminium-parylene (Al/P) filters.

Table II. Summary of PG1159 stars and observed X-ray count rates

Star	m_V	X-ray Count Rate [a] (cs^{-1})		Ratio Al/P:3Lx
		3Lx	Al/P	
PG1159-035	14.84	0.047 ± 0.004	< 0.0015	< 0.033
K1-16	15.09	0.019 ± 0.0006	0.0019 ± 0.0006	0.10
PG1144+005	15.04	0.0124 ± 0.0016	< 0.0032	< 0.26
(Basile and Holberg, 1985)				
KPD0005+5106	13.32	0.048 ± 0.003	0.0039 ± 0.0013	0.081
(Downes et al, 1985)				
VV-47	16.00	< 0.0013		
(Kaler, 1978)				
PG0122+200	16.1	(Wesemael et al, 1985)		
PG1151-029	16.4	(Wesemael et al, 1985)		
PG1424+535	16.2	(Wesemael et al, 1985)		
PG1520+525	15.52	(Wesemael et al, 1985)		
PG1707+427	16.4	(Bond et al, 1984)		
PG2131+066	16.63	(Bond et al, 1984)		

[a] Count rates are averaged over several observations for K1-16 and PG1159-035. All upper limits are 3σ

4.2. X-ray Pulsations

PG1159-035 was seen to exhibit X-ray pulsations. These results have been reported by Barstow et al (1986) but are summarised here. Figure 4 shows the portion of the background corrected (40s time bins) EXOSAT data which coincides with the optical time series. By comparison, the EXOSAT data have a low signal-to-noise ratio. However, subtle correlations are still evident between the X-ray and optical data.

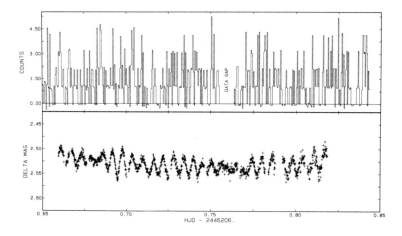

Figure 4. A comparison of the simultaneous portions of the EXOSAT (upper panel) and optical (lower panel) light curves for PG1159-035.

A region of the power spectrum of the entire 1985 EXOSAT data set is shown in Figure 5. Three strong peaks are present, ccorresponding to confidence levels above 99%. A simulation of PG1159-035 X-ray data, using poissonian counting statistics, has been performed incorporating intensity modulations at the observed periods. This has shown that the relative pulsation amplitudes, obtained by folding the data with respect to the observed periods, are consistent with the levels of the peaks in the observed power spectrum and demonstrates that 99% is an acceptable confidence level.

Two of the largest peaks in Figure 5 correspond to modes previously indentified in the optical power spectra of PG1159-035 (Winget et al, 1985); the 516 ± 1s and 540 ± 1s peaks. In addition, a new peak is seen at 524 ± 1s, with an amplitude comparable to those at 516s and 540s. Although the peak is not found in the optical data from 1979 to 1984, (with a semi-amplitude limit of $< 10^{-3}$) it does appear in the extensive March-May 1985 data set (Kepler et al, 1987). The X-ray and optical pulsations are in phase, indicating that the radiation travel time between the two regions where these fluxes arise is less than a few seconds.

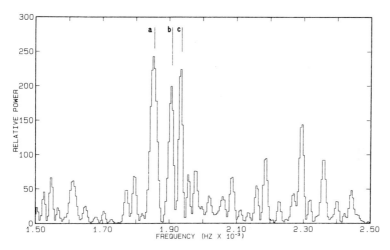

Figure 5. A power spectrum of the entire 16hr EXOSAT time series. The lines designated a, b, and c correspond to the locations of the 539, 524 and 516s optical periods.

Six more periods are present in the optical power spectrum that are undetected in the X-ray data. These absences may be due either to a relatively low sensitivity in the X-ray observation or physical selection of modes due to limb darkening of the X-ray flux. The amplitudes of X-ray and optical pulsations should have similar relative amplitudes, as they are both caused by temperature fluctuations and the instrument band passes are constant. Upper limits to the undetected X-ray amplitudes were determined by comparing simulated modulations with the observed power spectrum. Table III summarises the optical and X-ray amplitudes (and upper limits) and X-ray/optical amplitude ratios. Taking the ratios for the detected X-ray periods as representing the range expected the remaining upper limits provide some useful information. The ratio limits at 390, 424.4, 645.2 and 831.7s are close to the expected values and, if present in the X-ray data, would be below the instrument sensitivity. However, the 451.5 and 495.0s ratio limits are less than half the expected value and should have been detected by EXOSAT. The absence of these modes could be a result of selection by limb darkening. If so, detailed modelling of the behaviour of the emergent optical and X-ray fluxes when a star pulsates may shed light on actually what pulsation harmonics (denoted by l, k and m) are being excited. Kawaler (1986) indicates, from analysis of period spacings, that PG1159-035 is an l=1 or l=3 (or both) pulsator with high radial (high k) overtone modes. The analysis suggested above would be a useful independent test of his conclusion.

Table III Comparison of the amplitudes of optical and X-ray pulsations in PG1159-035

Period (s)	Optical semi-amplitude	X-ray amplitude	Ratio X-ray:Optical
390.0	0.002	< 0.05	< 25
424.4	0.0031	< 0.05	< 16
451.5	0.0066	< 0.05	< 8
495.0	0.0058	< 0.05	< 9
516.0	0.0092	0.15	16
524.93	0.0052	0.16	31
538.9	0.0083	0.17	20
645.2	0.0022	< 0.05	< 23
831.7	0.0036	< 0.05	< 14

4.3. Effective Temperatures

Determination of the effective temperatures of these stars from the X-ray photometry requires an appropriate model atmosphere to predict the emergent flux to be folded through the instrument response. Here there is some difficulty. First, as already discussed, their composition is poorly determined. Second, few high temperature WD model atmospheres bear even a passing resemblance to the constituents of these stars. The two closest models available are the pure He atmospheres of Wesemael (W, 1981) and the planetary nebulae nuclei of Hummer and Mihalas (H&M, 1970). Each has drawbacks. For W, although He/H is > 1, the real abundance of H may not be zero and CNO are not considered. H&M do incorporate CNO at solar abundances but He/H is too low (0.16). The C, N and O K absorption edges, above which energies these elements give their highest contribution to the atmospheric opacity, all lie outside the principal energy band of EXOSAT that is sensitive to PG1159 stars. In addition, the absorption cross section of C is similar to that of He at energies below its K edge. If the abundances of N and O are quite low the pure He atmosphere may represent a good approximation.

Using the pure He model (taking log g=8.0, although note that fluxes are insensitive to log g), count rates were predicted for each EXOSAT filter as a function of T and N_H, with the spectra normalised to the optical V magnitude of each star. From this data, curves of constant count rate can be generated in the T/N_H plane (Figure 6). For K1-16 and KPD0005+5106, where Al/P count rates were measured, single valued solutions are obtained where the filter curves meet - yielding T = 127000 ± 2000K, N_H = 1.7 ± 0.2 × 10^{20} cm^{-2} and T = 122000 ± 2000K, N_H = 1.2 ± 0.2 × 10^{20} cm^{-2} respectively. The estimate of T for KPD0005+5106 is somewhat higher than the 80000K derived from UV and optical data (Downes et al, 1987). The Al/P upper limit for PG1144+005 determines a range of solutions corresponding to T = 116000 − 13300K and N_H = 6.3 × 10^{19} − 3.6 × 10^{20} cm^{-2}. There is no consistent solution for PG1159-035, but taking the point of closest match between the Al/P and 3Lx data T is ≈ 130000K and N_H ≈ 1.9 × 10^{20} cm^{-2}. All the estimated columns are within the galactic values of Heiles (1975). A useful upper limit, of 133000K, to the effective temperature of VV-47 can be generated, from the galactic column limit (≈ 4.5 × 10^{20} cm^{-2}), provided the source has no intrinsic absorption.

Clearly the results of the He model should be treated with some caution, although a fairly consistent picture does arise. While the approximation may be appropriate for K1-

16, KPD0005+5106 and PG1144+005 this does not seem to be the case for PG1159-035. If the X-ray pulse amplitude is interpreted as a temperature change, the half amplitude determined with the He model is 600K compared to a value of 1100K implied by the optical amplitude, underlining the problem. PG1159-035 must have H, N or O in quantities that modify the flux significantly. Addition of H to the atmosphere will increase the emergent flux at a given temperature, lowering the temperature estimate whereas N or O would have the opposite effect.

Figure 6. T/N_H curves of constant count rate for the detected PG1159 objects. The solid line is for the 3Lx filter and the dashed line is Al/P. Arrows mark upper limits.

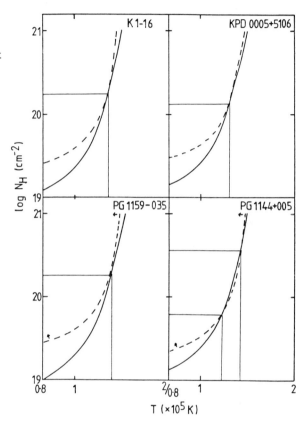

5. CONCLUSION

The presence of pulsations in WDs has been discussed in general terms incorporating a qualitative description of the theory of pulsation and the driving mechanisms responsible in these systems. Particular attention has been paid to the group of objects containing the hottest DVs - the He rich PG1159 objects - which are well suited to soft X-ray studies. Recent observations of these objects have been summarised, discussing the detection of soft X-ray pulsations (photospheric in origin) in PG1159-035 and measurement of effective temperatures. These objects are potentially extremely interesting in the context of white dwarf evolution but are not yet supported by appropriate theoretical models. Their development must be the next step in this field.

6. ACKNOWLEDGMENTS

The author would like to thank Dr J B Holberg for useful discussions during preparation of this review. The author acknowledges the financial support of the SERC through a Research Associateship and of this NATO ASI.

7. REFERENCES

Barstow, M.A., Holberg, J.B., Grauer, P.D. and Winget, D.E., 1986, *Ap.J.*, **306**, L25.
Basile, J. and Holberg, J.B., 1985, *Bull.A.A.S.*, **17**, 838.
Bond, H.E., Grauer, A.D., Green, R.F. and Liebert, J., 1984 *Ap.J.*, **279**, 751.
Cox, A.N., Starrfield, S.G., Kidman, R.B. and Pesnell, W.D., 1987, *Ap.J.*, **317**, 303.
Cox, J.P., 1980, **Theory of Stellar Pulsation**, *Princeton University Press*.
Downes, R.A., Liebert, J. and Margon, B., 1985, *Ap.J.*, **290**, 321.
Downes, R.A., Sion, E.M., Liebert, J. and Holberg, J.B., 1987, in press.
Grauer, A.D. and Bond, H.E., 1984 *Ap.J.*, **277**, 211.
Heiles, C., 1975, *Ap.J.Suppl.*, **20**, 37.
van Horn, H.M., 1975, in **Multiple Periodic Variable Stars**, ed. Fitch, 259.
Hummer, D.G. and Mihalas, D., 1970, *Mon.Not.R.astr.Soc.*, **147**, 339.
Kaler, J.B., 1978, *Ap.J.*, **266**, 947.
Kawaler, S.D., 1986, IAU Symp. No. 123, Advances in Helio and Asteroseismology, in press.
Kawaler, S., 1987, in IAU Colloquium 95, **The Second Conference on Faint Blue Stars**, in press.
Kepler, S.O., Winget, D.E., Grauer, A.D., Cropper, M., O'Donoghue, D., Holberg, J.B. and Barstow, M.A., 1987, in preparation.
Koester, D., Weidemann, V. and Vaudair, G., 1983, *Astron.& Astr.*, **123**, L11.
Liebert,J., Wesemael,F., Hansen,C.J., Fontaine,G., Shipman,H.L., Sion,E.M., Winget,D.E., and Green,R.F., 1986, *Ap.J.*, **309**, 241.
McGraw, J.T., Starrfield, S.G., Liebert, J.and Green, R.F., 1979, in IAU Colloquium 53, **White Dwarfs and Variable Degenerate Stars**, ed. H.M. van Horn and V. Weidemann, 377.
Robinson, E.L., 1979, in IAU Colloquium 53, **White Dwarfs and Variable Degenerate Stars**, ed. H.M. van Horn and V. Weidemann, 343.
Sion, E.M., 1986, *Publs.astr.Soc.Pac.*, **98**, 821.
Sion, E.M., Liebert, J. and Starrfield, S.G., 1985, *Ap.J.*, **292**, 471.
Starrfield, S., Cox, A.N., Kidman, R.B. and Pesnell, W.D., 1984, *Ap.J.*, **281**, 800.
Unno, W., Osaki, Y., Ando, H. and Shibahashi, H., 1979, **Non-Radial Oscillations of Stars**, University of Tokyo Press.
Warner, B. 1975, in **Multiple Periodic Variable Stars**, ed. Fitch, 247.
Wesemael, F., 1981, *Ap.J.Suppl.*, **45**, 177.
Wesemael, F., Green, R.F. and Liebert, J., 1985, *Ap.J.Suppl.*, **58**, 379.
Winget, D.E., van Horn, H.M., Tassoul, M., Hansen, C.J. and Fontaine, G., 1983, *Ap.J.*, **268**, L33.
Winget, D.E., Kepler, S.O., Robinson, E.L., Nather, R.E. and O'Donoghue, D., 1985, *Ap.J*, **292**, 606.
Winget, D.E., Robinson, E.L. and Nather, R.E., 1982, *Ap.J.*, **262**, L11.
Winget, D.E., Robinson, E.L., Nather, R.E. and Balachandrar, S., 1984, *Ap.J.*, **279**, L15.

RADIATION FROM GAS ENVELOPES AROUND Be STARS

Krishna M.V. Apparao and S.P. Tarafdar
Tata Institute of Fundamental Research
Homi Bhabha Road
Bombay 400005
India

Be stars [1] are in the mass range 5-20 $M_\odot$ and have a surface temperature between 10,000 - 30,000 °K. They show large rotational velocities - between 100 and 500 km.s^{-1}. These stars sometimes have emission lines. These emission lines build up in intensity, persist for a while and disappear gradually. The star then becomes a B star with normal absorption lines. This phenomenon repeats in a quasi-periodic fashion with periods of a few years to a few tens of years. The emission lines are mostly Balmer lines, but sometimes also contain He I and low excitation metal lines. The appearance of the emission lines is interpreted as due to formation of a gas envelope around the Be star - the gas envelope is in the equatorial region in the form of a ring. The formation of the ring is related to the high rotation velocity - destabilisation forces [2] lead to ejection of a gas ring from the equatorial regions.

The line emission comes from the gas envelope. The density of hydrogen atoms is estimated to be between 10^{10} - 10^{13} cm^{-3}. Optical depths for absorption of UV radiation is large and case B situation applies. Also due to the high density in the envelope, collisional excitation becomes important.

BALMER LINE EMISSION FROM Be STARS [3]

The Lyman continuum absorption by the gas in the envelope forms an ionized region (Stromgren Sphere or H II region). If we assume a gas ring separated from the Be star, this H II region is confined to a thin inner portion ($\lesssim 10^{10}$ cm) compared to the dimension of the ring ($\sim 10^{12}$ cm). The calculated energy in H_α emission

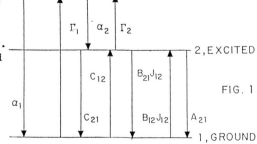

FIG. 1

from the H II region for various spectral types is shown in Table 1. The Lyman continuum values are obtained using Kurucz atmosphere calculations. The density used is 10^{12} cm^{-3}. Table 1 also shows the highest observed value of H_α energy for each spectral type. It is seen that the calculated values are lower than the observed values. It is sometimes suggested [6] that, due to the high rotation of Be stars, there could be an underestimate of the spectral types by about 1.5 numbers. There could also be errors in distance estimation. All these errors cannot amount to more than a factor of ten in the observed values. It is seen that the discrepancy between calculated and observed values persists for later Be spectral types.

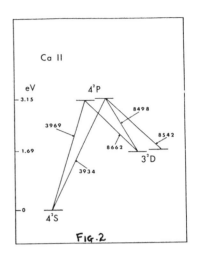

FIG. 2

It is earlier suggested [7] that absorption of Balmer continuum can enhance the ionization. The Lyman photons emitted during recombination can keep an excited (level 2) population of hydrogen. Balmer photons from the star can then be absorbed and ionize from the excited level. We [3] have calculated this ionization enhancement by considering the relevant radiation transfer equations together with a two stream approximation. The calculated H_α emission from the enhanced ionized region (produced by absorption of Lyman and some Balmer photons) is given in Table 1. It is seen that even this process is inadequate to account for the observed H_α emission for the later types (beyond B5). It seems additional ultraviolet photons, other than from the stellar surface emission are needed. The number of UV photons of energy greater than 13.6 eV needed to account for the observed H_α emission are 2×10^{44} s^{-1} and 8×10^{44} s^{-1} for the B8 and B5 stars respectively, corresponding to energies of 4×10^{33} and 1.6×10^{34} ergs s^{-1}; these values are upper limits. A possible origin of these UV photons is coronal emission. Note also the required energy is in the range of observed X-ray emission in a number of Be star X-ray source binaries.

CALCIUM LINE EMISSION FROM Be STARS [8]

Several Be stars are observed [9] to emit the infrared Calcium II triplet λ8498, λ8542, λ8662. The energy in the lines to that in the continuum is about 1. The observed intensity ratio of the lines is 1:1:1, while the expected recombination ratio is 1:9:5. Also the expected Ca II H and K lines are not observed. The above facts are explained on the basis of a large optical depth for these lines which

Table 1

H$_\alpha$ Line Emission from Be Stars

Temperature °K	Spectral Type	Lyman photons LOG Q_L s^{-1}	Balmer photons* LOG Q_B s^{-1}	Observed+ H$_\alpha$ Emission ergs s^{-1}	Calculated H$_\alpha$ emission ergs s^{-1}	
					Lyman photons	Lyman + Balmer photons
25000	B1	45.78	48.37	2.5 x 10^{34}	8.2 x 10^{33}	2.7 x 10^{34}
20000	B3	44.24	47.83	4 x 10^{33}	2.4 x 10^{32}	5.8 x 10^{32}
16000	B5	42.42	47.12	8 x 10^{32}	3.6 x 10^{30}	8.3 x 10^{30}
13000	B8	40.71	46.53	5 x 10^{32}	7.0 x 10^{28}	1.7 x 10^{29}

* Thompson (1984), using Kurucz atmosphere.

+ Ashok et al. (1984), see text.

is obtained in the envelopes around Be stars. The ratio of Calcium infrared line emission energy to that in the continuum is observed to be 1. However, the Ca II emission, which we calculated, from the H II region produced by the Be star is insufficient to explain the observed values by orders of magnitude (see Table 2). We have suggested [4] that the emission occurs beyond the H II region, which we call the C II region. The ultraviolet photons from the Be star ionize C I atom (ionization potential 11.26 eV) and Ca II atoms (ionization potential 11.87 eV). The electrons for recombination of Ca III ions are from ionization of C I. The extent of the C II region is simple to calculate and is given in Table 2 (in brackets). In the case of early type of stars the region calculated is larger than the typical dimension of the gas ring. We have adopted a size of 5×10^{12} cm for the extent of the ring.

We have considered the radiation transfer of the Ca II lines (see the energy level diagram) and calculated the emission intensity of the infrared lines. All the parameters are known except the collisional deexcitation C_1 rate of level 2. We have considered the line emission for $C_1 = 0$ and $C_1 = 10^3$ and the values of the ratios of the emission in the lines to that in the continuum are given in Table 2. The values for $C_1 = 10^3$ agree with observed values of about 1.

EFFECT OF PRESENCE OF COMPACT OBJECTS [10]

Several Be stars are found to be binaries and strong X-ray emission indicates the presence of a compact object like a white dwarf or a neutron star. The compact object accretes matter from the gas

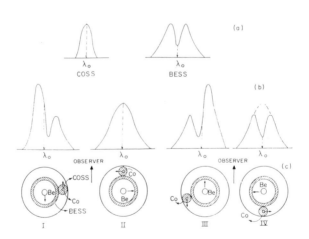

Fig. 3

TABLE 2

C II Region and Ca II triplet strength

Spectral Type	Temperature °K	Radius (cm)	Extent of H II region* (cm)	Dimension (cm) of C II region		Ratio of Ca II triplet strength to continuum from		
				Density bound	(Ionization bound)	H II region	C II region Density bound	
							$C_1=0$	$C_1=10^3$
B1	25000	4.41+11**	1.34+9	5.0+12	(3.1+13)	2.8-3	2.8	2.8
B3	20000	3.22+11	5.32+7	5.0+12	(1.8+13)	1.5-4	1.4	1.4
B5	16000	2.36+11	8.05+5	5.0+12	(8.1+12)	5.9-6	0.6	1.0
B8	13000	1.97+11	1.57+4	4.7+12	(4.7+12)	2.2-7	0.2	1.1

*Thickness of the H II region, assuming it is at a distance of 10^{12} cm from the star.

**a+n means a x 10^n

envelope and emits X-rays. The X-radiation ionizes the gas around and forms its own H II region. In the case of the X-ray source A 0538-66 in the Large Magellanic Cloud the intensity of X-ray emission is about 10^{39} ergs s^{-1} during its peak and the whole gas envelope around the Be star is ionized except that shadowed by the Be star. The H II region produced by the compact object (COSS) emits its own line emission. The observed line emission is the combined emission from the Be star H II region (BESS) and COSS and leads to the observed so-called V/R variation.

The emission from BESS is in the form of a double hump emission (see Fig. 3). It is observed that at times the violet hump is larger than the red hump, which gradually decreases and then the red hump increases. The ratio of emission in the violet to that in the red hump (V/R), is found to vary in a quasi periodic fashion. This can be explained, if we consider the combination of BESS and COSS emission (Fig. 3). The COSS region follows the compact object in its binary motion around the Be star. When the compact object approaches the observer the COSS emission is shifted towards the violet and when it is receding from the observer it is shifted to the red. Thus COSS emissions alternately contribute to the violet and red humps causing the V/R variation.

REFERENCES

1. V. Doazan, B stars with and without Emission Lines, Eds. A. Underhill and V. Doazan, NASA-456, 1982.
2. K.M.V. Apparao, H. Antia and S.M. Chitre, Astr. Astrophys.
3. K.M.V. Apparao and S.P. Tarafdar, Ap. J. (In press)
4. R.I. Thompson, Ap.J., 283, 165, 1984
5. N.M. Ashok et al., M.N.R.A.S., 153, 471, 1984
6. A. Slettebak, Sp. Sci. Rev., 23, 541, 1979
7. J. Scargle et al., Ap. J., 224, 527, 1978
8. K.M.V. Apparao and S.P. Tarafdar, submitted to Astr. Astrophys.1987
9. R.S. Polidan, I.A.U. Symposium No.70, 'Be and Shell Stars', Ed. A. Slettebak, D'Reidel Publishing Co., Holland, 1976.
10. K.M.V. Apparao and S.P. Tarafdar, Astr. Astrophys. 161, 271, 1986

5. SUPERNOVA REMNANTS AND THE HOT INTERSTELLAR MEDIUM

X-RAY OBSERVATIONS OF HOT THIN PLASMA IN SUPERNOVA REMNANTS

B. Aschenbach
Max-Planck-Institut für Physik und Astrophysik
Institut für extraterrestrische Physik
8046 Garching, W-Germany

ABSTRACT. An overview is given of X-ray observations of supernova remnants carried out on the Einstein, Tenma und EXOSAT satellites as well as from a few sounding rocket experiments. Our current interpretation of images, energy spectra and the first few spectrally resolved images is discussed.

1. INTRODUCTION

In 1963, only one year after the discovery of the first non-solar cosmic X-ray source, the Crab Nebula became the first object identified with an X ray emitting supernova remnant, i.e. the debris from a stellar catastrophic explosion. Until 1976 X-rays from additional 7 supernova remnants had been detected, including Cas-A, Tycho, IC443, Puppis-A, the Cygnus Loop and Vela (Gorenstein and Tucker, 1976). They are found at distances ranging from 500 pc to 3 kpc and they are observed as extended sources with angular diameters from a few arcminutes like Cas-A to ~ 5 degrees like the Vela remnant, and even more in case of the galactic loop structures. These angular diameters correspond to linear diameters from a few parsecs to about 50 pc. With X-ray luminosities of ~ 10^{34} erg/s to ~ 10^{38} erg/s supernova remnants represent the second brightest class of galactic X-ray sources following the high luminosity binaries.

As of today 40 galactic X-ray supernova remnants are known. Outside the Galaxy 32 remnants have been found in the Large Magellanic Cloud and 6 in the Small Magellanic Cloud, respectively (Matthewson et al., 1985). It was the Einstein observatory with its imaging and spectroscopic capability through which substantial progress has been made. An extensive compilation of papers dealing with the Einstein observations is contained in the proceedings of the IAU Symposium No. 101 (Danziger and Gorenstein, 1983). Since then further contributions have come from the EXOSAT observatory (Aschenbach, 1985), the Tenma satellite and a few sounding rocket experiments. During these years X ray instrumentation has evolved to such a state that detailed studies of images, energy spectra and time variability of individual remnants can be made.

2. IMAGES

High quality images became available first from the Einstein observatory with an angular resolution of typically 10 arcsec. The brighter remnants have subsequently been imaged by the smaller EXOSAT telescopes with less resolution. These images have been used to support the view of at least two main classes of remnants, which are the Crab like remnants and the shell-type remnants. In his review Seward (1985) defines the Crab like remnants as extended objects with a non-thermal X-ray spectrum and some evidence for a central energy source like an active pulsar. Clearly, this class of remnants is intimately connected to neutron star research and a review on this topic has been given by Helfand and Becker (1984). More relevant in the context of this conference on hot thin plasma is the second class of remnants which is that of shells with a thermal spectrum. Unlike the Crab-like remnants most likely energized by a central compact object, the shell type remnants emit radiation from an optically thin plasma, which has been heated to X-ray temperatures by a shock wave. This shock wave may either be the blast wave associated with the stellar explosion (Heiles 1964), which interacts with the ambient interstellar or circumstellar material, or it may be a reverse shock wave (McKee 1974), which propagates inwards from the decelerated blast wave and raises the temperature of the stellar ejecta. In both cases the X-ray emitting shell (or shells) is expected to be perfectly circular and to have no spatial structure in surface brightness except a radial gradient for a spherically symmetrical explosion and homogeneous media. The X-ray images, however, have revealed quite the opposite with a great deal of structure present in all remnants. The young remnants Cas-A, Kepler, Tycho and SN1006 are still maintaining an approximate circular shape whereas older ones like Puppis-A are hard to be reconciled with a shell at all (see figure 1).

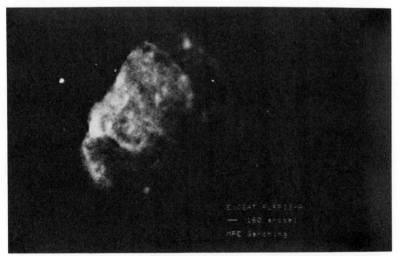

Fig. 1. EXOSAT soft X-ray image of Puppis-A.

Since X-ray emission scales with the square of density it is obvious to consider density variations within or across the remnant as the dominant source for the observed X-ray brightness structures. Large scale uniform density changes like the interstellar density gradient perpendicular to the galactic plane have been made responsible for the asymmetry in galactic latitude, observed in Puppis-A for instance (Petre et al. 1982). The increased density towards the galactic plane gives rise to a higher X-ray surface brightness in the same direction as well as a smaller extent due to the increased slowing-down of the blast wave. If a remnant has grown to such a size the blast wave has passed quite a number of individual cool, high density interstellar clouds of different sizes, as well as their warm, medium density envelopes, which are embedded in the general hot, low density intercloud medium (McKee 1981), and these structures will directly be imaged in X-rays if heated to the required temperatures.

The detailed X-ray appearance of clouds in a low density medium depends on the physics of the interaction with the shock wave. Cowie and McKee (1977) and McKee (1981) have considered mass evaporation and hydrodynamic ablation to increase the density in the adjacent intercloud medium thus raising the X-ray emission. An alternative approach to increase the X-ray emission near dense clouds has been proposed by Hester and Cox (1986). Led by an analysis of the X ray and optical emission of the Cygnus Loop, they have proposed an additional compression of the down-stream X-ray emitting plasma by reflected or bow shocks around dense clouds. It is interesting to note (McKee 1981), that evaporation dominated remnants are affected on a large scale by the mass transfer from the clouds to the intercloud medium. In this case pressure and density will increase radially inwards contrary to the classical Sedov solution. Finally, non-thermal contributions to the X-ray emission may be not negligible in regions of electron acceleration and magnetic field amplification associated with the shock waves (Reynolds and Chevalier 1981).

3. ENERGY SPECTRA

The thermal origin of the emission from shell type supernova remnants was for the first time established unambiguously by the detection of the Fe-K emission line in the spectrum of Cas-A (Serlemitsos et al. 1973). However, the clear detection of additional emission lines at lower energies from other atomic species was severely hampered by the low spectral resolution of the collimated proportional counter detectors. This changed with the advent of the solid state spectrometer (SSS) used in the focus of the Einstein observatory X-ray telescope. With an improved resolution of about 160 eV over the nominal 0.5-4.5 keV detection band, numerous emission lines from highly ionized Mg, Si, S, Ar, and Ca have been measured in the spectra of young supernova remnants including Cas-A, Kepler and Tycho (Holt 1983). Additional lines from highly ionized N, O and Ne, as well as from Fe XVII have been discovered with the Einstein Bragg Focal Plane Crystal Spectrometer (FPCS) in the older remnant Puppis-A. With a resolving power of 100-1000, the forbidden, intercombination and recombination lines of the He-like triplets from O VII and Ne IX have been resolved for the first time (Winkler et al. 1981). The spectral survey of emission lines from Puppis-A, which is the most detailed result of non-solar X-ray spectroscopy so far, is displayed in figure 2. Winkler et al. (1983) have pointed out in a very clear way how the various line ratios can be used to perform detailed plasma diagnostics in determining electron temperature, ion population, ionization temperature and ionization time.

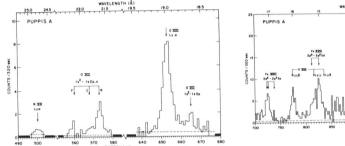

Fig. 2. Einstein FPCS spectra of Puppis-A (from Winkler et al. 1981).

The Einstein energy band extended up to about 4.5 keV and therefore the Fe-K line emission as well as the high energy continuum, whose existence which was known from previous experiments, could not be covered simultaneously. Energy spectra over the 2-10 keV range with a spectral resolution better than the collimated proportional counter by typically a factor of 2 have become available from the non-imaging gas scintillation proportional counters (GSPC) flown on board of the EXOSAT and Tenma satellites.

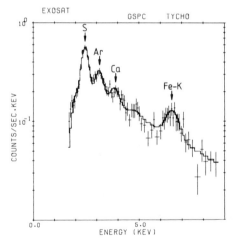

Fig. 3. EXOSAT GSPC spectrum of Tycho. The fit assumes a thermal bremsstrahlung continuum of kT=6.5 keV and emission lines of S, Ar, Ca and Fe superimposed (from Smith et al. 1987).

Figure 3 shows the EXOSAT GSPC spectrum of Tycho (Smith et al. 1987). The emission lines from transitions of He-like S, Ar, Ca and Fe ions are clearly resolved. The spectrum above about 5 keV is dominated by the Fe-K line and the continuum, the latter of which has been used to determine the electron temperature. Similiar spectra are available for Cas-A, Kepler, RCW 103 and W49B and a detailed account of the EXOSAT spectra has been given by Smith (1987). Tsunemi et al. (1986) have published the

Tenma GSPC energy spectra of Cas-A and Tycho which extend to some lower energies to include the Si-K lines.

The emission from shell type remnants is considered to originate from an optically thin plasma, which has been heated to X-ray temperatures by shock waves. Model fits to the measured spectra attempt to determine X-ray temperature, total emission measure and elemental abundances, from which the density of the pre-shocked interstellar medium, the total swept-up mass and the explosion energy of the supernova can be derived using some 3-dimensional geometry and hydrodynamical solution for the expansion (see for instance Gorenstein and Tucker, 1976). If the contributions from the interstellar medium and the stellar ejecta can be disentangled, density and mass of the ejecta can be determined. In this way a link from the remnant to the progenitor star and the supernova type can be established. Early fits to the spectra have been based on the following assumption:

1. the emitting plasma is in collisional equilibrium ionization;
2. there is thermal equilibrium between ions and electrons implying the same temperature for both;
3. the electrons have a Maxwellian velocity distribution;
4. the plasma parameters including temperature, abundances, ionization stages, etc. are homogeneous over the entire remnant, at least over the field of view of the instrument;
5. the hydrodynamical evolution of the remnant can be described by the self similarity Sedov solution, and
6. the remnant expands into an ambient homogeneous medium.

The general results of these early analyses, which hold for almost any well studied remnant, can be summarized as follows:

1. The spectrum is well described by the superposition of the emission from a two component plasma characterized by a low temperature of about 0.2 - 0.5 keV and a high temperature of a few keV;
2. a third, very high temperature component of about 30 keV has been suggested to be present in Cas-A (Pravdo and Smith 1979) which is, however, inconsistent with the EXOSAT data (Jansen et al. 1987). For Tycho, Pravdo and Smith derive a similar high temperature component which could not be confirmed by EXOSAT because of inadequate sensitivity (Smith et al. 1987);
3. elemental abundances from oxygen burning nucleosynthesis products including Si, S, Ar and Ca are largely overabundant compared with solar or cosmic values, whereas Fe tends to be less than solar;
4. even young remnants contain a large X-ray emitting mass. For instance, Fabian et al. (1980) have found more than 15 solar masses in Cas-A, and Reid et al. (1982) estimate 15 solar masses for Tycho as well.

The validity of each of the six assumptions listed above has been questioned very early on mainly from theoretical arguments. Now, there is growing observational evidence supporting more or less substantial modifications of the assumptions with subsequent revision of the parameters derived, the amount of which depends on age and environment of the individual remnant.

As pointed out by Gorenstein et al. (1974) the plasma in supernova remnants may not be in collisional ionization equilibrium because the time scale to ionize the heavy elements to the equilibrium level by electron collisions is of the order of $10^4/n_e$ years (Canizares 1984). For electron densities ne of about 1 - 10 cm^{-3}, this is large compared with the age of young remnants and has the effect that the heavy elements are underionized compared with their equilibrium population. Direct observational evidence for non equilibrium ionization (NEI) has been obtained from the analysis of the high resolution FPCS spectra of regions in Puppis-A (Canizares et al. 1983, Winkler et al. 1983, Fischbach et al. 1987) and in Cas-A (Markert et al. 1987). The NEI conditions are derived from the weakness of the forbidden lines relative to the resonance lines of OVII and NeIX (see figure 2). Vedder et al. (1986) have recently published the FPCS spectrum taken in the northern bright region of the Cygnus Loop. From the forbidden to resonance line ratio of OVII they conclude that even in a remnant as old as the Cygnus Loop at least sections exist where collision equilibrium ionization has not been reached. However, Gabriel et al. (1985) have pointed out that fast electrons in an otherwise thermal plasma can mimic NEI conditions because they tend to excite preferentially the resonance line, and therefore a weak forbidden to resonance line ratio may not be conclusive. They compute that 20 keV electrons having a proportion of 1 % of the total number of electrons are sufficient to explain the OVII lines as observed in Puppis-A.

Further evidence for NEI conditions in Tycho and Cas-A has been obtained from the EXOSAT (Jansen et al. 1987, Smith et al. 1987) and Tenma (Tsunemi et al. 1986) observations of the Fe-K line. Both experiments have measured consistently a line energy significantly lower than the value expected from CEI which is based upon a temperature derived from the high energy continuum. Applying NEI models both experiments agree that Tycho is substantially more underionized than Cas-A, which is plausible from the higher density in Cas-A.

Less direct but nonetheless evidence for NEI is deduced from the fact that NEI models fit the observed X-ray spectra better than CEI models. NEI models have been constructed by numerous authors to describe time dependent ionization (Itoh 1977, 1979, Gronenschild and Mewe 1982, Shull 1982, Hamilton et al. 1983, Hamilton and Sarazin 1984, Nugent et al. 1984). The gross effect of NEI is an enhanced emission from lower ionization stages, which mimic a separate low temperature CEI plasma in addition to the temperature indicated by the high energy continuum. This is the reason that early X-ray spectra could be approximated by two temperature CEI models. However, with increasing better spectral resolution and broader energy coverage the fits became increasingly poorer. In order to explain the strong line emission from heavy elements observed in young remnants, a significant overabundance compared to solar values had to be adopted. Furthermore high emission measures implying high densities were needed to explain the high level of soft X-ray emission, which in turn led to high masses for the remnants. NEI models have recently been applied to some remnants with the result that the CEI based estimates about the total X-ray emitting mass have been refined. Using NEI emission to model the surface brightness distribution observed with the Einstein HRI, Gorenstein et al. (1983) derive a total of about 4 solar masses for Tycho shared equally the ejecta and swept up matter. This has to be contrasted with the result of about 15 solar masses which Reid et al. (1982) deduced from the

Einstein IPC image and a CEI model. An even lower value of only 0.6 solar masses for the X-ray emitting mass of Tycho has been presented by Tsunemi et al. (1986), which they conclude from a NEI analysis of the Tenma energy spectrum. Although the mass has come down significantly the heavy elements are still a factor of 6-15 overabundant compared with solar values including iron. The Tenma spectrum of Cas-A has also acceptably been fitted with a single component NEI model with a remarkable low value of 2.4 solar masses and element abundances very close to solar values. The Tenma analysis, however, is in conflict with the EXOSAT results analyzed by Smith et al. (1987) and Jansen et al. (1987) who claim that the spectra cannot be explained by a single component but a two component NEI model. It is interesting, that the Tenma results, for the first time, are clearly in line with respect to element abundances and mass with what is currently predicted from supernova explosion models advocating a type I for Tycho and type II for Cas-A.

The NEI models discussed above assume that the hydrodynamic evolution of the remnant can be described by the self similarity solutions of the Sedov type. Hughes and Helfand (1985) have instead used a numerical hydrodynamic shock code into which the time dependent ionization equations have been incorporated. Thus, the time dependent ionization structure can be computed simultaneously in both the blast wave heated ambient plasma as well as the reverse shock heated ejecta. Interestingly, they have found that the surface brightness distributions and the spectra of Kepler's supernova remnant as measured with the Einstein observatory instruments can be equally good reproduced if the emission originates predominately either from the heated interstellar medium with the remnant in the Sedov phase or from the heated ejecta.

In summary, in recent years progress in spectral modelling by involving NEI in particular and better fits have been obtained with generally lower masses and lower elemental abundances have been obtained compared with CEI models. However, there seems to be no conclusion on the number of temperature components present in young remnants. It is striking that the analysis of relatively narrow band spectra like that by Tsumeni et al. for Tycho and Cas-A and that by Hughes and Helfand for Kepler favour a dominant single component. In contrast to this, the analysis of broad band spectra including the Fe-K line complex like that of Jansen et al. for Cas-A and Smith et al. for Tycho and that of Hamilton et al. for Tycho as well clearly require two NEI components which are associated with plasma heated by the blast wave and the reverse shock.

4. SPECTRALLY RESOLVED IMGAGES

Spectrally unresolved images obtained with the Einstein HRI or IPC, or the EXOSAT CMA instrument, or even the earlier collimated scanning counters demonstrated a great deal of spatial structure to be present in many remnants. Structure like this may originate from different kinds of variations across the remnant, such as those of emission measure, i.e. density and depth of line of sight within the remnant, temperature, ionization structure, and even interstellar absorption column density towards remnants of significant extent. Clearly, these effects are important to be considered when determining elemental abundances and X-ray emitting mass, and therefore a spectrum spatially integrated over the whole remnant or over substantial portions of the remnant is of limited information (see for instance Brinkmann and Fink 1987).

Spectral variations across the remnant of Cas-A were first discovered in the Einstein IPC data, which show significantly different pulse height spectra in the south-west and the north- east section (Murray et al. 1979). On a sounding rocket flight, the first spectrally resolved image of Cas-A has been obtained with an imaging X-ray telescope and a position sensitive proportional counter in focus (Aschenbach, 1985). Assuming CEI emission, a temperature map has been derived, which shows that the faint southwest region has a temperature in excess of 4 keV, whereas the bright northeast part is between 0.3 and 1 keV. The overall temperature distribution shows two distinct peaks at 0.5 keV and 5.4 keV. The hot component forms an almost complete shell along the outermost boundary at a radial distance of about 3 arcmin from the centre. The interior is rather uniform at the low temperature level. Assuming two spherical shells for the emission region, density and pressure maps have been constructed and the remnant is not found in pressure equilibrium, but with the highest pressure occurring along the outer boundary. Interestingly, the interior which seems to be in pressure equilibrium is separated from the outermost annulus by an about 30 arcsec wide pressure minimum, indicating a deceleration of the blast wave.

The EXOSAT imaging telescopes have been used to take very long exposures of Cas-A and to measure the spectral variations with high statistical accuracy, but with proportional counter type resolution (Jansen et al. 1987). They also find significant CEI temperature variations with the highest temperatures occurring in an outer annulus about 3 arcmin from the centre. Analysing the temperature distribution in radial sectors, the high temperature component is most pronounced in the smooth and faint regions whereas in the clumpy and bright regions the high temperature component appears reduced.

The Einstein IPC, the sounding rocket and the EXOSAT PSD results largely agree on the gross temperature distribution and demonstrate two temperature components in Cas-A one of which is apparently associated with the blast wave indicated by its high temperature, its outermost location and the underlying smooth and faint brightness distribution. As expected the low temperature component is associated with the reverse shock, which has heated the clumpy ejecta located more inside the remnant. So even in a remnant as young as Cas-A significant emission from heated ejecta has been found and this has independently been established by the remnant integrated spectra like those from the SSS, EXOSAT and Tenma as well as the spectrally resolved, but low energy images.

The first spectrally resolved images of the middle-aged remnant Puppis-A have been reported by Pfeffermann et al. (1980) from a sounding rocket experiment. The CEI temperature map with a resolution of 3 arcmin shows a rather uniform temperature of $(2-5) \cdot 10^6$K for the interior, which agrees fairly well with the results from the Einstein high resolution FPCS CEI analysis (Winkler et al. 1981 a, b). In a subsequent paper, however, Canizares et al. (1983) showed that the spectrum taken with the 3 arcmin by 30 arcmin aperture of the FPCS cannot be explained by a CEI plasma but by a still ionizing plasma of an electron temperature in excess of $5 \cdot 10^6$K. The MPI rocket experiment revealed a second high temperature component of more than 10^7K which has been found earlier in spatially unresolved counter spectra. Unlike in Cas-A this component is not associated with the outer periphery but shows up in some pixels at the north eastern rim, along some filaments in the interior, but predominantly in the faint western parts, which is plausible since the region is presumably of lower density of the interstellar medium and thus heated to a higher temperature by the blast wave. Furthermore it was found that the area of the bright eastern knot was by far the coolest part of the remnant, although the statistics were not sufficient to resolve the temperature of the knot itself. Significant spectral variations have also been detected among the 8 fields in Puppis-A observed with the 6 arcmin wide aperture of the SSS (Szymkowiak 1985). These variations are most obvious in the equivalent line widths from the He-like ions of heavy elements and in the soft X-ray part of the spectrum below 1 keV. The attempt to fit two temperature CEI models to the data failed and even fits with a single component NEI plasma are not convincing. Also, the pointings of the high resolution FPCS to various different fields in Puppis indicate spectral variations and possibly different temperatures (Fischbach et al. 1987).

A significant improvement in counting statistics and angular resolution has become possible with the EXOSAT absorption filter spectroscopy. Since the broad band transmission of the Lexan and boron filters used, is X-ray energy dependent, the ratio of the two X-ray fluxes depends on the source spectrum. In case of a CEI plasma of cosmic abundance the filter ratio scales approximately with log T and thus by dividing the two images a CEI temperature map of Puppis-A has been produced with a resolution of 1-arcmin and better (Aschenbach 1985). The map shows a great deal of temperature structure on scales even as small as the angular resolution limit and with temperatures between $5.9 \leq \log T < 7.1$. Beyond the upper limit the filter ratio is insensitive to temperature. There are 3 distinct regions forming the coolest parts within the remnants, which include the two bright eastern and northern knots and the outer section of the southeast elongated patch. It is interesting to note that the CEI X-ray temperature of the bright eastern knot is not uniform but varies between 6.0 and 6.4 for log T. Teske and Petre (1987) have reported optical CCD images of the region of the eastern knot in the forbidden red and green coronal iron lines, and under the assumption that the knot, which is supposed to be an interstellar cloud, is in CEI at a temperature of $\log T = 6.35$ they derive a minimum mass of about 0.1 solar masses with a density in the range of 23 to 49 cm^{-3}. The EXOSAT results show that there is significantly cooler gas in the knot region, and thus they do not support Teske's and Petre's view that the visible FeXIV emission is from cooler inclusions in a still hotter medium.

Spectrally resolved images have been taken also from the old Vela supernova remnant and the Cygnus Loop with the Einstein IPC. For the Cygnus Loop, Ku et al. (1984) have pointed out that CEI best fits show lower temperatures along the limb compared with the centre, and that temperatures anticorrelate with intensity as expected for approximate pressure equilibrium. This view has been supported by Charles et al. (1985) by a detailed spectral study of two 1° wide fields, located at the southern and western boundary of the remnant. Temperature and emission measure variations on scales as small as 4 arcmin have been found, favouring the presence of an inhomogeneous cloudy interstellar medium. Similar results have been obtained for the Vela remnant by Kahn et al. (1985), although the two remnants differ significantly in their overall morphological appearance. However, the observations do not preclude pressure variations as large as a factor of 10.

5. FUTURE PROSPECTS

Since the early years of proportional counters much progress has been made in supernova remnant research due to the availability of high resolution imaging, high resolution spectroscopy and broad band energy coverage. High resolution imaging made visible the cloudy interstellar medium and the clumpy ejecta, and it made possible the first comparative investigations between X-ray, optical and radio morphology; in particular the first detailed observational studies of shocked clouds and possible thermal evaporation. The Einstein FPCS and SSS have shown that non-equilibrium ionization may be present in many remnants, even as old as the Cygnus Loop. However, it seems that the present generation of NEI models does not explain the broad band, high energy spectra as observed by EXOSAT for instance, unless multi NEI models with a sufficient number of free parameters are used. In order to better understand the images and spectra, hydrodynamical shock codes coupled with time dependent ionization are needed with an improved knowledge of the electron energy distribution which determines the ionization structure. These model calculations should take into account the results from the supernova explosion simulations including velocity, density and elemental abundances of the debris to predict in detail the effects of the progenitor star. Similarly, the ambient medium has to be considered into which the remnant expands, including a stellar wind of the progenitor star, and a multi component medium.

A key issue for further progress from an observational point of view is the future availability of spectrally resolved images which cover the energy region from 0.3 to about 10 keV. An energy band as broad as this is needed to disentangle the emission from the ejecta and the interstellar medium and to determine independently elemental abundances for each component.

At present there are 5 missions being planned which carry imaging telescopes, i.e. the German ROSAT (1990) and Spectrosat (1993) missions, the Italian SAX (1992) mission, NASA's AXAF (1995) and XMM (1998) of the European Space Agency. The numbers in parenthesis give the currently envisaged launch dates. Except ROSAT, all other four missions await final approval. ROSAT will take images with high angular resolution (5 arcsec), and extremely high contrast. This will allow to search for compact objects within remnants at an significantly increased level of sensitivity compared with Einstein. With a throughput about 8 times greater than the Einstein HRI and a comparable background level, remnants can be studied to lower surface brightness within and outside our galaxy. Spectrally resolved images with an angular resolution of 20 arcsec and a spectral resolving power of about 2.5 will be taken with the position sensitive proportional counter in the energy band 0.1 - 2.2 keV. This spectral resolution is insufficient to resolve individual lines and to perform detailed plasma diagnostics but it will establish spectral variations across many remnants at an unprecedented level of angular resolution, and it will constrain the non-thermal component in Crab-like and composite remnants. Spectrosat will be a ROSAT follow-up modified by a transmission grating which will increase the resolving power to about 50-100.

At present, it looks as if SAX will be the first mission which will carry imaging telescopes working up to 10 keV, with a spatial resolution of about 1 arcmin and a resolving power of about 10 at the Fe-K line. Very few years before the turn of the millenium, hopefully, the two great observatories AXAF and XMM will be launched into orbit. Both observatories cover the energy band from 0.1 to 10 keV. With AXAF sub-arcsecond imaging will become possible and very high spectral resolution as well, although at moderate throughput. XMM will perform low to high spectral resolution observations with high throughput but at the expense of angular resolution (30 arcsec).

6. REFERENCES

Aschenbach, B., Sp.Sc.Rev. **40**, 447 (1985).
Brinkmann, W. and Fink, H.H., IAU Coll. No. **101**, Roger, R.S. and Landecker, T.L., eds., Cambridge Uni. Press (1987).
Canizares, C.R., Proc. of the Intern. Symp. on "X-ray Astronomy '84", Oda, M. and Giacconi, R., eds., Bologna, 275 (1984).
Canizares, C.R., Winkler, P.F., Markert, T.H., Berg, C., "Supernova Remnants and their X-ray Emission", IAU Symp. No. 101, Danziger, J. and Gorenstein, P., eds., (D. Reidel), 205 (1983).
Charles, P.A., Kahn, S.M., McKee, C.F., Ap.J. **295**, 456 (1985).
Cowie, L.L. and McKee, C.F., Ap.J. **211**, 135 (1977).
Danziger, J. and Gorenstein, P., "Supernova Remnants and their X-ray Emission", Proc. of the IAU Symp. No. 101, Venice (D. Reidel) 1983.
Fabian, A.C., Willingale, R., Pye, J.P., Murray, S.S., Fabbiano, G., M.N.R.A.S. **193**, 175 (1980).
Fischbach, K.F., Canizares, C.R., Markert, T.H., Coyne, J.H., IAU Coll. No. **101**, Roger, R.S. and Landecker, T.L., eds., Cambridge Uni. Press (1987).
Gabriel, A.H., Acton, L.W., Bely-Dubau, F., Faucher, P., Proc. ESA Workshop on "Cosmic X-Ray Spectroscopy Mission", ESA **SP-239**, 137 (1985).
Gorenstein, P., Harnden, Jr., F.R., Tucker, W.H., Ap.J. **192**, 661 (1974).
Gorenstein, P., and Tucker, W.H., Ann. Rev. Astronphys. **14**, 373 (1976).
Gorenstein, P., Seward, F., Tucker, W., "Supernova Remnants and their X-ray Emission", IAU Symp. No. **101**, Danziger, J. and Gorenstein, P., eds., (D. Reidel), 1, 1983.
Gronenschild, E.H.B.M. and Mewe, R., Astron. Astrophy. Suppl. **48**, 305 (1982).
Hamilton, A.J.S. and Sarazin, C.L., Ap.J. **284**, 601 (1984).
Heiles, C., Ap.J. **140**, 470 (1964).
Helfand, D.J. and Becker, R.H., Nature **307**, 215 (1984).

Hester, J.J. and Cox, D.P., Ap.J. **300**, 675 (1986).
Holt, S.S., "Supernova Remnants and their X-ray Emission", IAU Symp. No. **101**, Danziger, J. and Gorenstein, P., eds., (D. Reidel), 17 (1983).
Hughes, J.P. and Helfand, D.J., Ap.J. **291**, 544 (1985).
Itoh, H., Publ. Astron. Soc. Japan **29**, 813 (1977).
Itoh, H., Publ. Astron. Soc. Japan **31**, 541 (1979).
Jansen, F.A., Smith, A., Bleeker, J.A.M., de Korte, P.A.J., Peacock, A., White, N.E., submitted to Ap.J. (1987).
Kahn, S.M., Gorenstein, P. Harnden, Jr., F.R., Seward, F.D., Ap.J. **299**, 821 (1985).
Ku, W.H.-M., Kahn, S.M., Pisarski, R., Long, K.S., Ap.J. **278**, 615 (1984).
Markert, T.H., Blizzard, P.L., Canizares, C.R., Hughes, J.P., IAU Coll. No. **101**, Roger, R.S. and Landecker, T.L., eds., Cambridge Uni. Press (1987).
Matthewsen, D.S. Ford, V.L., Tuohy, I.R., Milk, B.Y., Turtle, A.J., Helfand, D.J., Ap.J. Suppl. **58**, 197 (1985).
McKee, C.F., Ap.J. **188**, 335 (1974).
McKee, C.F., "Supernova: A Survey of Current Research", Proc. NATO ASI Cambridge, Rees, M. and Stoneham, R.J., eds., (D. Reidel), 433 (1981).
Murray, S.S., Fabbiano, G., Fabian, A.C., Epstein, A., Giacconi, R., Ap.J. **234**, L69 (1979).
Nugent, J.J., Pravdo, S.H., Garmire, G.P., Becker, R.H., Tuohy, I.R., Winkler, P.F., Ap.J. **284**, 612 (1984).
Petre, R., Canizares, C.R., Kriss, G.A., Winkler, P.F., Ap.J. **258**, 22 (1982).
Pfeffermann, E., Aschenbach, B., Bräuninger, H., Heinecke, N., Ondrusch, A., Trümper, J., Bull. Am. Astron. Soc. **11**, 789 (1980).
Pravdo, S.H. and Smith, B.W., Ap.J. **234**, L195 (1979).
Reid, P.B., Becker, R.H., Long, K.S., Ap.J. **261**, 485 (1982).
Reynolds, S.P. and Chevalier, R.A., Ap.J. **245**, 912 (1981).
Serlemitsos, P.J., Boldt, E.A., Holt, S.S., Ramaty, R., Brisken, A.F., Ap.J. **184**, L1 (1973).
Seward, F.D., Comm. Astroph. XI, 1, 15 (1985).
Smith, A., Davelaar, J., Peacock, A., Taylor, B.G., Morini, M., Robba, N.R., submitted to Ap.J. (1987).
Smith, A., IAU Coll. No. **101**, Roger, R.S. and Landecker, T.L., eds., Cambridge Uni. Press (1987).
Shull, J.M., Ap.J. **262**, 308 (1982).
Szymkowiak, A.E., NASA Techn. Memor. 86169 (1985).
Teske, R.G. and Petre, R., Ap.J. **314**, 673 (1987).
Tsunemi, H., Yamashita, K., Masai, K., Hayakawa, S., Koyama, K., Ap.J. **306**, 248 (1986).
Vedder, P.W., Canizares, C.R., Markert, T.H., Pradhan, A.K., Ap.J. **307**, 269 (1986).
Winkler, P.F., Canizares, C.R., Clark, G.W., Markert, T.H., Petre, R., Ap.J. **245**, 574 (1981a).
Winkler, P.F., Canizares, C.R., Clark, G.W., Markert, T.H., Kalata, K., Schnopper, H.W., Ap.J. **246**, L27 (1981b).
Winkler, P.F., Canizares, C.R., Bromley, B.C., "Supernova Remnants and their X-ray Emission", IAU Symp. No. **101**, Danziger, J. and Gorenstein, P., eds., (D. Reidel), 245 (1983).

THE HOT INTERSTELLAR MEDIUM : OBSERVATIONS

R. ROTHENFLUG
Service d'Astrophysique (DPHG/IRF)
C.E.N. Saclay (France)

ABSTRACT. The observational evidence that substantial parts of the interstellar medium are filled up with hot gas is presented. The soft X-ray experiments showed that the vicinity of the sun consists of a hot bubble ($T\sim10^6$ K, $n_e \sim 10^{-2}$ cm^{-3}), remnant of at least one supernova explosion. OVI absorption line measurements in the UV range by Copernicus depicted the near ($d < \sim 2$ kpc) galactic disk as partially made of hot ($T \sim 3.10^5$ K) gas regions (~ 6 per kpc). Such gas, mixed with gas at lower temperatures, probably extends as far as 3 kpc above the galactic plane, as shown by IUE measurements of SiIV, CIV and NV.

1. INTRODUCTION

Several phases of gas coexist in the interstellar medium (ISM). In the solar neighbourhood, half the mass lies in molecular hydrogen, the other half in atomic hydrogen. This atomic hydrogen itself is divided into cold clouds (~80 K) and a more diffuse warm compoment (~8000 K). In volume, the warm component may occupy a fraction as high as 50% of the interstellar medium (Kulkarni and Heiles, 1987 and references therein).
    Hot phases of the ISM were revealed by the observations of highly ionized species of high Z elements. UV absorption lines of SiIV, CIV, NV and OVI characteristic of hot components ($\sim 10^5$ to several 10^5 K) were detected up to several kpc. X-ray lines of even more ionized species provide the bulk of the soft X-ray background emission ($\sim 10^6$ K) coming from more local regions.
    In this paper I present a review of the observations concerning the hot ISM, starting from the sun vicinity up to more distant regions. Section II is devoted to the local hot ISM as measured by soft X-ray experiments. Section III deals with the galactic disk and section IV with the galactic corona.

2. THE HOT LOCAL ISM

2.1. Soft X-ray observations.

The sky has now been almost entirely scanned in the soft X-ray range by experiments using proportional counters. These detectors have a very poor energy resolution : typically E/ΔE~2 for E<1 keV and the detected X-rays are discriminated by different windows. So their results are presented as broad band maps. The figure 1 compares the efficiencies of several experiments devoted to the soft X-ray background (from Arnaud and Rothenflug,1986) and depicts the energy ranges of the different bands.

All sky maps in the B band (130-188 eV), C band (160-284 eV) and M bands (440-930 eV; 600-1100 eV) were made by the Wisconsin group with a $6.5°$ angular resolution (Fried et al,1980; McCammon et al,1983) using rocket experiments. A C band map was obtained with an angular resolution of $2.9°$ FWHM by an experiment put in the SAS 3 satellite (Marshall and Clark,1984). Data for the Be band (80-110 eV) exist for parts of the sky (Bloch et al, 1986) (see §II.3). Two spectroscopic experiments concern particular regions of the sky (see §II.2).

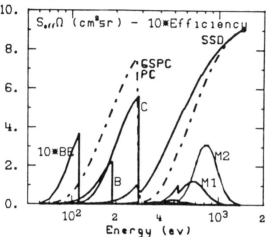

Figure 1: Efficiencies of experiments on the soft X-ray background. PC: proportional counters (Hayakawa et al,1978). GSPC: gas scintillating proportional counters (Inoue et al, 1979). SSD: solid state detector (Rocchia et al,1984). Effective area-solid angle products versus energy of the B, C, M_1, M_2 bands (McCammon et al,1983) and of the Be band (Bloch et al,1986).

Maps of the diffuse emission in galactic coordinates for the B and the C bands are shown in McCammon et al(1983). The C map obtained by SAS 3 presents the same general characteristics than the Wisconsin C map. The most important features of the soft X-ray background in these B and C bands are:
        1. The B and C maps appear very similar, with the exceptions of some features appearing in the C map and which are associated to particular objects (North Polar Spur,Eridanus,etc: see below).
        2. The flux in the galactic plane is approximately the same in all longitudes for the B and C bands.
        3. Again for the B and C bands,the observed flux is generally higher by a factor of two to three at high latitudes. A large scale anticorrelation with HI column density exists.

2.2.Evidence for a thermal emission

The explanation of the soft X-ray background as thermal emission from an interstellar gas was proposed very soon. The B/C band ratios and the spectra lead to temperatures around 10^6 K. At such temperatures, the bulk of the

X-ray emission comes from lines, unresolved with proportional counters. Two experiments were performed by two different groups with detectors having $E/\Delta E \sim 4$ at 600 eV, thus better than usual proportional counters and measured soft X-ray spectra in particular directions.

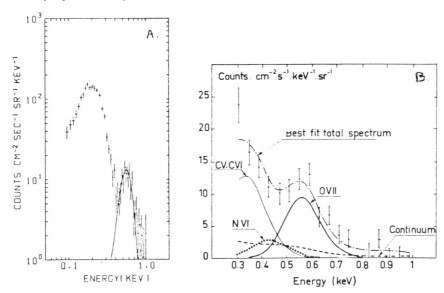

Figure 2 : A. Oxygen line at 580 eV observed by a GSPC in the direction of the Hercules hole (Inoue et al,1979).
B. Spectrum of the soft X-ray background obtained with a SSD above 300 eV (Rocchia et al, 1984).

1. A Japanese group observed the Hercules Hole region ($20° \times 20°$ around $l^{II} \sim 80°$ and $b^{II} \sim 40°$) with a gas scintillating proportional counter (Inoue et al,1979). This region presents a very low column density in HI ($\sim 1.7 \ 10^{20}$ H atoms/cm^2). The observed spectrum revealed a clear peak at about 570 eV which can be identified with the OVII emission line (see figure 2a). Emission from continua alone can be rejected at better than 99%. Their spectrum is in agreement with a thermal emission at a temperature around $1.4 \ 10^6$ K.

2. A collaborative experiment between the Smithsonian Astronomical Observatory and the Saclay group observed the soft X-ray background with solid state detectors (Rocchia et al,1984). Their experiment scanned the region roughly delimited by $b^{II} > 10°$, $0 < l^{II} < 180°$ with a ~ 1 ster field of view. In figure 2b, we reproduce the spectrum they obtained in regions of the sky outside the North Polar Spur. The feature around 570 eV was attributed to the OVII emission line. There is also a strong excess at low energies attributed to a blend of CV(300eV) and CVI(360 eV) emission lines. Their temperature determination was $1.14 \pm 0.08 \ 10^6$K (at the 90% confidence level).

These two experiments confirm the idea that the emission of the soft X-ray background comes from an interstellar gas at a temperature around 10^6 K. To derive an emission measure from the measured fluxes, one must use the computation of the X-ray emission made with a particular plasma model: it appears that differences exist among the different models. The widely used model of Raymond and Smith (1977) leads to systematically lower emission measures than the Kato (1976) model (see Arnaud and Rothenflug(1986) for a more complete discussion): with the Kato model, Rocchia et al(1984) gave an emission measure around $1.2 \cdot 10^{-2}$ cm^{-6} whereas McCammon et al(1983) obtained between 2.10^{-3} and 5.10^{-3} cm^{-6} with the Raymond and Smith model. Finally, one must keep in mind that the B and C bands measure in fact emission coming from different lines, thus from different elements: at 10^6 K, magnesium lines dominate in the B band and silicon lines in the C band. Thus the conclusions derived from band analysis depend on the assumed abundances.

2.3. The local contribution

The soft X-rays of the B and C bands are attenuated by their photoelectric absorption with the cold interstellar matter. For X-rays below the absorption edge of carbon (284 eV), about one third of the absorption is due to hydrogen, almost all the remainder is due to helium (Brown and Gould,1970; Morrison and McCammon,1983). The mean free paths are respectively :

- 1. 10^{19} HI atoms/cm^2 for the Be band.
- 6. 10^{19} HI atoms/cm^2 for the B band.
- 13. 10^{19} HI atoms/cm^2 for the C band.

In the direction of the galactic plane, these values imply that in the case of the C band, the emission comes from the nearest ~300 pc. Using the shadowing of nearby clouds, one can restrict this distance to ~100 pc (Arnaud et al,1981).

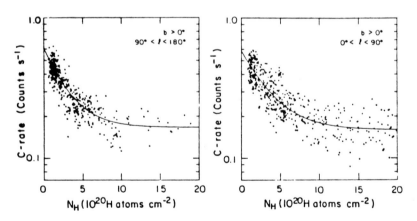

Figure 3 : Examples of correlations between the C band rates and the HI column densities from the SAS 3 survey (Marshall and Clark,1984).

At high galactic latitudes, the observed correlation between the soft X-ray fluxes and the HI column density prompted many workers (Bowyer and Field,1969; Bunner et al,1969; Marshall and Clark,1984) to propose the interstellar absorption as origin for the variations. Some examples of such correlations obtained with the SAS 3 satellite are shown in figure 3 (Marshall and Clark,1984).It was proposed that this kind of data can be explained by a model with two components, an unabsorbed local component and a distant component absorbed by N_H with an effective cross-section σ_{eff}. From the analysis of the measured correlations one concludes that, in this picture, the effective cross sections must be one third of the photoelectric cross section for the B band and two thirds for the C band.

A possible explanation of such small effective cross sections is that variations of gas column density reduce the predicted absorption (Bowyer and Field,1969; Bunner et al,1969; Marshall and Clark,1984; Jakobsen and Kahn,1986). These variations could be produced by a clumping of interstellar matter in clouds. One finds that a random distribution of clouds of constant column density $N_c = 1.3\ 10^{20}$ HI atoms reduces the cross sections to values very near the required values for both the B and C bands.

However the clumping hypothesis can be tested directly by 21 cm emission measurements: the atomic hydrogen represents the bulk of the material in the line of sight and its clumping would produce fluctuations measurable by 21cm studies. To rule out clumping, it must be shown that the required fluctuations do not exist on any angular scale. Jahoda et al (1985,1986) studied column density fluctuations down to 10' resolution and concluded that the required clumping does not exist on any angular scale: the existing fluctuations constrain the cross-section to values greater than ~90% of the photoelectric cross section for the B and C bands.

Moreover there is a general tendency for higher values of C/B ratios to occur in regions of higher intensities: this is the contrary of what would be expected if the bulk of variations was due to absorption (Fried et al,1980).

Finally the measurement in the very soft X-ray range (the Be band) reported by Bloch et al(1986) brings the strongest argument for a local origin of the soft X-ray background.

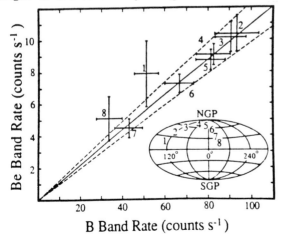

Figure 4 : Observed Be band rates versus the B rates from the Wisconsin sky survey. The solid line is the best fit slope. The dashed lines correspond to a 99% confidence interval if the ratio is assumed to be constant (Bloch et al,1986).

The scan path of this experiment is depicted in the insert of figure 4. Taking into account the distribution of the local cold interstellar matter (see the review by York and Frisch, 1984), the mean Be band free path implies that this emission must be produced within surely less than ~100 pc.

The ratio of the Be to B band rates is almost constant over $120°$ of the sky (see figure 4). This strongly suggests a common origin for the two band emissions. Moreover, its constancy means that there is no more than about 5.10^{18} HI atoms/cm^2 between the observer and the bulk of the Be and B band emissions (Bloch et al,1986). This is necessary to avoid that variations in the HI column density lead to variations in the Be/B ratios. With such a low value, the Be band emission must come from the nearest ~50 pc.

This discussion leads me to conclude that **the bulk of the soft X-ray emission is probably local.** The observed large scale correlation with HI column densities could be explained as a displacement effect (Sanders et al, 1977): where the column density gets large, there is less room left for the hot gas and its emission measure is reduced.

2.4. An old supernova remnant

The dimension of ~50 pc and the emission measures deduced from the X-ray measurements (with $T \sim 10^6$ K) lead to a very low density for the local hot medium: $n_e \sim 2.10^{-2}$ cm^{-3}. The pressure is about 4.10^4 K.cm^{-3} and the thermal energy ~ several 10^{50} ergs, reminiscent of values found for old supernova remnants (SNR).

Cox and Anderson (1982) devised a model where the emission of an old SNR expanding in a finite pressure medium accounts for the observed mean B and C fluxes. Arnaud and Rothenflug (1986) made an attempt to constrain this kind of model with the spectroscopic results and found that both spectral results and the mean Be flux can be explained by a depletion of high Z elements. If such SNR models really apply, the hot plasma is limited to a thin shell and must expand in a cavity.

An alternate idea presented the sun as embedded in an old superbubble in a radiative phase (Innes and Hartquist,1984): the initial energy required to explain the present day thermal energy represents the energy released by an entire OB association as winds and supernova explosions. This idea of a very old bubble was also discussed by Cox and Snowden(1986). Tomisaka(1986) studied the hydrodynamics evolution of a superbubble formed by sequential explosion of supernovae. (for a discussion of the local ISM, see also the recent review of Cox and Reynolds,1987).

2.5. The soft X-ray background in the M band.

The background in the M band (0.5-1.2 keV) seems highly isotropic (Nousek et al, 1982). At high galactic latitude, an important contribution of the extragalactic X-ray background (EXBG) detected at higher energies (E>2 keV) is expected. This contribution strongly decreases at low galactic latitudes because of the absorption by the interstellar matter in the galactic disk. Once this contribution is removed, the remaining M band flux then increases toward the galactic plane and can be only explained by a component with a large scale height.

The SNR model of Cox and Anderson (1982) cannot produce such an M flux. The situation was improved in that respect by Arnaud and Rothenflug (1986): using up-dated values for the atomic physics and constraining their model with the spectroscopic observations, they accounted for a larger part of the M flux. It remains that such SNR models encounter great difficulties to explain why the M flux will increase toward the galactic plane and the B and C fluxes will have an opposite behaviour.

One can try then to build a model for which in addition to the EXBG component and to the local hot medium contribution, there exist one or several other components (Nousek et al,1982; Sanders et al,1982). In addition, measurements at $1°$ resolution made by the Einstein IPC (Kahn and Caillaud,1986) revealed a dip on top of a narrow component in the direction of the galactic plane: it is explained as due to the absorption of a galactic component probably linked with the galactic ridge seen at higher energies (see below).

Thus in spite of its isotropy, the M flux contains at least 4 components: the absorbed EXBG, a contribution of the local hot medium, the galactic ridge and a component with a large scale height. Among the possible origins of this last component, X-ray emission from stars, and specially M dwarfs have been proposed (Rosner et al,1981) although their contribution was recently estimated to be less than 10% (Caillault et al,1986). A contribution from the galactic halo cannot be excluded, but would be very difficult to sort out.

2.6.Other local hot cavities.

Several enhancements show up in the soft X-ray maps. They are explained as emissions from local (d<500 pc) hot cavities. I give below in table 1 a list of such bubbles, with their very approximate distances and locations in galactic coordinates.

TABLE 1

Hot local bubbles

NAME	l^{II}	b^{II}	Dist pc	Radius pc	Temp. K	REF.
North Polar Spur	330°	20°	200	100	3.10^6	Iwan(1980)
Eridanus Hot Spot	205°	-40°	200	52	2.10^6	Naranan et al(1976)
Monoceros	204°	10°	300	50	2.10^6	Nousek et al(1981)
Lupus Loop	330°	15°	500		3.10^6	Winkler et al(1979)

3. HOT GAS IN THE GALACTIC DISK

3.1. The Galactic Ridge.

Because of the interstellar absorption, observations with soft X-rays are limited to the nearest hundreds of parsec. At higher energies (E>2 keV), the emission from a Galactic Ridge was observed by different experiments (Bleach et al,1972; Warwick et al,1985). Recent spectral observations reported by Japanese groups (Koyama et al,1986) discovered an intense feature around 6.7 keV due to iron K line emission: the measured value of the line energy proves the thermal nature of the emission. The temperature is found to vary from region to region in the range 5 to 10 10^7 K. Explained in terms of interstellar gas emission, such media would have a very high pressure (~10^5 K cm^{-3}), more reminiscent of pressures found in young SNR. Moreover, the thermal velocity exceeds the escape velocity from the galactic plane. The origin of this component must be probably searched in yet unresolved sources like young SNR or nebulae around star formation regions (Koyama et al,1986), rather than truly diffuse processes.

Such diffuse processes seem to be best detected through absorption of UV lines in the spectra of hot stars.

3.2. UV absorption lines.

Several ions were used to trace hot gas in the interstellar medium (see table 2). The ionization potential of OV and the temperature of maximum abundance reveal the importance of OVI to the search for a hot ISM. With increasing ionization potential, photoionization in the warm gas appears as less likely for the production of ions and seems highly unlikely for OVI (Savage,1987). The column POTENT. in table 2 corresponds to the energy needed to produce the ion from the ionization stage below.

From an UV absorption spectrum, one can first estimate the ion column density, using the adopted oscillator strength and the measurement of the equivalent width. In addition, the measurements provide absorption line velocities and line widths (for a complete discussion and formula, see for instance Cowie and Songaila, 1986 or Savage,1987). The line widths or shapes give a direct information on the temperature if the line broadening is thermal: a Doppler broadened line produced by an ion of mass number A at a temperature T will have a full velocity width at half maximum intensity :

$$FWHM\ (km.s^{-1}) = 0.215\ (T/A)^{0.5}$$

The FWHM thermal broadenings expected for each ion at the temperature of maximum abundance are given in table 2.

The satellite Copernicus was launched in 1972 and operated until 1980: it had a resolution of 13 km.s^{-1} between 912 and ~1400 A and was the only instrument making possible OVI line absorption studies. The International Ultraviolet Explorer (IUE) is still operating : it is limited to the range 1150-3300 A with a resolution of 25 km.s^{-1}. However the IUE has obtained spectra of objects 500 times fainter than the faintest objects observed by Copernicus. In counterpart, Copernicus had the spectral resolution to determine reliable temperature informations.

TABLE 2

UV absorption lines

IONS	λ(A)	POTENT. eV	T_{max} (K)	Th. Br. km/s
Si IV	1403. 1394.	33.5	$0.6\ 10^5$	10
CIV	1551. 1548.	47.9	$1.0\ 10^5$	20
NV	1243. 1239.	77.5	$2.0\ 10^5$	26
OVI	1038. 1032.	113.9	$3.0\ 10^5$	29

3.3. The OVI coronal gas.

The first measurement of an OVI absorption line by Copernicus was reported as early as 1973 by Rogerson et al, followed up by the more extensive works of Jenkins and Meloy (1974) and York (1974,1977). The most complete OVI observational results are discussed in the two papers of Jenkins (1978,a,b). He analyzed the observations towards 72 O or B stars. As these young stars are preferentially found in clusters at low galactic latitudes, the sky coverage was not homogeneous. Because of the sensitivity of the Copernicus satellite, the farthest star is at 3.2 kpc and most of them lie below 1 kpc.

Jenkins derived for each observation the column density of OVI, the velocity centroid of the line and its velocity variance. As OVI can be produced in bubbles around hot stars (Weaver et al,1977), he tested the hypothesis of a circumstellar origin : neither the column densities nor the velocities can be correlated with star characteristics, ruling out this hypothesis. A loose correlation between column densities and star distances brings some support to the interstellar origin (Jenkins,1978c) (see figure 5). One can derive from the analysis of the large fluctuations of this last correlation that there exist about 6 OVI regions per kiloparsec, each having an OVI column density of $\sim 10^{13}$ cm^{-2}. The filling factor of the OVI gas is found to be less than 20%. The average OVI density lies around 2.10^{-8} cm^{-3}.

There is a lower cut-off in the distribution of the velocity variances at ~ 100 km^2.s^{-2}. Interpreted as an internal velocity dispersion, it corresponds to a thermal broadening for oxygen ions at $T=2.10^5$ K (a temperature where the OVI ionic abundance is appreciable). Large column densities have larger than usual variances, suggesting out that one is viewing sometimes individual components, sometimes a superposition of narrow components. The dispersion of random bulk velocities for the entire population equals 26 km.s^{-1}, implying that these regions were not heated recently by supernova shocks (Cowie and Songaila,1986).

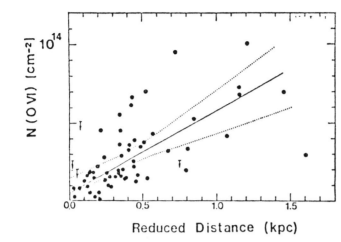

Figure 5: Observed OVI column densities as a function of reduced distances. The true distances toward stars away from the galactic plane were foreshortened to compensate for an overall trend toward lower OVI densities at high z (Jenkins,1978c).

3.4. High temperature gas from other ions.

The IUE satellite pursued the study of the hot interstellar gas, with the absorption lines of SiIV,CIV,NV, characteristic of lower temperatures than OVI (see table 1). The production of SiIV and CIV by photoionization cannot be totally excluded, specially around hot stars and it appeared to be the case in a large fraction of the observations.

Among others, Jenkins(1981) made an attempt to sort out what quantities of these ions are relevant to the interstellar medium. Savage and Massa(1987) presented the data obtained for 67 distant stars, mostly B0 or cooler to minimize the contribution of HII regions. Their lines of sight avoid directions of obvious nebulosities and their results represent the low density and interarm gas. In the galactic plane, they suggested the following densities:

$n(SiIV) = 2.10^{-9}$ cm^{-3}, $n(CIV) = 7.10^{-9}$ cm^{-3}, $n(NV) = 3.10^{-9}$ cm^{-3},

within a factor of two from the values of Jenkins(1981).

3.5. Clues for the origin.

McKee and Ostriker(1977) suggested that the OVI ions were produced in conductive interfaces between cold clouds and a hot substrate (T~10^6 K). This hot substrate could fill as much as 80% of the interstellar medium. This idea received some support when Cowie et al(1979) found a correlation between the kinematics of the OVI lines and those of other strong UV lines from lower ionization stages (SiIII,NII).

However this scenario implies a mean free path between conductive interfaces of ~10 pc if one uses the cloud spectrum of McKee and Ostriker (1977) and this seems contrary to the value of 6 regions per kpc derived by Jenkins (Cox,1986). Ballet, Arnaud and Rothenflug (1986) have carried out calculations of ion production in evaporative interfaces including the effects of the non-equilibrium ionization and have shown that these effects enhance the column densities per cloud. Using reasonable values for the cloud filling factor and radius, they derived mean densities of ions much higher than observed. Obviously the role of conductive interfaces around clouds needs to be clarified.

The hot substrate in which the clouds are supposed to be embedded would also produce measurable OVI column densities: 2.10^{13} cm^{-2} for 1 kpc, with $T=10^6$ K and $n_e=2.10^{-3}$ cm^{-3}. Such a component could be quite broad (100-200 km.s^{-1}) because of the turbulence (McKee and Ostriker,1977) and thus very difficult to observe. From a search in the Copernicus archives, Cowie and York failed to find such broad lines: their non-detection lead to an upper limit of 5.10^{13} cm^{-2} per kpc (quoted by Cowie and Songaila, 1986). Such a hotter component was searched through FeX and FeXIV absorption line studies in the optical (Hobbs,1984,a,b): the derived upper limits around 10 mÅ for the line equivalent width correspond to column densities of FeX less than 10^{17} cm^{-2}. If iron is assumed both to be present at its standard abundance and to be entirely in the form of FeX, such column density leads to a density less than 0.5 cm^{-3} for the hot gas. This upper limit does not really constrain the models.

An alternate idea was suggested by Cox(1986) : OVI ions are produced in very old bubbles (age~10^6 years) generated by supernovae. They would have the correct column densities per object and would be spaced with the correct mean free path between them. In such situations, the gas happens to be in a cooling process, isobaric, isochoric or some combinations of the two in time. Highly ionized ions thus persist to lower temperatures than would be found in equilibrium situations (Jenkins,1981;Edgar and Chevalier,1986).

TABLE 3

Ion densities

DENSITIES	SiIV	CIV	NV	Ref.	
Gal. plane	0.1	0.23	0.15	Meas.	
Isochoric	0.015	0.17	0.072	Ed.C.	Ed.C.: Edgar, Chevalier(1987).
Isobaric		0.036	0.043	Ed.C.	
Evap. Cl.		0.125	0.043	Bal.	Bal.: Ballet et al(1986).

One can then compare the measured ion densities in the interstellar medium with densities expected in different possible cooling situations, as Jenkins(1981) did. This is given in table 3 by comparison with OVI. Clearly CIV and NV can be produced by the same gas which produced OVI, but some additional mechanism (probably photoionization) is needed for SiIV ions in the galactic disk.

4. THE GALACTIC HALO

Spitzer first suggested in 1956 the existence of hot gas in the galactic halo measurable through UV absorption line studies. We now have measurements with UV absorption lines and probably UV emission lines indicating that hot gas fills part of the galactic halo.

4.1. UV absorption lines.

The OVI line absorption studies with Copernicus were limited to several hundreds of parsec above the galactic plane. Using hot stars in the LMC and the SMC as target stars, Savage and de Boer (1979,1981) revealed with IUE the existence of strong UV absorption lines, coming from the galactic halo and belonging to a wide range of ionization stages up to SiIV,CIV. Other studies used halo stars in our galaxy (Pettini and West,1982; de Boer and Savage,1984).

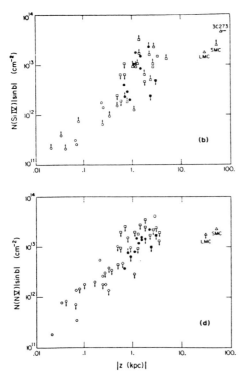

Figure 6: N(ion) sin b versus z plotted for measurements of SiIV and NV towards the 67 stars of Savage and Massa (1987).

Savage and Massa (1987) presented the UV absorption lines for 67 galactic stars up to z~4 kpc. One of their most important results is the detection of NV along 14 lines of sight (see figure 6). This detection implies very probably the presence of hot gas: 77 eV are needed to convert NIV to NV by photoionization and sources of such UV spectra seem very difficult to find (Savage and Massa,1985).

Highly ionized gas is more extended away from the galactic plane than is HI or SiII. The z distribution suggests a scale height of about 3 kpc. For SiIV and CIV, the data indicate a 2-3 times enhancement for z>1 kpc over the exponential which best represents the data below that. The ratios of ion densities appear constant from the disk to the halo: the values

n(SiIV)/n(CIV)~0.29
n(NV)/n(CIV)~0.32

are acceptable for both.

However the existence of numerous upper or lower limits may hide the presence of large fluctuations.

Computations of ionic fractions in coronal plasmas at equilibrium (Arnaud and Rothenflug,1985) immediately indicate that a single temperature cannot reproduce the two ratios at the same time. As CIV could be at least in part produced by photoionization, the measured value of the ratio n(NV)/n(CIV) leads to a gas temperature greater than $1.5 \; 10^5$ K.

4.2. UV emission lines.

In a hot plasma, an emission line intensity varies as $\exp(-E_k/kT)/\sqrt{T}$ (where E_k is the excitation energy). For the UV lines we are interested in the E_k values lie around several eV. Measurements of UV line emission would mean then that $kT \sim E_k$ and rule out photoionization for which kT would be $\ll E_k$.

UV lines from SiIV and CIV have been claimed to be detected (Feldman et al,1981) at a level of $\sim 10^4$ photons.cm^{-2}: the low resolution and the possible presence of atmospheric lines produced above the sounding rocket rend these intensities highly uncertain. New measurements of these lines were performed with the UVX Berkeley experiment on board one of the last flights of the shuttle (Martin and Bowyer,1986). Lines from CIV and OIII were detected in 4 out of 8 directions. The CIV line intensity exhibits a correlation with the galactic latitude, which favors an extraterrestrial origin. The CIV line intensity varies from ~600 to ~3000 ph.cm^{-2}.

4.3. Galactic fountain or photoionization?

Both the fact that NV be present in the halo and the measurement of the CIV emission line favor models producing hot gas in the halo. In the Galactic fountain model (Shapiro and Field,1976; Bregman,1980), buoyant disk gas flows away from the galactic disk, cools and returns as condensations.

In their model for cooling of a hot galactic corona, Edgar and Chevalier(1986) showed that the observed values of NV column densities can be reproduced with reasonable assumptions, but the predicted value for SiIV falls short by a large factor. They also predict a value for the CIV line emission within a factor of three of the observed value.

On the other hand, photoionization models (Chevalier and Fransson,1984; Hartquist,Pettini, and Tallant,1984; Bregman and Harrington,1986) predict the observed amount of SiIV. CIV can be produced by both models. The observations of SiIV, CIV and NV thus required probably a combination of the two kinds of model.

Acknowledgements

I wish to thank M. Arnaud and J. Ballet for their very useful comments and their friendly advices.

REFERENCES

Arnaud M.,Rocchia R.,Rothenflug R.,Soutoul A.,1981, Proceedings of the 17th ICRC, Vol.1, p.131, (Ed. CEA/Saclay)
Arnaud M.,Rothenflug R., 1985, Astron. Astrop. Suppl. Ser. <u>60</u>, 425

Arnaud M.,Rothenflug R., 1986, Adv.Space Res., Vol.6,No.2, p119
Ballet,J.,Arnaud M.,Rothenflug R., 1986, Astron.Astrop. 161, 12
Bleach,R.D.,Boldt,E.A.,Holt,S.S.,Schartz,D.A.,Serlemitsos,P.J., 1972, Ap.J.(Letters) 174, L101
Bloch,J.J.,Jahoda,K.,Juda,M.,McCammon,D.,D.N.,Sanders,W.T.,Snowden,S.L. 1986, Ap.J.(Letters) 308, L59
Bowyer,C.S.,Field,G.B., 1969, Nature 223, 573
Bregman,J.N., 1980, Ap.J. 236, 577
Bregman,J.N.,Harrington,J.P., 1986, Ap.J. 309, 833
Brown,R.L.,Gould,R.J.,1970, Phys.Rev.D 1, 2252
Bunner,A.N.,Coleman,P.C.,Kraushaar,W.L.,McCammon,D.,Palmieri,T.M., Shilepsky,A.,Ulmer,M., 1969, Nature 223, 1222
Burrows,D.N.,McCammon,D.,Sanders,W.T.,Kraushaar,W.L., 1984, Ap.J. 287, 208
Caillault,J.P.,Helfand,D.J.,Nousek,J.A.,Takalo,L.O., 1986, Ap.J. 304, 318
Chevalier,R.A.,Fransson,C., 1984, Ap.J.(Letters) 279, L43
Cowie,L.L.,Jenkins,E.B.,Songaila,A.,York,D.B., 1979, Ap.J. 232, 467
Cowie,L.L.,Songaila,A., 1986, Ann. Rev. Astron. Astrophys., 24, 499
Cox,D.P., 1986, in Proceedings of the Meudon Workshop on Model Nebulae
Cox,D.P.,Anderson,P.R., 1982, Ap.J. 253, 268
Cox,D.P.,Reynolds,R.J., 1987, Ann. Rev. Astron. Astrophys., to be published
Cox,D.P.,Snowden,S.L., 1986, Adv.Space Res., Vol.6,No.2
deBoer,K.S.,Savage,B.D., 1984, Astron. Astrop. 136, L7
Dupree,A.K.,Raymond,J.C., 1983, Ap.J.(Letters) 275, L71
Edgar,R.J.,Chevalier,R.A., 1986, Ap.J.(Letters) 310, L27
Feldman,P.D.,Brune,W.H.,Henry,R.C., 1981, Ap.J.(Letters) 249, L51
Fried,P.M.,Nousek,J.A.,Sanders,W.T.,Kraushar,W.L., 1980, Ap.J. 242, 987
Hartquist,T.W.,Pettini,M.,Tallant,A., 1984, Ap.J. 276, 519
Hayakawa,S.,Kato,T.,Nagase,F.,Yamashita,K.,Tanaka,Y., 1978, Astron. Astrop., 62, 21
Hobbs,L.M., 1984a, Ap.J. 280, 132
Hobbs,L.M., 1984b, Ap.J. 286, 252
Innes,D.E.,Hartquist,T.W., 1984, M.N.R.A.S., 209, 7
Inoue,H.,Koyama,K.,Matsuoka,M.,Ohashi,T.,Tanaka,Y.,Tsunemi,H., 1979, Ap.J. 227, L65
Iwan,D.A., 1980, Ap.J. 239, 316
Jahoda,K.,McCammon,D.,Dickey,J.M.,Lockman,F.J., 1985, Ap.J. 290, 229
Jahoda,K.,McCammon,D.,J.M.,Lockman,F.J., 1986, Ap.J. 311, L57
Jakobsen,P.,Kahn,S.M., 1986, Ap.J. 309, 682
Jenkins,E.B., 1978a, Ap.J. 219, 845
Jenkins,E.B., 1978b, Ap.J. 220, 107
Jenkins,E.B., 1978c, Comments Astrophys. Vol.7,N°4,p.121
Jenkins,E.B., 1981, in The Universe at UV Wavelengths, ed R.D.Chapman, (Greenbelt:NASA CP2171),p.541
Jenkins,E.B.,Meloy,D.A., 1974, Ap.J.(Letters), 193, L121
Kahn,S.M.,Caillault,J.P., 1986, Ap.J. 305, 526
Kato,T., 1976, Ap.J.Sup. 30, 397
Koyama,K.,Makishima,K.,Tanaka,Y.,Tsunemi,H., 1986, P.A.S.J. 38, 121
Kulkarni,S.,Heiles,C., 1987, in Interstellar Processes, p87
  D.J.Hollenbach and H.A.Thomson,Jr.,eds,(D.Reidel publishing company).

McCammon,D.,Burrows,D.N.,Sanders,W.T.,Kraushaar,W.L., 1983, Ap.J. 269, 107
McKee,C.F.,Ostriker,J.P., 1977, Ap.J. 215, 213
Marshall,F.J.,Clark,G.W., 1984, Ap.J. 287, 633
Martin,Bowyer, 1986, Adv.Space Res., Vol.6,No.2
Morrison,R.,McCammon,D., 1983, Ap.J. 270, 119
Naranan,S.,Shulman,S.,Friedman,H.,Fritz,G., 1976, Ap.J. 208, 718
Nousek,J.A.,Cowie,L.L.,Hu,E.,Linblad,C.J.,Garmire,G.P.,1981,Ap.J. 248, 152
Nousek,J.A.,Fried,P.M.,Sanders,W.T.,Kraushaar,W.L., 1982, Ap.J. 258, 83
Pettini,M.,D'Odorico,S., 1986, Ap.J. 310, 700
Pettini,M.,West,K.A.,1982, Ap.J. 260, 561
Raymond,J.C.,Smith,B.W., 1977, Ap.J.Sup. 35,419
Rocchia,R.,Arnaud,M.,Blondel,C.,Cheron,C.,Christy,J.C.,Rothenflug,R., Schnopper,H.W.,Delvaille,J.P., 1984, Astron. Astrop. 130,53
Rogerson,J.B.,York,D.G.,Drake,J.F.,Jenkins,E.B.,Morton,D.C.,Spitzer,L., 1973, Ap.J.(Letters), 181, L110
Rosner,R., et al, 1981, Ap.J. 249, L5
Sanders,W.T.,Burrows,D.N.,Kraushaar,W.L.,McCammon,D., 1982, IAU Symposium 101, Supernova Remnants and their X-ray emission, J.Danziger and P.Gorenstein ed.,(Dordrecht,Reidel), p.361
Sanders,W.T.,Kraushaar,W.L.,Nousek,J.A.,Fried,P.M., 1977, Ap.J. 217, L87
Savage,B.D., 1987, in Interstellar Processes, p123, D.J.Hollenbach and H.A.Thomson,Jr.,eds,(D.Reidel publishing company).
Savage,B.D.,Massa,D., 1985, Ap.J.(Letters) 295, L9
Savage,B.D.,Massa,D., 1987, Ap.J. 314, 380
Savage,B.D.,deBoer,K.S., 1979, Ap.J.(Letters) 230, L77
Savage,B.D.,deBoer,K.S., 1981, Ap.J. 243, 460
Shapiro,P.R.,Field, 1976, Ap.J. 207, 460
Spitzer,L.Jr., 1956, Ap.J. 124, 20
Tomisaka,K., 1986, Adv.Space Res., Vol.6,No.2, p
Warwick,R.S.,Turner,M.J.L.,Watson,M.G.,Willingale,R., 1985, Nature, 317, 218
Weaver,R.,McCray,R.,Castor,J.,Shapiro,P.,Moore,R.,1977, Ap.J. 218, 377
Winkler,P.F.,Hearn,D.R.,Richardson,J.A.,Behnken,J.M., 1979, Ap.J. (Letters) 229, L123
York,D.G., 1974, Ap.J.(Letters), 193, L127
York,D.G., 1977, Ap.J. 213, 43
York,D.G.,Frisch,P.C., 1984, in proceedings of the IAU Colloquium 81, "Local Interstellar Medium", NASA Conf. Publ. 2345

THEORY OF SUPERNOVA REMNANTS AND THE HOT INTERSTELLAR MEDIUM

R. A. Chevalier
Department of Astronomy
University of Virginia
P. O. Box 3818
Charlottesville, VA 22903 U.S.A.

ABSTRACT. The three major classes of supernova, Types Ia, Ib, and II, can probably be identified with the explosions of white dwarfs, massive stars without their hydrogen envelopes, and massive stars with their hydrogen envelopes, respectively. The Type Ib supernovae and massive Type II supernovae are likely to have progenitor stars that strongly affect their surroundings through stellar winds and photoionizing radiation. The Type Ib events may have a particularly complex environment because their progenitors may go through both fast and slow wind phases. Type Ia supernovae may interact more directly with the interstellar medium. Self-similar solutions are particularly useful in describing the hydrodynamic interaction of the expanding supernova with the surrounding gas. In most Type II supernovae, the supernova first interacts with the red supergiant wind, giving radio and X-ray emission. Type Ib supernovae are probably related to oxygen-rich remnants, so these objects are expected to have a particularly complex interaction. Type Ia remnants can have a simpler interaction and X-ray spectroscopy is a powerful tool to investigate the properties of the supernova. Hot gas created by supernovae fills at least 20% of the interstellar medium. It is possible that it fills most of the medium and that there is a circulation rate of about $10 M_\odot$ yr^{-1} through the hot gas. This model predicts a large column density of O VI in the galactic corona. The alternative model with a small filling factor of hot gas implies nonthermal support of the galactic corona.

1. INTRODUCTION

Recent research on supernova remnants has shown that their properties are closely related to the nature of the progenitor stars and their presupernova evolution. The progenitor star affects its surroundings through mass loss and photoionizing radiation; the region thus affected is called the circumstellar medium. This review will thus deal with supernovae themselves as well as their remnants. Some of this material was already covered in Chevalier (1988); the proceedings of I.A.U. Colloquium 101 (Landecker and Rogers 1988) give a detailed overview of supernova remnants and the interstellar medium.

Since the emphasis here is on hot gas, energy input by a central pulsar will not be considered. Section 2 discusses the expansion of supernovae, with an emphasis on the composition structure. The expected nature of the circumstellar medium is covered in section 3. Hydrodynamic features of the interaction between supernovae and their surroundings are discussed in section 4; self-similar solutions are especially useful in this area. Interactions with circumstellar gas and interstellar gas are presented in sections 5 and 6 respectively. Some properties of radiative shock waves are discussed in section 7. Section 8 deals with the hot interstellar medium and section 9 gives conclusions.

2. SUPERNOVAE

Supernovae are observationally divided into two major classes depending on whether hydrogen lines are absent (Type I) or present (Type II) in their spectra. It has recently been realized that there are two categories of Type I events. The Type Ia events are associated with an old stellar population and have strong Fe line emission in their late spectra while the Type Ib supernovae are associated with a young stellar population and have lines of O and other intermediate elements in their late spectra (Kirshner and Oke 1975; Gaskell et al. 1986; Filippenko and Sargent 1986).

The most successful theoretical model for Type Ia supernova is the carbon deflagration of a white dwarf (Chevalier 1981; Sutherland and Wheeler 1984; Nomoto, Theilemann, and Yakoi 1984). This model can generally reproduce the light curves, early time spectra, and late time spectra (Axelrod 1980) of the supernovae. The models of Nomoto et al. (1984) are particularly detailed with regard to the composition and density structure and their model W7 has been used to model the observed spectra of Type Ia supernovae (Branch et al. 1985). The result of this work is that mixing of the intermediate element layers with velocities above 8000 km s^{-1} significantly improves the spectral fit. The density structure of the gas in the free expansion phase is complex because of the partial incineration of the gas. The process that mixes the gas may also smooth some of the dense features in the density profile.

The mechanism that is responsible for the probable mixing is not known. However, the two-dimensional carbon burning calculations of Muller and Arnett (1986) are suggestive of the complex motions that can accompany the propagation of a burning front. They find that the burning creates hot bubbles that are Rayleigh-Taylor unstable and result in a corrugated burning front. It is not clear to what extent the numerical resolution plays a role in the calculated structure or whether the flow becomes fully turbulent.

The observed characteristics of Type Ib supernovae make an interpretation as the explosions of massive stars that have lost their hydrogen envelope attractive (Wheeler and Levreault 1985; Chevalier 1986). These are Wolf-Rayet stars. The deduction that SN1985f ejected at least 5 $M_\odot$ of oxygen (Begelman and Sarazin 1986) is particularly suggestive of this interpretation. Schaeffer, Casse, and Cahen (1987) have shown that the light curves of exploding Wolf-Rayet stars are roughly consistent with those of Type Ib supernovae. The hydrogen envelope can be lost either through massive single star evolution or through mass transfer in a close binary system. Blaauw (1985) has estimated that 18% of early B type stars are in close binaries. One-dimensional models of Wolf-Rayet star explosions show the ejection of layers of heavy elements (Ensman and Woosley 1987). However, the line profile of [OI]λ6300 in the spectrum of SN1985f implies the ejection of oxygen over a broad velocity range of 0 to > 3000 km s^{-1} (Filippenko and Sargent 1986; Fransson 1986a). The origin of this mixing is unknown, but it may be related to the inhomogeneous ejecta observed in the Cassiopeia A supernova remnant. There, some fast moving knots of heavy element gas are observed with heavier elements having higher velocities than lighter elements (Chevalier and Kirshner 1979). This gradient is opposite to that expected in the models.

The explosions of Wolf-Rayet stars are expected to have a range of progenitor masses. The uniformity of Type Ib supernovae, including the radio regime (Panagia, Sramek, and Weiler 1986), is thus a possible problem for the massive star model. The possibility that mass loss from massive stars drives the core toward a common structure is not anticipated in stellar evolution theory.

If a massive star explodes with its hydrogen envelope, the result is a Type II supernova. This is the least controversial of the various supernova models. The explosion is generally expected to occur in a red supergiant envelope, although the recent supernova 1987a shows the possibility of a blue supergiant progenitor. The available evidence points to the explosion of the B3 Ia star Sk-69 202 in this case (Kirshner et al. 1987; Woosley, Pinto, and Ensman 1987). The reason for the explosion of a blue supergiant is still controversial. Hillebrandt et al. (1987), Arnett (1987), and Woosley (1987) suggest that the primary factor is the low metallicity in the LMC (Large Magellanic Cloud), but mass loss is still a possibility. In the evolutionary models of Hillebrandt et al. (1987) and Arnett (1987), the star remains blue throughout its life. However, the presence of red supergiants in the LMC with relative numbers comparable to those in our Galaxy (Humphreys and Davidson 1979) suggests that an earlier red supergiant phase is likely. The models of Woosley (1987) with restricted convection do have such a phase. His model with the preferred progenitor mass, 20 $M_\odot$, evolves from a red supergiant phase to a blue supergiant explosion in 20,000 years. The existence of a red supergiant phase is very important for the remnant evolution because of the mass loss properties in this phase. The evolution leading to Wolf-Rayet stars is still controversial (Chiosi and Maeder 1986), but mass loss due a red supergiant phase is quite plausible (Maeder 1981). Typical lifetimes of Wolf-Rayet stars are (3-8) x 10^5 years.

If Type Ib supernovae have Wolf-Rayet star progenitors and Type II supernovae have progenitors with hydrogen envelopes, it is plausible that most Type Ib events come from more massive stars than do SN II and the relative rates of these supernovae give information on the range of progenitor masses. Van den Bergh, McClure, and Evans (1987) estimate that the SN Ib rate is 0.36 times the SN II rate (see also Branch 1986). For an assumed initial mass function like that of Kennicutt (1984), van den Bergh (1987) derives a lower mass limit of about 8-12 $M_\odot$ for SN II and a lower limit of about 20-30 $M_\odot$ for SN Ib. These numbers are for a Hubble constant in the range 50-75 km s^{-1} Mpc^{-1}. The mass limit for the SN Ib is in rough agreement with expectations for Wolf-Rayet stars (Chiosi and Maeder 1986), although it is somewhat low. The implication is that early B stars generally become SN II, while O stars generally become SN Ib (van den Bergh 1987).

3. CIRCUMSTELLAR ENVIRONMENTS

Our understanding of the progenitors of SN Ia is not sufficiently well developed to be able to predict their circumstellar environment. Iben and Tutukov (1984) have considered a number of paths that potentially lead to a SN Ia explosion through the evolution of binary stellar systems. Many of the paths involve mass loss from the binary at some point in the evolution, but the evolution timescale is sufficiently long and the progenitor velocities are likely to be sufficiently high that the exploding star may expand into the ambient interstellar medium. An exception is if the progenitor is a white dwarf accreting mass from the wind of a companion star.

The SN II and SN Ib have massive star progenitors which are known to lose mass during their evolution and to emit ionizing radiation. For the main sequence evolution, the mass 15 $M_\odot$ is a critical stellar mass above which the circumstellar effects are large and below which they decrease rapidly. This is because the rate of emission of ionizing photons drops dramatically for stars later than type B0 (e.g. Panagia 1973). For the massive stars, values of $r_s n_H^{2/3}$, where r_s is the radius of the Stromgren sphere and n_H is the density, are in the 10's of pc cm^{-2}, while for the later-type stars the values drop to a few pc cm^{-2} (see Table I). For O-type main sequence stars, strong winds are clearly

TABLE I. MAIN SEQUENCE EVOLUTION

Mass ($M_\odot$)	Spectral Type	Lifetime (10^6 yr)	L_{wind} (1.) (erg s^{-1})	$r_{wind} n_H^{1/5}$ (pc cm$^{-3/5}$)	$r_s n_H^{2/3}$ (2.) (pc cm^{-2})
100	O4 V	3.4	2.1×10^{37}	100	138
60	O6 V	4.2	5.3×10^{36}	86	82
20	O9 V	10.3	1.1×10^{35}	68	40
15	B0.5 V	11.1	2.4×10^{34}	53	10
10	B2 V	21.1	...		2.9

(1.) Based on Abbott 1982.
(2.) Based on Panagia 1973.

present while for the B stars, effects of a wind are usually unobservable (Abbott 1982). An exception is the Be stars, which are observed to have winds with velocities 600 to 1100 km s^{-1} and mass loss rates of 10^{-11} to 3×10^{-9} M$_\odot$ yr^{-1} (Snow 1981). The wind luminosity is much lower than that of the O-type stars. Table I shows the radii of the bubbles that can be created by the winds assuming that adiabatic theory applies (Weaver et al. 1977). McKee, Van Buren, and Lazareff (1984) have examined the propagation of a wind bubble in a cloudy medium. They find that during the main sequence lifetime of an O4 to B0 star, a region of radius

$$R_h = 53 \, n_m^{-0.3} \text{ pc,}$$

where n_m is the average density, is made homogeneous by the photoionizing radiation. A wind bubble expands adiabatically out to this radius, but suffers radiative losses at larger radius so that R_h may characterize the final bubble radius.

The mass range for SN II discussed in the previous section covers both O4-B0 stars with large bubbles and the early B stars which are likely to have smaller regions affected by winds. The nature of the interstellar medium is likely to be important for the early B stars. In our scheme for SN Ib, they are expected to be in large bubbles created by the main sequence wind and photoionizing radiation. Some Wolf-Rayet stars may be the result of the binary evolution of lower mass stars, in which case an extended cavity may not be present.

During the next evolutionary phase for a massive star, the red giant or supergiant phase, the star has a slow wind with velocity 5-50 km s^{-1} and a mass loss rate of $10^{-7} - 10^{-4}$ M$_\odot$ yr^{-1} (Zuckerman 1980). The total duration of the red giant phase is about 10% of the main sequence lifetime. The rate of mass loss may evolve during this phase; the highest rates of mass loss have been observed in OH/IR stars at the tip of the red giant branch with M < -6. The duration of the OH/IR phase may be about 5×10^5 yr and the wind velocity is about 15 km s^{-1} (e.g. Herman 1985). However, most OH/IR stars may have initial masses in the range 2-5 M$_\odot$, which is below the range of interest for supernovae.

Most Type II supernovae are expected to explode at this point, but extreme mass loss can complicate the evolution. Loops can occur in the HR diagram (Chiosi and Maeder 1986); while the star is relatively blue, a faster, lower density wind is expected. Loss of the hydrogen envelope leads to a blue Wolf-Rayet star; these stars have typical mass loss rates of $10^{-5} - 10^{-4}$ M$_\odot$ yr^{-1} and wind velocities of 1000 - 2000 km s^{-1} (Chiosi and Maeder 1986). This is the expected immediate environment of a Type Ib supernova. Massive stars with their hydrogen envelopes but low metallicity may also explode as blue stars (see the discussion of SN 1987a in section 2).

The interaction of the fast wind from the blue star with the red supergiant wind creates a shocked, cool shell of the dense wind and a hot shell of shocked fast wind (McCray 1983; Chevalier and Imamura 1983). Ring nebulae have been observed around Wolf-Rayet stars. These typically have radii of 3-10 pc, velocities of 30-100 km s^{-1}, and densities of 300-1000 cm^{-3} (Chu et al. 1983). Some of the nebulae show evidence for abundance enhancements of N and He (Kwitter 1981,

1984). Such enhancements are expected in 15 and 25 $M_\odot$ red supergiants if stellar winds have removed at least 9 $M_\odot$ of the star (Lamb 1978). Many of the wind blown nebulae are asymmetrically distributed about the Wolf-Rayet star (Chu et al. 1983). A model in which this is due to stellar motion is attractive, but is not tenable if the fast wind only interacts with the red supergiant wind because both winds have the same space motion. Interaction with the interstellar medium is needed (e.g. Bandiera 1987). Asymmetries in the winds may also be a factor.

4. HYDRODYNAMIC EVOLUTION

The result of the supernova explosion is a radial flow in free expansion. The velocity field is given by $v = r/t$ where t is the age of the explosion and the density by $\rho = Bt^{-3}f(v)$ where B is a constant and f(v) is a function that depends on the initial hydrodynamic evolution. The pressure in the expanding gas is negligible because of adiabatic expansion. The pressure and velocity of the ambient medium can generally be neglected because of the high initial supernova velocities and only the density distribution is relevant. Because of these simplifications, self-similar solutions can be very useful in delineating the major features of the interaction. These solutions are calculated for one-dimensional flows. Although two-dimensional flows are expected to be self-similar when the ambient or supernova density can be separated into radial and angular functions, they are not easily calculated because partial differential, not ordinary differential, equations are involved. For spherically symmetric flows, the ambient medium is generally taken to be of the form $\rho \propto r^{-s}$, where s = 0 (interstellar medium) or 2 (circumstellar medium). This circumstellar medium results from a constant velocity wind with a constant mass loss rate.

The first case is a point explosion in a power law medium. Sedov (1959) obtained an analytic solution for this case and noted that the shock radius increases as $t^{2/(5-s)}$. For s=0, the shocked gas is concentrated at the shock front. For s=2, the solution is particularly simple, with velocity $v \propto r$, density $\rho \propto r$, and pressure $p \propto r^3$. These solutions assume adiabatic postshock flow. In a young supernova remnant, heat conduction will tend to flatten the temperature profile if conduction is not impeded by magnetic fields. Korobeinikov (1956) and Solinger, Rappaport, and Buff (1975) discussed isothermal blast waves. The transport of heat in to the shock reduces the shock compression ratio to 2.38 from 4. During the early phases, there is unlikely to be energy equipartition between ions and electrons. Electron heat conduction occurs on a faster timescale than proton conduction so that the electrons may be isothermal with the ions adiabatic. Cox and Edgar (1983) have investigated self-similar solutions in this case and for s = 0 find a shock compression of 3.24. They assume that ion-electron energy equilibrium is achieved in the collisionless shock front. Even if the magnetic field geometry is favorable for the action of heat conduction, it is quite possible that plasma instabilities reduce the mean free paths of the ions and electrons below the Coulomb interaction values. This has the effect of reducing heat conduction.

The other class of self-similar solutions for young remnants includes the shocked supernova ejecta gas as well as the shocked ambient medium. Chevalier (1982) and Nadyozhin (1985) found solutions for the interaction of a power law ejecta profile ($\rho \propto t^{-3} v^{-n}$) with a power law ambient medium. Dimensional analysis shows that the shock waves and contact discontinuity expand as $t^{(n-3)/(n-s)}$. For n = 5, the expansion law is the same as that for the Sedov solution, so for $n \leq 5$ the flow should approach that for a point explosion. For s=0, the density at the contact discontinuity drops to 0 for both the shocked supernova ejecta and the ambient medium, while for s=2, the density at the contact discontinuity becomes infinite for both media. The density ratio between the shocked supernova gas and the shocked ambient medium increases for larger values of n.

Self-similar solutions with reverse shock waves and the isothermal gas have not been found. Bedogni and d'Ercole (1987) have carried out numerical computations of the above case with heat conduction included and two-fluid flow. They find that the flow becomes complex with reverse shocks forming close to the contact discontinuity and thermal conduction driving a broad inner shocked region. Smooth self-similar flows may not exist in this case.

The reverse shock solutions do assume that the expanding supernova gas is smoothly distributed. Some remnants, like Cas A, indicate that the ejecta may be clumpy. Hamilton (1985) has investigated the interaction of clumpy ejecta with an ambient gas for cases similar to the reverse shock solutions discussed above. In order to preserve self-similar flow, certain assumptions, such as undecelerated clump motion, were necessary. Ablation of the clumps was allowed. If the clumps interacted strongly with the ambient medium, the solution for smooth flow was recovered. For weaker interaction, the clumps moved out ahead of the shock front in the ambient medium. This type of behavior is qualitatively expected.

The above reverse shock solutions assume a relatively steep power law density profile. Hamilton and Sarazin (1984a) have found self-similar solutions for the initial phases of a reverse shock in a medium with a flat density profile (n < 1). The solutions apply to the time when the distance between the reverse shock wave and the edge of the "freely expanding" ejecta is much less than the radius so that the flow is approximately planar. For uniform ejecta, the reverse shock propagates as $z \propto t^{(5-s)/2}$, where z is the distance to the edge of the ejecta if it continued in free expansion. As opposed to the steep power law case, the shocked supernova ejecta have a density peak at the contact discontinuity for both s = 0 and s = 2. The solutions for the shocked ambient medium resemble those for the steep power law case, although the flow is not exactly self-similar.

For the transition from the early reverse shock flow and for more general density distributions than power-laws in radius, numerical hydrodynamic calculations are needed. Because of the steep density profiles present in the early flows, many computational zones are needed in the one-dimensional calculations to reproduce the self-similar solutions (Jones and Smith 1983; Hamilton and Sarazin 1984a). A number of computations have been carried out on the interaction with circumstellar matter (Fabian, Brinkmann, and Stewart 1983; Itoh and Fabian 1984; Dickel and Jones 1985). The general expectations for the expansion of a massive star are that it will first interact with the dense wind ejected in the red supergiant phase, it will then approach free expansion in the bubble created by the fast wind lost in the main sequence phase, and will finally interact with the swept up wind bubble shell. The interaction with the interstellar medium takes place in a late evolutionary phase. If Type Ib supernovae have Wolf-Rayet star progenitors, the dense wind does not occur close to the stellar surface, but may be present further out from an earlier evolutionary phase. In this case, there is likely to be interaction with a shell of swept up red supergiant wind. Type Ia supernovae may interact more directly with the interstellar medium.

Multidimensional calculations are needed for studying the supernova remnant interaction with clouds (see McKee 1988 for a recent review). Although it was not intended to represent supernova expansion, Woodward's (1976) calculation of flow past a cloud has a particularly accurate treatment of the cloud boundary. His calculation demonstrated the growth of Rayleigh-Taylor and Kelvin-Helmholtz instabilities along the cloud boundary. However, the calculation stopped before the shock wave in the cloud had completely transversed the cloud. After this initial phase, the ram pressure of the flow continues to accelerate the cloud. McKee, Cowie, and Ostriker (1978) calculated cloud motion in this phase on the assumption of a constant cloud cross section. Instabilities are likely to modify the cross section and may break up the cloud (Nittman, Falle, and Gaskell 1982). The final evolution of clouds is still not known.

The overall evolution of supernova remnants in their late radiative phases in a uniform medium has been described by numerical computations. Cioffi, McKee, and Bertschinger (1987) have recently obtained an analytic description of the remnant evolution in these late phases. Some particular topics regarding radiative shock waves are discussed in section 7.

5. CIRCUMSTELLAR INTERACTION

There is excellent evidence for interaction with a dense circumstellar wind for the Type II supernovae SN1979c and SN1980k (see reviews by Chevalier 1984a and Fransson 1986b). The evidence includes radio emission from the interaction region for both supernovae (Weiler et al. 1986), infrared dust echoes for both supernovae (Dwek 1983), thermal X-ray emission from the interaction region in SN1980k (Canizares et al. 1982), and the ultraviolet line emission from highly ionized atoms in SN1979c (Fransson et al. 1984). The radio emission is a particularly good diagnostic because the early absorption of the radio emission can be interpreted as free-free absorption by the preshock gas and an estimate of the circumstellar density is obtained. Lundqvist and Fransson (1987) have made a detailed study of the temperature and ionization of the circumstellar gas in the radiation field of the supernova and have been able to reproduce detailed features of the radio light curves. They derive mass loss rates of 12×10^{-5} $M_\odot$ yr^{-1} and 3×10^{-5} $M_\odot$ yr^{-1} for a wind velocity $v_w = 10$ km s^{-1} for SN1979c and SN1980k respectively.

There are presently 5 radio supernovae with fairly extensive data, including the rising part of the radio light curve. They are SN1979c, SN1980k, SN1983n (Sramek, Panagia, and Weiler 1984), SN1986j (Rupen et al. 1987), and SN1987a (Turtle et al. 1987). Table II lists the supernova type, the time of optical depth at 20 cm, t_{20}, and the circumstellar density given in terms of the presupernova mass loss rate divided by the wind velocity. SN1986j was probably not observed near maximum light and the Type II designation given here is based solely on the presence of hydrogen line emission. Rupen et al. (1987) have suggested a Type V designation. Of the 5 supernovae, only SN1979c and SN1980k show clear evidence for circumstellar interaction outside of

TABLE II. RADIO SUPERNOVAE

Supernova	Type	t_{20} (days)	$\dot{M}/v_w$ ($M_\odot$ yr^{-1})/(km s^{-1})	
1979c	II	950	1×10^{-5}	a
1980k	II	190	3×10^{-6}	a
1983n	Ib	30	5×10^{-7}	b
1986j	II	1600	2×10^{-5}	c
1987a	II	2	1×10^{-8}	d

a. Lundqvist and Fransson, 1987
b. Chevalier 1984b; Sramek, Panagia, and Weiler 1984
c. Chevalier 1987
d. Chevalier and Fransson 1987

radio wavelengths. For SN1983n and SN1987a this is attributable to the low circumstellar density and for SN1986j to the late discovery.

The results show that the winds around SN1979c, 1980k, and 1986j are consistent with the dense slow winds expected around red supergiant stars. The density around the Type Ib event SN1983n is considerably lower, but it roughly consistent with the value expected around a Wolf-Rayet star. A wind velocity of 1000 km s^{-1} and $\dot{M} = 10^{-4}$ $M_\odot$ yr^{-1} leads to a value of $\dot{M}/v_w$ that is a factor of 5 below the estimated value. SN1987a had an even earlier turn-on and was a faint radio supernova, but the estimated value of $\dot{M}/v_w$ is roughly consistent with the density expected around a B3 Ia star like the Sk-69 202 progenitor star (Chevalier and Fransson 1987). The observational estimate is again a factor of a few larger than the expected value. If there is clumping in the circumstellar wind, the observational estimates are reduced.

In the circumstellar interaction model for the radio emission, the radio luminosity at a given age should be correlated with the density of circumstellar material. This is observed. It appears that the circumstellar interaction does give information on the properties of the supernova progenitor.

An exciting development is the possibility of resolving radio supernovae with very long baseline interferometry (VLBI) techniques. The expansion of SN1979c has been measured (Bartel et al. 1985; Bartel 1986) and Bartel (private communication) has estimated that if the expansion follows $R \propto t^m$, then $m = 0.9 \pm 0.1$. The radius and the expansion law are consistent with circumstellar interaction. SN1986j, which is currently the brightest radio supernova, has also been resolved by VLBI observations (Bartel, Rupen, and Shapiro 1987). Measurements of the expansion of the radio source should allow the age of the supernova to be estimated.

During the next phase of evolution the supernova may approach free expansion in a low density wind bubble. When SN1979c enters this phase, an accelerated rate of decline of the radio emission is expected. One way to identify supernova remnants in this evolutionary stage would be to search for pulsar nebulae which show little or no evidence for interaction with a surrounding medium. This may be the explanation for "Crabs without shells" which are about equal in number to the Crabs with shells (Helfand and Becker 1987). Of course the Crab Nebula itself is lacking a shell.

Of the young supernova remnants, Cassiopeia A is the most likely to be related to Type Ib supernovae. It has fast moving oxygen-rich gas and is interacting with dense nitrogen-rich circumstellar gas; Fesen, Becker, and Blair (1987) have suggested that the progenitor was a Wolf-Rayet WN star. The problems with the Type Ib identification are that Cas A has recently been found to have fast-moving hydrogen-rich gas (Fesen et al. 1987) and the supernova was probably too faint to be a typical Type Ib event, even if it was observed by Flamsteed in 1680 (Ashworth 1980). Although the presence of hydrogen would appear to rule out a Type I supernova, it is perhaps possible that the hydrogen would not have been detectable spectroscopically near maximum light. If the progenitor was a Wolf-Rayet star, it is likely that the slowly moving, nitrogen-rich gas is in a shell (see section 3). This has not generally been taken into account in discussions of the dynamics of Cas A.

A remnant with oxygen-rich ejecta which is likely to be interacting with circumstellar gas is N132D in the Large Magellanic Cloud. Optical studies of this remnant show a faint outer shell of radius 40 pc, an inner disk of radius 16 pc, and a ring of expanding oxygen-rich ejecta at a radius of 3 pc (Lasker 1978, 1980). X-ray emission from the remnant has approximately a 16 pc radius (Mathewson et al. 1983). The inner oxygen knots give an age of only 1300 years, which implies that the X-ray emitting gas has been expanding in a low density medium if it is a normal supernova (Hughes 1987). Hughes (1987) suggests that the cavity was created by an HII region. A fast stellar wind may also play a role. The present X-ray remnant is then the result of interaction with the swept up red supergiant wind. The faint outer shell may be a remnant of the stellar wind and photoionizing radiation while the progenitor star was on the main sequence. This remnant is then a particularly good example of the type of circumstellar environment expected for a Wolf-Rayet star.

An interesting recent suggestion is that Kepler's supernova remnant is the result of a Type Ib supernova. An analysis of the X-ray emission indicates that the initial stellar mass was $> 7 M_\odot$ (Hughes and Helfand 1985), although this is rather uncertain. Bandiera (1987) has argued that the progenitor star was a runaway Wolf-Rayet star from the galactic plane. Proper motion studies of the dense optical knots do imply a high space velocity (van den Bergh and Kamper 1977) and the asymmetry of the supernova remnant may be due to the interaction of presupernova mass loss with an ambient medium (Bandiera 1987).

There is evidence that some larger supernova remnants are interacting with circumstellar gas. The Cygnus Loop, with a radius of 20 pc, is interacting with density $\gtrsim 5$ cm^{-3} over much of its surface area, yet the appearance of the remnant shows circular symmetry. The X-ray properties imply that the remnant has been expanding in a low density medium until recently (Charles, Kahn, and McKee 1985). The implication is that the remnant is interacting with a spherical shell of radius 20 pc. Since this radius is smaller than the bubbles expected around O stars, Charles, Kahn, and McKee (1985) suggest that the progenitor star was an early B star.

6. INTERSTELLAR INTERACTION

The physical properties of young supernova remnants are probably best studied by X-ray spectroscopy and there have been a number of recent theoretical studies in this area. The first case to be examined in detail was the X-ray emission from a self-similar Sedov blast wave (Gronenschild and Mewe 1982; Hamilton, Sarazin, and Chevalier 1983). The properties of the emission are determined by two parameters, e.g. n_o^2 E and t where n_o is the ambient density, E is the total energy, and t is the age. The most important property of the flow is that the gas is underionized compared to equilibrium values because ionization timescales can be longer than the hydrodynamic timescales. Since underionization can favor line emission, the X-ray luminosity from a nonequilibrium flow may be a factor of 10 higher than that from an equivalent flow assumed to be in ionization equilibrium.

It is unknown to what extent electrons are heated in collisionless shock fronts so that the amount of heating is often a parameter in theoretical studies (e.g. Hamilton, Sarazin, and Chevalier 1983). Even if collisionless heating does take place in the shock, electrons may be released by ionization in the postshock flow which have not been subject to this heating. Itoh (1984) noted that some fast shocks appear to be moving into a partially neutral medium and that the electrons released from the neutrals in the postshock flow are only subject to Coulomb heating. Hamilton and Sarazin (1984c) noted a similar process for the postshock ionization and heating of a heavy element gas. In either case, it is necessary to take into account two populations of electrons.

Hamilton and Sarazin (1984b) found that the X-ray emission from a variety of self-similar flows can be estimated without carrying out detailed calculations for each case. Two important parameters are an ionization time, τ, which is weighted by a Boltzmann factor and an emissivity parameter, ε, which is a function of radius. Two supernova remnants that have similar values of τ versus ε through the remnant belong to the same structural type. The two basic types are the Sedov type, which approaches a hot, low density medium in the postshock flow, and a type which approaches a cold, high density medium in the post-shock flow. The interaction of a steep power law density profile with an $s = 0$ medium is of the first type. The interaction of a steep profile with an $s = 2$ medium and the reverse shock wave for uniform ejecta are of the second type. Two remnants of the same type that have similar values of average τ, average temperature, and total emissivity are expected to produce similar X-ray spectra. The fact that detailed spectra have been calculated for Sedov blast waves makes this method quite useful.

A more general way to calculate X-ray spectra is to solve the time dependent ionization equations along with a numerical hydrodynamic computation (Itoh and Fabian 1984; Nugent et al. 1984; Hughes and Helfand 1985). Hughes and Helfand (1985) have developed a useful matrix method that speeds up the calculation of the ionization equations.

As discussed in the previous section, the explosions of massive stars are likely to interact with circumstellar matter during their early phases. Direct interaction with the interstellar medium is most likely for Type Ia supernovae and Tycho's supernova (SN 1572) and SN1006 may belong to this class. Detailed modeling of the X-ray spectra of these remnants has been carried out by Hamilton, Sarazin, and Szymkowiak (1986a,b). They find that in both cases, the spectra are best fit by models of the second type discussed above, i.e. models with cool dense ejecta. A range of temperatures is needed to give an approximate power law continuum and to produce emission that approximates ionization equilibrium. Hamilton et al. concentrate on models in which constant density ejecta with a sharp edge expand into the interstellar medium. For both supernova remnants, the models can accomodate $\gtrsim 0.5 \, M_\odot$ of Fe as expected in a Type Ia supernova because the Fe is either unshocked or is at low density. The presence of cold Fe in SN1006 appears to be confirmed by the presence of broad ultraviolet Fe absorption in the direction of the Schweizer-Middleditch star (Wu et al. 1983; Fesen et al. 1987).

The X-ray spectra of Tycho and of SN1006 are very different in that Tycho shows strong line emission while SN1006 does not. Hamilton et al. attribute this to a low density surrounding SN1006 so that its "ionization age" is less than that of Tycho and the ionization has not yet proceeded to the stage which gives X-ray line emission. Kirshner, Winkler, and Chevalier (1987) have recently confirmed this hypothesis by measuring the Balmer line emission from two remnants and using it to estimate shock velocities. SN1006 has a higher shock velocity even though it is an older remnant, which implies it is expanding into a low density medium.

The models of Hamilton et al. appear to be very promising but they do not allow for the presence of a steep outer power law component to the density profile that is expected from supernova modeling (see section 2). A related problem may be that the expansion rate of the supernova remnants in the models are larger than is indicated by optical and radio observations. A power law region with n somewhat greater than 5 would lead to greater deceleration of the outer shock front. Possible resolutions of these problems are that there is an outer power law region but with a relatively small amount of mass or that clumping of the ejecta plays an important role.

7. RADIATIVE SHOCK WAVES

The general agreement between radiative shock wave models and the optical emission from most supernova remnants strongly argues that we are observing emission from shock waves (McKee and Hollenbach 1980; Raymond 1984 and references therein). The shock wave emission can then be used to determine basic physical properties such as the shock velocity, the preshock density, and the elemental abundances. For example, Blair, Kirshner, and Chevalier (1981) used optical observations of the supernova remnants in M31 to determine the abundance gradient in the galaxy. Raymond et al. (1987) have extensively examined a radiative shock wave in the Cygnus Loop and deduced that the postshock thermal pressure was substantially less than the total shock ram pressure, $\rho_o v_{sh}^2$, which suggests that some form of nonthermal pressure comes to dominate in the postshock flow.

The striking morphology of optical supernova remnants has been interpreted in a number of ways. For example, Cowie, McKee, and Ostriker (1981) and Fesen, Blair, and Kirshner (1982) suggested that the shock emission from the Cygnus Loop is from small clouds (size < 1 pc) that have been overrun by a blast wave in an intercloud medium. However, detailed studies of the surface brightness variations around a filament together with high-resolution spectra of individual lines have shown that the emission is from a wavy sheet that is observed edge on in places (Hester, Parker, and Dufour 1983; Hester 1987). The presence of a wavy sheet suggests the operation of an instability. In fact, the shell built up by a radiative shock wave with an internal pressure is overstable (Chevalier and Theys 1975; Vishniac 1983). The properties of the instability may be modified if magnetic pressure dominates in the cool layer. An argument against the instability is that the H-alpha, nonradiative shock emission from the Cygnus Loop also appears to be from a wavy sheet.

Another instability can operate in the cooling region of a shock front. Chevalier and Imamura (1982) showed that a radiative shock front with a cooling function of the form $\Lambda \propto T^\alpha$ is overstable to planar perturbations if $\alpha \lesssim 0.5$. For a realistic model of an interstellar shock wave, it is necessary to put together time-dependent hydrodynamic and non-ionization equilibrum codes. Innes, Giddings, and Falle (1987a,b) and Gaetz, Edgar, and Chevalier (1987) have done this and found that instability occurs for $v_{sh} \gtrsim 140$ km s^{-1}. While this is larger than the values of v_{sh} generally inferred from observations of radiative shock waves, the instability can cause the shock velocity to vary over a range of velocities that overlaps with the observed range. There are some shock emission regions where the steady-state shock models do not provide a good fit to the relative line intensities and some type of time-dependent behavior is likely (Raymond et al. 1980, 1987; Fesen et al. 1982). This is generally interpreted as being due to a recent onset of cooling, but the cooling instability could also play a role. Regions where a cooling shock wave is just forming are the most likely to show the instability; the numerical calculations of Falle (1981) did show oscillations at just this time.

8. HOT INTERSTELLAR MEDIUM

The basic theoretical point on the hot interstellar medium is that supernova remnants create large volumes of hot gas, which has a relatively long cooling time. The supernova remnants can overlap so that hot gas created by one supernova is re-heated by a subsequent supernova; Cox and Smith (1974) showed that this is a plausible occurrence and that a significant volume of the interstellar medium contains hot gas. McKee and Ostriker (1977) argued that a medium with a density of 0.1-0.3 cm^{-3} and temperature of about 10^4K would be transformed into a hot ($3 \times 10^5 - 10^6$K) low density (3×10^{-3} cm^{-3}) medium by the action of supernovae. Since the supernova remnants then occupy most of the interstellar medium, they determine the interstellar pressure and the evolution of other supernova remnants. There are some uncertainties in this argument. One is that Type II supernovae are likely to be strongly correlated in space in OB associations, which can limit the total volume that they affect. A more serious uncertainty is the total interstellar pressure. Cox (1988) has suggested that the nonthermal pressure is several times the thermal pressure, based on the total weight of halo gas. A high interstellar pressure stops the expansion of hot gas at a relatively small radius and the filling factor of hot gas is reduced.

McKee and Ostriker (1977) developed a model for how supernova remnants would evolve in a low density medium containing clouds. The main effect is that the outer shock front would be moving into low density gas and that gas would be released by thermal evaporation of clouds inside the remnant. Thus X-ray emission would peak well inside the outer shock front. Observations of older supernova remnants have not shown this effect; the X-ray emission appears to peak at the shock front, as expected for a substantial pre-shock density. For the remnants of Type II supernova, the absence of evidence for expansion in a low density medium can be attributed to interaction with circumstellar matter (section 5 and McKee 1988). The Cygnus Loop is a possible example. However, Type Ia supernovae might be expected to interact directly with the interstellar medium. SN1572 appears to be interacting with gas of density 0.3 cm^{-3} and SN1006 with gas of density about 0.02 cm^{-3} (e.g. Kirshner, Winkler, and Chevalier 1987). In both cases, the presence of optical hydrogen line emission implies that the remnants are expanding into partially neutral gas. SN1572 is at a galactic altitude of 60 pc and SN1006 at an altitude of 500 pc. The deduced densities at these heights are comparable to those found in the neutral hydrogen study of Lockman (1984) if the gas is in fact diffuse.

There is little direct observational evidence on the hot interstellar medium. There is a soft X-ray background, but it is likely to be local to the solar neighborhood (Cox and Reynolds 1987). The observations of OVI by the Copernicus Observatory give information on gas with a temperature of about 3×10^5K. The ionization potential of O^{+4} is sufficiently high that collisional ionization is much more likely than photoionization. Jenkins (1978) analyzed the observations and found that gas with OVI occupies about 20% of the volume of the interstellar medium. Since the OVI is expected to be present at a lower temperature than that of the typical hot interstellar medium, it appears in an intermediate phase. One theory for the formation of OVI involves conductive interfaces around interstellar clouds or circumstellar shells (Weaver et al. 1977; Cowie 1987). Cowie (1987) estimates that 10 $M_\odot$ yr^{-1} of gas must be evaporated in order to produce the observed column densities. There is some question whether heat conduction operates efficiently in the interstellar medium because of magnetic and plasma effects. OVI must be produced in shock heated gas, but the gas velocities would be too large to be compatible with observations of OVI line profiles. Another possibility is that the hot gas comes into pressure equilibrium with the interstellar medium and radiatively cools, producing OVI. The observed column densities can be produced with a cooling mass flux of less than 1 $M_\odot$ yr^{-1} (Edgar and Chevalier 1986).

If hot gas does have a large filling factor, it will naturally expand into the galactic halo, creating a galactic corona. Spitzer (1956) argued for the presence of such a corona in order to provide pressure balance for high altitude clouds. Shapiro and Field (1976) noted that hot gas produced by supernovae might cool in the halo, creating a galactic fountain. Chevalier and Oegerle (1979) found that a mass circulation rate of about 10 $M_\odot$ yr^{-1} is needed in order to produce the pressure balance suggested by Spitzer. There are a number of attractive features of a model with such a circulation rate. (1) The cooling gas can produce clouds which fall to the galactic plane and are observed as high-velocity clouds (Bregman 1980). (2) The cooling gas produces a column density of NV that is in approximate accord with absorption line observations (Edgar and Chevalier 1986; Savage and Massa 1987). (3) The observed emission line strength of CIV (Martin and Boyer 1986) is reproduced.

A different kind of model is one in which halo gas has nonthermal support by cosmic rays or magnetic fields. This model was originally proposed on the basis of observations of synchrotron radiation from outside the galactic disk (e.g. Badhwar and Stephans 1977). Lockman (1984) has observed an HI layer extending to 1 kpc from the galactic disk. If this gas is truly diffuse and is not in high-velocity clouds, it requires nonthermal support. As noted above, SN1006 may be expanding into such a medium. The cool high altitude gas is photoionized by stars and by the extragalactic background radiation. Column densities of CIV and SiIV produced in such a model are compatible with absorption line observations (Pettini and West 1982; Chevalier and Fransson 1984; Fransson and Chevalier 1985), but NV may be underproduced by a factor of 10. However, there is considerable uncertainty in the far ultraviolet-soft X-ray extragalactic background radiation, so photoionization cannot be definitely ruled out. An attractive feature of the photoionization model is that the extragalactic ionizing radiation penetrates down to a height of 1 kpc from the galactic place. The observations of SiIV, CIV, and NV suggest that the densities of these ions increase at this distance from the plane (Pettini and West 1982; Savage and Massa 1987).

Both of these models have attractive features and some combination of the two may be indicated. However, the hot model does require a large filling factor for the hot gas, which would be a problem for the nonthermal support model. I suspect that only one interpretation is correct. A good test of the hot model is the prediction of a substantial column density, about 6×10^{14} cm^{-2}, of OVI in the corona (Edgar and Chevalier 1986). Photoionization is not a plausible source of this ion.

9. CONCLUSIONS

Although it often seems as if a separate physical picture is needed for each supernova and supernova remnant, some general trends in the interpretation of these objects are becoming clear. For massive stars, mass loss plays a crucial role both for the supernova explosions and their remnants. Type Ib events may be closely related to Type II supernovae, but have lost their hydrogen envelopes in presupernova evolution. Radio supernovae give good evidence for the expansion of massive explosions into the nearby circumstellar wind. The further evolution may be related to a wind bubble and its associated shell. Current observations of Type Ia supernovae are consistent with direct interaction with the interstellar medium. They are not observed as radio supernovae (Weiler et al. 1986) and the remnants of SN1572 and SN1006 appear to be interacting with the interstellar medium.

With regard to hydrodynamical modeling, spherically symmetric models for the interaction of a supernova with an ambient medium have now been calculated in considerable detail. However, there is the expectation of hydrodynamic instabilities and the observational data show evidence for clumpiness and mixing. It will eventually be important to clarify the three-dimensional evolution of supernovae and their remnants.

The study of supernova remnants has reached the state where such detailed modeling is warranted. On the other hand, the study of the hot interstellar medium is still in its infancy. We do not yet have a well-accepted model for how it permeates the galactic disk and halo or for the dominant physical processes that determine its evolution. More observational data, particularly of ultraviolet resonance lines, are necessary.

This work was supported in part by NSF grant AST-8615555 and NASA grant NAGW-764.

REFERENCES

Abbott, D. C. 1982, *Ap. J.*, **263**, 723.
Arnett, W. D. 1987, *Ap. J.*, **319**, 136.

Ashworth, W.B. 1980, J. Hist. Astr., 11, 1.
Axelrod, T. S. 1980, Ph.D. thesis, University of California, Santa Cruz.
Badhwar, G. D. and Stephans, S. A. 1977, Ap. J., 212, 494.
Bandiera, R. 1987, Ap. J., 319, 885.
Bartel, N. 1986, Highlights of Astr., 7, 655.
Bartel, N., Rogers, A. E. E., Shapiro, I. I., Gorenstein, M. W., Gwinn, C. R., Marcaide, J. M., and Weiler, K. W. 1985, Nature, 318, 25.
Bartel, N., Rupen, M. and Shapiro, I. I. 1987, IAU Circ. 4292.
Bedogni, R. and d'Ercole, A. 1987, Astr. Ap., submitted.
Begelman, M. C. and Sarazin, C. L. 1986, Ap. J. (Letters), 302, L59.
Blaauw, A. 1985, in Birth and Evolution of Massive Stars and Stellar Groups, W. Boland and H. van Woerden, eds. (Dordrecht: Reidel), p. 211.
Blair, W. P., Kirshner, R. P., and Chevalier, R. A. 1981, Ap. J., 254, 50.
Branch, D., Doggett, J. B., Nomoto, K., and Thielemann, F. K. 1985, Ap. J., 294, 619.
Branch, D. 1986, Ap. J. (Letters), 300, L51.
Bregman, J. N. 1980, Ap. J., 236, 577.
Canizares, C. R., Kriss, G. A., and Feigelson, E. D. 1982, Ap. J. (Letters), 253, L17.
Charles, P. A., Kahn, S. M., and McKee, C. F. 1985, Ap. J., 295, 456.
Chevalier, R. A. 1981, Ap. J., 246, 267.
Chevalier, R. A. 1982, Ap. J., 258, 790.
Chevalier, R. A. 1984a, Ann. N.Y. Acad. Sci., 422, 215.
Chevalier, R. A. 1984b, Ap. J. (Letters), 285, L63.
Chevalier, R. A. 1986, Highlights of Astr., 7, 599.
Chevalier, R. A. 1987, Nature, in press.
Chevalier, R. A. 1988, in Landecker and Rogers 1988.
Chevalier, R. A. and Fransson, C. 1984, Ap. J. (Letters), 279, L43.
Chevalier, R. A. and Fransson, C. 1987, Nature, 328, 44.
Chevalier, R. A. and Imamura, J. N. 1982, Ap. J., 261, 543.
Chevalier, R. A. and Imamura, J. N. 1983, Ap. J., 270, 554.
Chevalier, R. A. and Kirshner, R. P. 1979, Ap. J., 233, 154.
Chevalier, R. A. and Oegerle, W. R. 1979, Ap. J., 227, 398.
Chevalier, R. A. and Theys, J. C. 1975, Ap. J., 195, 53.
Chiosi, C. and Maeder, A. 1986, Ann. Revs. Astr. Ap., 24, 329.
Chu, Y.-H., Treffers, R. R., and Kwitter, K. B. 1983, Ap. J. Suppl., 53, 937.
Cioffi, D. F., McKee, C. F., and Bertschinger, E. 1987, Ap. J., submitted.
Cowie, L. L. 1987, in Physical Processes in the Interstellar Medium, eds. D. J. Hollenbach and H. A. Thronson, Jr. (Dordrecht: Reidel), p. 245.
Cowie, L. L., McKee, C. F., and Ostriker, J. P. 1981, Ap. J., 247, 908.
Cox, D. P. 1988, in Landecker and Rogers 1988.
Cox, D. P. and Edgar, R. J. 1983, Ap. J., 265, 443.
Cox, D. P. and Reynolds, R. J. 1987, Ann. Rev. Astr. Ap., 25, 303.
Cox, D. P. and Smith, B. W. 1974, Ap. J. (Letters), 189, L105.
Dickel, J. R. and Jones, E. M. 1985, Ap. J., 288, 707.
Dwek, E. 1983, Ap. J., 274, 175.

Edgar, R. J. and Chevalier, R. A. 1986, Ap. J. (Letters), 310, L27.
Ensman, L. and Woosley, S. E. 1987, in preparation.
Fabian, A. C., Brinkmann, W., and Stewart, G. C. 1983, in IAU Symposium 101, Supernova Remnants and Their X-Ray Emission, I. J. Danziger and P. Gorenstein, eds. (Dordrecht: Reidel), p. 83.
Falle, S. A. E. G. 1981, M.N.R.A.S., 195, 1011.
Fesen, R. A., Becker, R. H., and Blair, W. P. 1987, Ap. J., 313, 378.
Fesen, R. A., Blair, W. P., and Kirshner, R. P. 1982, Ap. J., 262, 171.
Fesen, R. A., Wu, C. -C., Leventhal, M., and Hamilton, A. J. S. 1987, preprint.
Filippenko, A. V. and Sargent, W. L. W. 1986, A. J., 91, 691.
Fransson, C., et al. 1984, Astr. Ap., 132, 1.
Fransson, C. 1986a, Highlights of Astr., 7, 611.
Fransson, C. 1986b, in Radiation Hydrodynamics in Stars and Compact Objects, D. Mihalas and K. H. A. Winkler, eds. (Berling: Springer), p. 141.
Fransson, C. and Chevalier, R. A. 1986, Ap. J., 296, 35.
Gaetz, T. J., Edgar, R. J., and Chevalier, R. A. 1987, Ap. J., submitted.
Gaskell, C. M., Cappellaro, E., Dinerstein, H. J., Garnett, D. R., Harkness, R. P. and Wheeler, J. C. 1986, Ap. J. (Letters), 306, L77.
Gronenschild, E. H. B. M. and Mewe, R. 1982, Astr. Ap. Suppl., 48, 305.
Hamilton, A. J. S. 1985, Ap. J., 291, 523.
Hamilton, A. J. S. and Sarazin, C. L. 1984a, Ap. J., 281, 682.
Hamilton, A. J. S. and Sarazin, C. L. 1984b, Ap. J., 284, 601.
Hamilton, A. J. S. and Sarazin, C. L. 1984c, Ap. J., 287, 282.
Hamilton, A. J. S., Sarazin, C. L., and Chevalier, R. A. 1983, Ap. J. Suppl., 51, 115.
Hamilton, A. J. S., Sarazin, C. L., and Symkowiak, A. E. 1986a, Ap. J., 300, 698.
Hamilton, A. J. S., Sarazin, C. L., and Symkowiak, A. E. 1986b, Ap. J., 300, 713.
Helfand, D. J. and Becker, R. H. 1987, Ap. J., 314, 203.
Herman, J. 1985, in Mass Loss from Red Giants, eds. M. Morris and B. Zuckerman (Dordrecht: Reidel), p. 215.
Hester, J. J. 1987, Ap. J., 314, 187.
Hester, J. J., Parker, R. A. R., and Dufour, R. J. 1983, Ap. J., 273, 219.
Hillebrandt, W., Hoflich, P. Truran, J. W., and Weiss, A. 1987, Nature, 327, 597.
Hughes, J. P. 1987, Ap. J., 314, 103.
Hughes, J. P. and Helfand, D. J. 1985, Ap. J., 291, 544.
Humphreys, R. M. and Davidson, K. 1979, Ap. J., 232, 409.
Iben, I. and Tutukov, A. V. 1984, Ap. J. Suppl., 32, 351.
Innes, D. E., Giddings, J. R., and Falle, S. A. E. G. 1987a, M.N.R.A.S., 224, 179.
Innes, D. E., Giddings, J. R., and Falle, S. A. E. G. 1987b, M.N.R.A.S., 226, 67.
Itoh, H. 1984, Ap. J., 285, 601.
Itoh, H., and Fabian, A. C. 1984, M.N.R.A.S., 208, 645.
Jenkins, E. B. 1978, Ap. J., 220, 107.

Jones, E. M. and Smith, B. W. 1983, in IAU Symposium 101, Supernova Remnants and Their X-Ray Emission, I. J. Danziger and P. Gorenstein, eds. (Dordrecht: Reidel), p. 83.
Kennicutt, R. C. 1984, Ap. J., 277, 361.
Kirshner, R., Nassiopoulas, G. E., Sonneborn, G., and Crenshaw, D. M. 1987, Ap. J., in press.
Kirshner, R. P. and Oke, J. B. 1975, Ap. J., 200, 574.
Kirshner, R. P., Winkler, P. F., and Chevalier, R. A. 1987, Ap. J. (Letters), 315, L135.
Korobeinikov, B. P. 1956, J. Acad. Sci. Soviet Union, 109, 271.
Kwitter, K. B. 1981, Ap. J., 245, 154.
Kwitter, K. B. 1984, Ap. J., 287, 840.
Lamb, S. A. 1978, Ap. J., 220, 186.
Landecker, T. and Rogers, R. 1988, eds. IAU Colloquium 101, The Interaction of Supernova Remnants with the Interstellar Medium (Cambridge Univ. Press: Cambridge), in press.
Lasker, B. M. 1978, Ap. J., 223, 109.
Lasker, B. M. 1980, Ap. J., 237, 765.
Lockman, F. J. 1984, Ap. J., 283, 90.
Lundqvist, P. and Fransson, C. 1987, Astr. Ap., in press.
Martin, C. and Bowyer, S. 1986, Bull. A. A. S., 18, 1036.
Mathewson, D. S., Ford, V. L., Dopita, M. A., Tuohy, I. R., Long, K. S., and Helfand, D. J. 1983, Ap. J. Suppl., 51, 345.
McCray, R. A. 1983, Highlights of Astr., 6, 565.
McKee, C. F. 1988, in Landecker and Rogers 1988.
McKee, C. F., Cowie, L. L., and Ostriker, J. P. 1978, Ap. J. (Letters), 219, L23.
McKee, C. F. and Hollenbach, D. J. 1980, Ann. Rev. Astr. Ap., 18, 219.
McKee, C. F. and Ostriker, J. P. 1977, Ap. J., 218, 148.
McKee, C. F., Van Buren, D., and Lazareff, B. 1984, Ap. J. (Letters), 278, L115.
Muller, R. and Arnett, W. D. 1986, Ap. J., 307, 619.
Nadyozhin, D. K. 1985, Ap. and Sp. Sci., 112, 225.
Nittman, J., Falle, S., and Gaskell, P. 1982, M.N.R.A.S., 201, 833.
Nomoto, K., Thielemann, F. -K., and Yokoi, K. 1984, Ap. J., 286, 644.
Nugent, J. J., Pravdo, S. H., Garmire, G. P., Becker, R. H., Tuohy, I. R., and Winkler, P. F. 1984, Ap. J., 284, 612.
Panagia, N. 1973, A. J., 78, 929.
Panagia, N., Sramek, R. A., Weiler, K. W. 1986, Ap. J. (Letters), 300, L55.
Pettini, M. and West, K. A. 1982, Ap. J., 260, 561.
Raymond, J. C. 1984, Ann. Rev. Astr. Ap., 22, 75.
Raymond, J. C., Black, J. H., Dupree, A. K., Hartmann, L., and Wolff, R. S. 1980, Ap. J., 238, 881.
Raymond, J. C., Hester, J. J., Cox, D., Blair, W. P., Fesen, R. A., and Gull, T. R. 1987, Ap. J., submitted.
Rupen, M. P., van Gorkom, J. H., Knapp, G. R., Gunn, J. E., and Schneider, D. P. 1987, A. J., in press.
Savage, B. D. and Massa, D. 1987, Ap. J., 314, 380.
Schaeffer, R., Casse, M., and Cahen, S. 1987, Ap. J. (Letters), 316, L31.

Sedov, L. 1959, Similarity and Dimensional Methods in Mechanics, Academic Press, New York.
Shapiro, P. R. and Field, B. G. 1976, Ap. J., 205, 762.
Snow, T. P. 1981, Ap. J., 251, 139.
Solinger, A., Rappaport, S., and Buff, J. 1975, Ap. J., 201, 381.
Spitzer, L., Jr. 1956, Ap. J., 124, 20.
Sramek, R. A., Panagia, N., and Weiler, K. W. 1984, Ap. J. (Letters), 285 (L59.
Sutherland, P. G. and Wheeler, J. C. 1984, Ap. J., 280, 282.
Turtle, A. J. et al. 1987, Nature, 327, 38.
van den Bergh, S. 1987, Ap. J., submitted.
van den Bergh, S. and Kamper, K. W. 1977, Ap. J., 218, 617.
van den Bergh, S., McClure, R. D., and Evans, R. 1987, Ap. J., 322, in press.
Vishniac, E. T. 1983, Ap. J., 274, 152.
Weaver, R., McCray, R., Castor, J., Shapiro, P. R. and Moore, R. T. 1977, Ap. J., 218, 377.
Weiler, K. W., Sramek, R. A., Panagia, N., van der Hulst, J. M., and Salvati, M. 1986, Ap. J. (Letters), 301, 790.
Wheeler, J. C. and Levreault, R. 1985, Ap. J., 294, L17.
Woodward, P. R. 1976, Ap. J., 207, 484.
Woosley, S. E. 1987, preprint.
Woosley, S. E., Pinto, P. A., and Ensman, L. 1987, Ap. J., in press.
Wu, C. -C., Leventhal, M., Sarazin, C. L., and Gull, T. R. 1983, Ap. J. (Letters), 269, L5.

6. GALAXIES AND GALACTIC HALOS

OBSERVATIONS OF GALAXIES AND GALACTIC HALOS

Ginevra Trinchieri
Arcetri Observatory
Largo E. Fermi 5
50129 Firenze ITALY

A good diagnostic for the presence of hot thin plasmas is their emission in the far UV and x-ray bands. For this reason, high quality x-ray observations of normal galaxies should provide us with the right tool to detect directly hot extragalactic gas in these systems.

The main results on the x-ray emission from normal galaxies are relatively recent and have been obtained from the data collected with the EINSTEIN Observatory instruments. Previous space missions were not sensitive enough to detect sources as faint as those associated with normal galaxies, which have typical fluxes of $\sim 10^{-12} - 10^{-13}$ ergs cm^{-2}s^{-1}, with the exception of the detection of bright sources in very nearby objects (e.g the Magellanic Clouds, Clark et al 1978). Moreover, sources associated with galaxies turn out to be rather complex, with a morphology that can be properly understood with the aid of high quality images. These were not available prior to the launch of the EINSTEIN Observatory.

This talk will review the main results from the analysis of the x-ray data for normal galaxies, and will concentrate on the observations of hot plasmas in these systems. Galaxies referred to are mainly nearby, relatively isolated and not active, i.e. their emission will be extended on scales comparable to their optical size and not dominated by a single (usually nuclear) source. Galaxies will be separated in two main classes according to their optical morphology, which is closely related to the origin and morphology of the x-ray emission.

SPIRAL AND IRREGULAR GALAXIES

Only for a small sample of spiral and irregular galaxies do we have good x-ray data which can be analyzed in detail. Many more objects have been observed in x-rays, however, and these can be studied statistically, to obtain a more complete picture of the origin and nature of the x-ray emission in late type galaxies. The results of the analysis of the x-ray emission of late type galaxies indicate that:

- spiral and irregular galaxies, including the Milky Way galaxy, seem to be fairly homogeneous in their x-ray properties,
- the x-ray emission of these systems is primarily due to the integrated emission of discrete sources,
- a bright, dominant nucleus is very rarely observed, except for nuclear starburst activity,

- and finally, more relevant for the topic discussed in this school, diffuse, hot thin gas is very difficult to observe in these objects, and the evidence for such a component is still controversial.

Due to the limited spatial resolution and sensitivity of the x-ray detectors, only the brightest sources (typically those with $L_x \geq 10^{37}$ ergs s^{-1} for most objects observed) can be individually detected. These are often identified with close accreting binary systems (a main sequence star orbiting around a collapsed, compact object), or with a supernova remnant. The observations of nearby galaxies, such as M31 (Van Speybroeck et al. 1979), M33 (Long et al. 1981a) and the Magellanic Clouds (Long et al. 1981b, Seward and Mitchell 1981) are all excellent examples. A significant fraction of the emission is usually observed as an extended source, roughly coextensive with the optical image. This component is also likely to be due to many, lower luminosity, individual sources that are not detected individually (see Fabbiano and Trinchieri 1985, 1987 and references therein). Figure 1 gives an example of a good quality x-ray map for a nearby spiral galaxy (M33).

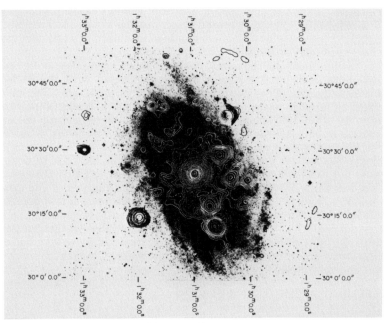

Figure 1. The x-ray map of M33 (IPC data) superposed on an optical image. The x-ray data have a resolution of ~ 45", or 150 pd at the galaxy's distance (from Trinchieri, Fabbiano and Peres 1988).

The complexity of the x-ray emission from these systems, and the presence of many different components, makes the detection of hot, diffuse gas a rather difficult task. With the exception of plasma related to supernova remnants, stellar coronae or starburst activity, the evidence for a truly diffuse hot gas in late type galaxies is still marginal, in spite of the efforts to detect it. Bregman and Glassgold (1982) have looked at NGC 3628 and NGC 4244, both edge-on systems, to detect a hot corona in or above the galactic plane. McCammon and Sanders (1984) have searched for

hot gas in the plane of the face-on object M101. In all three cases, the resulting
x-ray upper limits are over an order of magnitude lower than the supernova energy
deposition estimated for these systems.

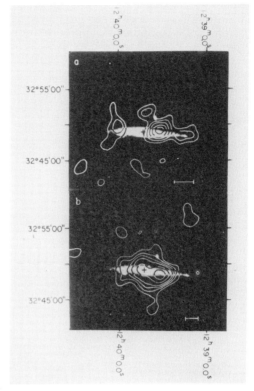

Figure 2. The x-ray image of NGC 4631 in two different energy bands, (0.2-0.8 keV) top, (0.8-4.0) bottom, superposed on the optical image (from Fabbiano and Trinchieri 1987).

There are however two cases in which a soft extended component in the plane is observed, for which one of several possible explanations is that it derives from diffuse, hot thin plasma. The first one, NGC 4631, is an edge-on system shown in Figure 2 in two different energy bands (within the EINSTEIN band). With the exception of the bright source west of the nucleus, the figure indicates the presence of two components distributed quite differently in the plane of the galaxy. The soft component in particular has an x-ray luminosity $L_x \sim 5 \times 10^{39}$ ergs s^{-1}, and it seems to extend further out towards the edges of the disk than does the harder component. The apparent lack of emission from the nucleus or the inner disk region is most likely due to absorption by the intervening material in the galaxy's disk, not important at higher energies. The soft emission is unlikely to be due to stars or supernova remnants; it might be related to the star formation activity in the many HII regions distributed in the disk, or it could instead be emission from diffuse, hot gas in the galaxy's plane (Fabbiano and Trinchieri 1987).

In M33 the evidence for a soft component comes from the results of the spectral analysis. As shown in Figure 1, the region of the galactic disk east of the nucleus is clearly void of bright sources, and the emission appears as a diffuse component in the disk. The spectrum of this 'disk component' is shown in Figure 3, compared to a single temperature bremsstrahlung model. The presence of material along the line of sight should reflect in a sizeable x-ray absorption at low energies. It is quite clear that there is instead an excess over the model, below ~ 1 keV. The observed spectral distribution would not be consistent with a single temperature model with a much lower value for kT: a detailed analysis indicates in fact a lower limit for kT of ~ 2 keV; see Trinchieri, Fabbiano and Peres (1988). The presence of two separate components, approximately one softer and the second one harder than about 1 keV, could instead explain the data. The hard component could be easily accounted for by the integrated contribution of lower luminosity compact sources, as discussed above. The soft component could instead be interpreted as due to the hot diffuse gas.

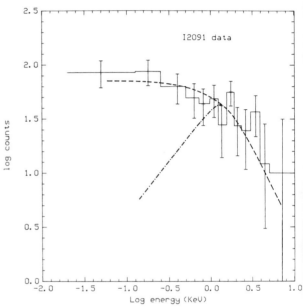

Figure 3. The spectral distribution of the counts from the disk of M33. A model spectrum (kT = 5 keV and no absorption) is indicated by the dashed line. The effect of absorption along the line of sight is shown by the dot-dashed line.

Unfortunately this interpretation is not completely satisfactory, as there are other components that will give a substantial contribution at energies below ~ 1 keV. The estimated x-ray luminosity of this component ranges from a few times 10^{37} to ~ 10^{38} ergs s^{-1}, depending upon the temperature assumed to convert the excess counts into an x-ray flux (kT=1 and 0.2 keV respectively for the above estimates). Coronal emission from the low mass stellar population (typically M stars) will already give a sizeable contribution, of $\geq 10^{37}$ ergs s^{-1}. This estimate does not include the emission from old supernova remnants, which could contribute substantially at low temperatures. It is therefore difficult to estimate the effective contribution of any diffuse hot gas to the soft emission observed in the disk of M33. More and more detailed x-ray observations of this or similar systems are needed before any more definitive statement about the presence and the intensity of the emission from hot diffuse gas in spiral disks can be made.

STARBURST COMPONENTS

The presence of active star formation can modify the above picture for normal galaxies. Star formation activity is in general associated with a higher x-ray emission: the observations of a sample of peculiar galaxies, whose colour and morphology suggest recent star formation, indicate that these objects have on average an x-ray luminosity and an x-ray-to-optical flux ratio f_x/f_b higher than normal galaxies (Fabbiano, Feigelson and Zamorani 1982).

The enhanced x-ray emission in these systems can usually be accounted for by an increased number of newly formed stars, and consequently by a higher fraction of massive binary systems and young supernova remnants (Fabbiano, Feigelson and Zamorani 1982; Stewart et al. 1982). In one case, NGC 5204, the observed emission can be directly related to the young stellar population of O and B stars, present as a result of a recent burst of star formation, and observed in the UV; the total x-ray luminosity can then be simply explained as the integrated contribution from the hot coronae of these stars (Fabbiano and Panagia 1983).

The above results are not exclusively applicable to entire galaxies, but could also explain enhanced emission from regions of star formations within normal galaxies, typically found around their nuclei (e.g. M83, Trinchieri, Fabbiano and Palumbo 1985; IC342, Fabbiano and Trinchieri 1987).

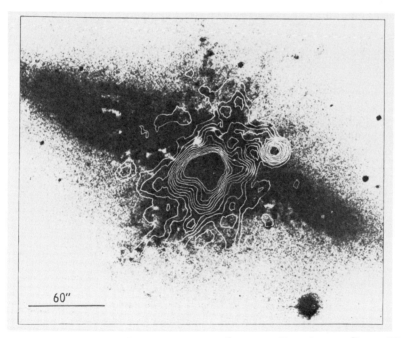

Figure 4. The x-ray map of M82 superposed on an H_α picture (from Watson, Stanger and Griffiths 1984).

Although in most cases the enhanced emission can be explained with a higher fraction of individual x-ray sources, a few examples esulate from such picture: the peculiar, extended features related to the starburst nuclei of M82 (Watson, Stanger and Griffiths 1982) and NGC 253 (Fabbiano and Trinchieri 1984).

The x-ray emission of M82 is shown in Figure 4. As in the x-ray observations of most spiral galaxies, there are point-like sources embedded in a more diffuse emission. In this particular case, however, this latter seems to extend above and below the galactic plane, and to correlate with the presence of H_α emission.

Watson et al. relate this phenomenon to the intense star formation activity that is taking place in the nuclear region of M82, and interpret it as emission due to hot gas, at a temperature of about ten million degrees. The gas has an x-ray luminosity of $L_x \sim 2 \times 10^{40}$ ergs s^{-1}, a mean density of $n_e \sim 0.2$ cm^{-3}, and a total mass of $\sim 10^7 M_\odot$. The estimated number of supernovae going off in the nuclear region would provide enough energy to heat the gas; in fact, only a fraction (~ 2 %) of the supernova energy input would be needed to explain the observed halo luminosity (Watson, Stanger and Griffiths 1984). The gas is most likely unbound to the galaxy, freely expanding outwards: the escape temperature from M82 is of $T \sim 2 \times 10^6$ K, which is also close to the minimum temperature that the gas should have to be detectable with the x-ray instrument. The analysis of the radial distribution of the halo component suggests independently that the gas is unbound: the density distribution n_e decreases with r, the distance from the nucleus, as r^{-2}, which is typical of free expanding winds. If the gas were instead bound to the galaxy, the steep density distribution would imply an uncomfortably large total mass for M82, unlikely to be true (see Fabbiano 1988 for details, and discussion on mass determination from the x-ray gas further on).

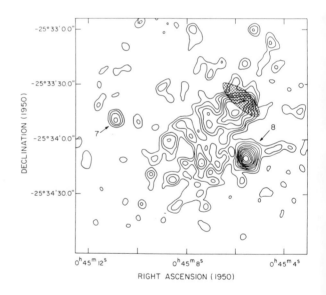

Figure 5. The x-ray map of the inner disk region of NGC 253 (data from the HRI on EINSTEIN). The shaded area indicates the nuclear radio source. Two point-like sources are also evident. (From Fabbiano and Trinchieri 1984).

A second example of gas outflow is given by the central region of NGC 253 shown in Figure 5. The emission is alongated in the south-east direction (which corresponds to the minor axis) and is asymmetric with respect to the nucleus. Fabbiano and Trinchieri (1984) explain this feature with emission from hot gas expanding outwards from the nucleus of NGC 253 and seen projected onto the galaxy's plane, in analogy with M82. In support of this, optical emission lines observed in the same region have also been interpreted as evidence for gas outflow. In this case, the lack of detection of symmetrical x-ray emission from the nucleus can simply be explained as due to obscuration from the intervening galactic plane of NGC 253 (Fabbiano and Trinchieri 1984).

It is evident that the inclination of a galaxy along the line of sight determines whether this kind of phenomenon is observable: if a galaxy is seen face-on, the gas outflowing from the nucleus will be seen projected onto the nuclear region, and its emission will become undistinguishable from any other emission in the nuclear region itself. Therefore, while it is likely that hot gas related to starburst activity is a widespread phenomenon, it is clear that it can only be observed under very favourable conditions, thus making its detection quite rare.

ELLIPTICAL AND S0 GALAXIES

The x-ray data from early type galaxies are not as detailed as those for spiral galaxies: the average distance for the E and S0 galaxies with good x-ray data is higher than for spirals, and consequently both resolution and sensitivity are on average lower. However, a great deal of new information about the properties of these systems has come out as a result of the study of their emission in the soft x-ray band; in particular it is quite relevant for the study of hot extragalactic gas. The observed x-ray properties of early type galaxies can be summarized as follows:

- all galaxies optically brighter than $\sim 10^9 \, L_\odot$ are found to be x-ray sources brighter than $\sim 10^{39}$ ergs s^{-1}, but they can have an x-ray luminosity as high as $\sim 10^{42}$ ergs s^{-1} (Figure 6).

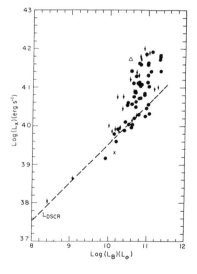

Figure 6. The observed relation between the total x-ray and optical luminosity for a sample of ~ 80 early type objects. The dashed line labelled DSCR indicates the expected contribution from the integrated emission from discrete x-ray sources as a function of the optical luminosity (adapted from Canizares, Fabbiano and Trinchieri 1987).

- the x-ray maps indicate that the emission is not dominated by a single, bright central source, but is usually roughly as extended as the optical emission and it can have a complex morphology (Figure 7).
- the spectral data of the brighter objects indicate a soft thermal spec-

trum with average temperature around $\sim 10^7$ K (although the full range of temperatures spans ~ 0.5 to 4×10^7 K).

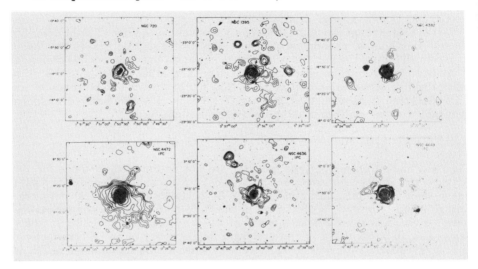

Figure 7. The x-ray maps of six well studied early type galaxies obtained from the IPC data (from Trinchieri, Fabbiano and Canizares 1986).

How do we explain the x-ray emission from these objects? Two components are likely to be present:
1) *discrete sources.* Point-like sources, most likely low mass binary systems, have been observed in bulges of spiral galaxies (in the Milky way, in M31) and in globular clusters. The similarity between spiral bulges and elliptical galaxies and the presence of globular clusters would then suggest a population of x-ray sources in early type systems. This component, however, could only dominate in objects of relatively low x-ray luminosity (Figure 6), while it would fail to explain the observed L_x of a large fraction of the optically luminous galaxies. Moreover, the expected x-ray luminosity from discrete sources should increase almost linearly with optical luminosity, $L_x \propto L_b$, as it is observed in late type galaxies (Fabbiano and Trinchieri 1985). It is evident form Figure 6 that a much steeper relation is needed to fit the data: $L_x \propto L_b^{\sim 1.6}$ (Trinchieri and Fabbiano 1985).
2) *emission from hot, diffuse gas.* Such a component will easily account for some of the observed properties of the x-ray emission in early type galaxies, such as high luminosities, distorted, complex morphologies and soft x-ray spectra.

The most striking example of a complex morphology is given by the tail observed in M86 (Forman et al. 1979), a galaxy in the Virgo cluster: the emission is peaked on the galaxy's nucleus, but there is an x-ray tail clearly not associated with obvious optical emission. Explaining such a feature with binary sources is very hard, while it is quite plausible that we are observing the effect of stripping caused by the motion of the galaxy in the cluster medium (Fabian, Schwarz and Forman 1980). Less striking distortions are observed also in other galaxies: NGC 4472 has a very asymmetric emission, that extends beyond the optical image in the south direction. In analogy with M86, this asymmetry could also be explained as the result of interaction with the ambient medium (NGC 4472 is also in the Virgo cluster), that seems to compress the gas in the northern sector.

The relatively low temperatures measured from the spectral analysis of the emission from bright objects rule out a dominant contribution from bright compact sources, which typically have harder spectra (both galactic binaries and sources in the M31 bulge), but would be consistent with emission from a diffuse, hot gas.

To explain the observed luminosities, bright early type galaxies should contain about $10^9 - 10^{10}$ $M_\odot$ of gas at the observed temperatures. This represents a large fraction of the gas shed by stars during their evolution. Previous attempts to detect interstellar gas in early type galaxies had failed to measure quantities as large as these, and models had been propounded to explain these failures (i.e. galactic winds or continuous star formation, see Faber and Gallagher 1976). Now some of these models will need to be revised.

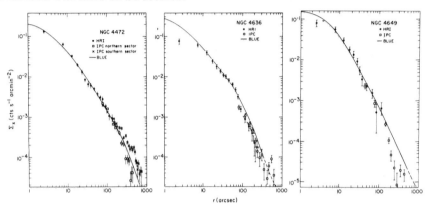

Figure 8. The azimuthally averaged x-ray surface brightness distribution compared to the optical surface brightness for three early type systems (from Trinchieri, Fabbiano and Canizares 1986).

What physical properties of the interstellar gas in early type galaxies can be derived from the x-ray data? Only for a few objects the x-ray data allow a detailed study of their x-ray emission. These can then be used to infer more general properties for the entire class of objects. Figure 8 shows the x-ray surface brightness distribution Σ_x for three well studied early type galaxies. The x-ray emission is clerly a smoothly decreasing function of radius, which can be simply parametrized with a 'king-type' model analogous to that used for the optical profiles, of the kind:

$$\Sigma(r) = \Sigma(0)[1 + (r/a_x)^2]^{-b}$$

a_x is the x-ray core radius, and is in general quite poorly determined by the present x-ray data (mostly obtained at a resolution of $\geq 40"$); b is in general easier to determine and is typically close to unity (Forman, Jones and Tucker 1985).

To first approximation, there is a striking similarity between the x-ray and the optical surface brightness distributions. Departures from this similarity are however observed both in the very inner region (e.g. NGC 4636, where the x-ray profile is depressed with respect to the optical one) and at the largest radii. In particular, the asymmetric morphology of NGC 4472 results in two different x-ray profiles, one steeper and one shallower than the optical one. A possible interpretation for this is that the original profile, originally similar to the optical one even at larger radii, has been perturbed by, for instance, the interaction with the outside medium as a result of the galaxy's motion. The steeper x-ray profile observed in NGC 4649 could similarly be the result of a succesful stripping in the outer regions of the galaxy, or alternatively, of the action of a partial wind in the outer regions only (see Trinchieri, Fabbiano and Canizares 1986).

The x-ray surface brightness profiles can be used to obtain the distribution of density and cooling time as a function of radius (see Fabian, this volume). Figure 9 shows the results for the same three galaxies discussed above. It is evident that the cooling times are quite short in the inner regions, and are in general shorter than a Hubble time throughout the entire body of the galaxies. Very short cooling times are also derived for other early type galaxies, even with lower quality x-ray data, under the simple assumption that they all have a similar shape for the x-ray surface brightness distribution (Canizares, Fabbiano and Trinchieri 1987). Galaxies with a detectable gaseous component should then have cooling flows (Nulsen Stewart and Fabian 1984) that probably extend over most of the galaxies' sizes. These cooling flows should be similar to those found in clusters of galaxies, although with relatively small flow rates, in the range 0.3 to ~ 1 $M_\odot$ yr^{-1}.

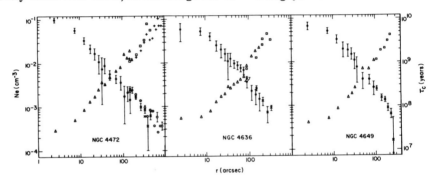

Figure 9. The distribution of density and cooling times as a function of galactic radius for three well studied early type galaxies (from Trinchieri, Fabbiano and Canizares 1986).

The results summarized above can then be used to determine whether models that have been propounded to explain the x-ray emission from early type galaxies can reproduce the observed quantities. In particular, they should help us understand important issues such as the heating mechanism for the gas. The most important sources of heat in early type galaxies are supernova explosions and gravitation processes. How these two mechanisms operate and their relative importance for the observed luminosity and distibution of the hot gas is addressed extensively by J Bregman (this volume, see also Trinchieri and Fabbiano 1985, Canizares, Fabbiano and Trinchieri 1987).

The hot gas in these systems can also be used to trace the gravitational potential and the total (gravitational) mass in early type systems. Under the simple assumption of a gas in hydrostatic equilibrium and of spherical symmetry, the total mass inside a radius r is given by (see Fabricant and Gorenstein, 1983):

$$M(r) = -\frac{kT}{G\mu m_H}\left(\frac{d\log\rho}{d\log r} + \frac{d\log T}{d\log r}\right) r$$

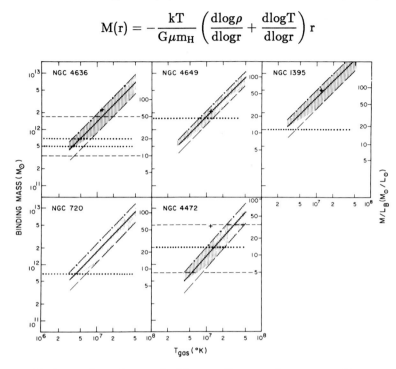

Figure 10. Estimated total mass inside a radius r and corresponding mass-to-light ratios as a function of temperature T and for different assumptions for the temperature gradient: $T \propto r^{0.5}$ (dashed line); isothermal (solid line); $T \propto r^{-0.5}$ (dot-dashed line). Horizontal lines indicate the estimates from optical data. Crosses are from Forman, Jones and Tucker (1985). (From Trinchieri, Fabbiano and Canizares 1986).

The observable quantities that appear in the formula, namely the temperature T, the temperature and density gradients at the radius r, are all *in principle* measurable with x-ray observations. Fabricant and Gorenstein have appied this formula to M87 and obtained a mass of $\sim 4 \times 10^{13}$ to 10^{14} $M_\odot$ within ~ 400 Kpc (see also Edge, this volume). However, for most galaxies:
- the present data can measure the density gradient, but not temperature gradients and the temperatures quoted are averaged over the entire galaxy's emission.
- the spherical symmetry - at least in NGC 4472 - is not strictly valid.
- the gas distribution in the outer regions might be modified by interaction with the external medium, or by partial winds.
Bearing in mind these limitations and under reasonable assumptions for the quantities not measured, total masses have been derived for a number of objects; examples are given in Figure 10. The uncertainties in these estimates are substantial, with mass-to-light ratios M/L ranging from ~ 10 to > 100. However, it is interesting to notice that the binding masses derived from the x-ray gas within the observed region are in good agreement with similar estimates based on optical data, and that all cases studied so far indicate M/L> 6, the canonical value expected from the

visible (stellar) population of early type galaxies. Lower limits to the total binding mass are also derived by Fabian et al. (1986) for the same galaxies. They assume the steepest temperature gradient allowed by a convectively stable gas and obtain M/L greater than ~ 20, with an average M/L ~ 75. These estimates all imply the presence of 'dark matter' in early type galaxies, and are therefore extreemely relevant for the determination of the total mass present in the universe.

The study of the properties of hot gas in other galaxies has only just begun, yet it has shown the relevance of the x-ray observations for this purpuse. It is however obvious that more and better quality data than currently available are needed, to investigate the presence and properties of hot thin plasma in other galaxies. We hope and expect that the next generation of satellites (see the reviews elsewhere in this volume) will give substantially improved data, thus enabling us to deepen and expand our understanding of the fundamental properties of galaxies.

4. REFERENCES

Bregman, J.N. and Glassgold, A.E. 1982, Ap.J., **263**, 564.
Canizares, C.R., Fabbiano, G. and Trinchieri, G. 1987, Ap. J., **312**, 503.
Clark, G., Doxsey, R., Li, F., Jernigan, J.G., and Van Paradijs, J. 1978, Ap. J. (Lett.), **221**, 137.
Fabbiano, G., Feigelson, E., and Zamorani, G. 1982, Ap. J., **256**, 397.
Fabbiano, G. and Panagia, N. 1983, Ap. J., **266**, 568.
Fabbiano, G. and Trinchieri, G. 1984, Ap. J., **286**, 491.
Fabbiano, G. and Trinchieri, G. 1985, Ap. J., **296**, 430.
Fabbiano, G. and Trinchieri, G. 1987, Ap. J., **315**, 46.
Fabbiano, G. 1988, preprint
Faber, S. and Gallagher, J. 1976, Ap. J., **204**, 365.
Fabian, A.C., Schwarz, J., and Forman, W. 1980, MNRAS, **192**, 135.
Fabian, A.C., Thomas, P.A., Fall, S.M., and White, R.E. III 1986, MNRAS, **221**, 1049.
Fabricant, D., and Gorenstein, P. 1983, Ap. J., **267**, 535.
Forman, W., Schwarz, J., Jones, C., Liller, W., and Fabian, A.C. 1979, Ap. J. (Lett.), **234**, L27.
Forman, W, Jones, C. and Tucker, W. 1985, Ap. J., **293**, 102.
Long, K.S., D'Odorico, S., Charles, P.A., and Dopita, M.A. 1981a, Ap. J. (Lett.), **246**, L61.
Long, K.S., Helfand, D.J., and Grabelsky, D.A. 1981b, Ap. J., **248**, 925.
McCammon, D., and Sanders, W.T. 1984, Ap. J., **287**, 167.
Nulsen, P.E.J., Stewart, G.C., and Fabian, A.C. 1984, MNRAS, **208**, 185.
Seward, F. and Mitchell, M. 1981, Ap. J., **243**, 736.
Stewart, G.C., et al. 1982, MNRAS, **200**, 61P.
Trinchieri, G. and Fabbiano, G. 1985, Ap. J., **296**, 447.
Trinchieri, G., Fabbiano, G. and Canizares, C.R. 1986, Ap. J., **310**, 637.
Trinchieri, G., Fabbiano, G. and Palumbo, G.G.C. 1985, Ap. J., **290**, 96.
Trinchieri, G., Fabbiano, G. and Peres, G. 1988, Ap. J., (in press).
VanSpeybroeck, L., et al. 1979, Ap. J., (Lett) **234**, L45.
Watson, M.G., Stanger, V., and Griffiths, R.E. 1984, Ap. J., **286**, 144.

A Theoretical Understanding Of Hot Gas Around Galaxies

Joel N. Bregman
National Radio Astronomy Observatory
Edgemont Road
Charlottesville, VA USA

1. INTRODUCTION

Within the past decade, it has become clear that some elliptical galaxies possess as much as 10^{10} $M_\odot$ of gas with a temperature of about 10^7 K (e.g., Forman, Jones, and Tucker 1985; Trinchieri, this volume). In addition, it has been known that million degree gas is common in our galactic disk (Nousek <u>et al</u>. 1982) and such gas might flow outward from the disk to surround the galaxy (Spitzer 1956). The production and evolution of hot gas in elliptical and spiral galaxies is considerably different, and they will be treated in turn. First, a number of relevant timescale and other quantities need to be discussed.

The primary cooling mechanism for gas in the environment of galaxies is optically thin line cooling, where C, O, and Fe are the most important elements. In a dilute gas, the cooling time can be significantly shorter than the dynamical time scales of the system. The instantaneous cooling time is proportional to $T/(\Lambda n)$, where Λ is the cooling function already discussed by Raymond (this volume) and has values of $6\times10^7/n_{-3}$ yr for T = 1×10^6 K, $8\times10^8/n_{-3}$ yr for T = 5×10^6 K and $13\times10^8/n_{-3}$ yr for T = 1×10^7 K, where n_{-3} is the density in units of 10^{-3} cm^{-3}. This can be compared to the sound crossing time of a region of size L, which is

$$t_s = 2.1\times10^7 \, (L/10 \text{ kpc}) \, (T/10^7 \text{ K})^{-0.5} \text{ yr.}$$

At low temperatures or high densities, the cooling time can be comparable to or shorter than the sound crossing time, which may be the case in spiral galaxies. At higher temperatures or in dilute gas, the sound crossing time is the shortest timescale, so the gas distribution has time to approach an equilibrium condition, which is representative of the properties of gas around elliptical galaxies. It is worth keeping in mind another timescale, the free fall time of a test particle in a galaxy, which is typically 1-3×10^7 yr.

There a correspondence between velocities and temperatures that is useful to remember. The characteristic velocity of gas is the sound speed, which is given by

$$c_s = (\gamma P/\rho)^{0.5} = 475 \, (T/10^7 \, K)^{0.5} \, km/s.$$

The velocity dispersion of stars that observers quote is usually the line of sight or one-dimensional velocity dispersion, which must be multiplied by $\sqrt{3}$ to obtain the three dimensional velocity dispersion. So a galaxy with an observed velocity dispersion (one-dimensional) of 300 km/s has an equivalent temperature of about 10^7 K.

2. GAS IN ELLIPTICAL GALAXIES

It has been well known for many years that elliptical galaxies are poor in the neutral (T < 5000 K) and photoionized gas (10^4 K) commonly seen in spiral galaxies. In elliptical galaxies, mass loss associated with the evolution of stars leads to about 1 $M_\odot$/yr of gas being deposited into the galaxy. If this gas were in the form of optically thin HI or as ionized gas (10^4 K), it would have been detected after accumulating for only a billion years. In order to render this gas unobservable, two theories were developed during the 1970s -- galactic winds and galactic stripping. In clusters with an intracluster medium, ram pressure stripping would remove gas from elliptical galaxies, while field galaxies drove the gas out through stellar winds (Gisler 1976; Takeda, Nulsen, and Fabian 1984; Gaetz, Salpeter, and Shaviv 1987). As a historical note it was believed that the lack of gas in ellipticals was a settled issue a decade ago. If this talk was being presented in 1977, I would have spoken of these theoretical explanations with strong conviction. However, in the past decade X-ray observations have shown that the interpretation of gas being physically removed from all elliptical galaxies is incorrect (Trinchieri and references therein, this volume). Many ellipticals have hot gas (10^7 K) that is gravitationally bound, and this gas is presumably the result of stellar mass loss. Not surprisingly, there are new theories that explain the behavior of this 10^7 K gas, and I will probably discuss these theories with the conviction that I might have had ten years ago for an entirely different model.

Before discussing specific models in detail, it is necessary to discuss the general properties of hot gas in the environment of elliptical galaxies. The temperature at which gas would have positive total energy and escape from the galaxy is the escape temperature (T_{esc}), which is about 2×10^7 K for giant ellipticals, $3-10 \times 10^6$ K for more ordinary elliptical galaxies, and significantly less for dwarf ellipticals. The source of mass deposition into the galaxy is by stars at the specific rate of 5×10^{-20} s^{-1} ($1/6 \times 10^{11}$ yr; Faber and Gallagher 1976); the rate per unit volume is proportional to the star density. This implies total mass deposition rates of 1-3 $M_\odot$/yr for giant ellipticals and 0.1 $M_\odot$/yr for more modest ellipticals. The detailed evolution of the stellar mass loss has not been calculated in great detail, but the following reasonable scenario has been given (Mathews and Baker 1971; Mathews and Loewenstein 1986; Sarazin and White 1986; Thomas et al. 1986).

Individual stars move through the galaxy at approximately the stellar velocity dispersion, which is typically 300-550 km/s. When a star loses a shell of gas in the planetary nebula phase, it may collide with the shell of gas from another star in this phase. The relative velocity of the gas shells is approximately the stellar velocity dispersion, which is enormous compared to the sound speed in the shell. Strong shocks are driven into the shells, which heat up to a temperature characteristic of the velocity dispersion of the stellar system, $4-10 \times 10^6$ K. The situation actually is a bit more complicated because some elliptical galaxies are filled with hot gas, leading to drag against a planetary nebula shell or a stellar wind. However, the relative velocity between the stellar ejecta and the hot gas in the galaxy, which is essentially stationary, will be reduced to zero through shocks and subsonic drag. If this heating occurs more rapidly than the radiative loss timescale, the temperature of the gas will be $4-10 \times 10^6$ K after it is thermalized. It is not entirely clear that all stellar mass loss is thermalized in such a manner, but it is likely that much of it is.

A potentially important heat source in ellipticals is supernovae. Until recently, the supernova rate in E and S0 galaxies was believed to be well known. Tammann (1974) based his estimate for the supernova rate on 16 type Ia events. If this rate is applicable, then heating by supernovae is 3-10 times more important than thermalization of stellar ejecta, and the gas would be driven above the escape temperature of the galaxy. However, van den Bergh (1987) has suggested that this rate is about a factor of three too high, and a more recent search for supernovae in early type galaxies by Tammann (1986) also indicates that true rate is at least a factor of 2.5 lower. The adjustment to the supernova rate might make it comparable to or less important than thermalization of stellar ejecta.

a) Galactic Winds

The theory of galactic winds were developed to describe the situation when the heating by supernovae raises the gas temperature above the escape temperature (Mathews and Baker 1971; Bregman 1978; MacDonald and Bailey 1981; White and Chevalier 1983; Mathews and Loewenstein 1986). For a spherical system, the equations that are solved are:

$$\frac{\partial \rho}{\partial t} + \nabla \cdot (\rho u) = \alpha \rho_*$$

$$\rho \frac{\partial u}{\partial t} = - \frac{\partial P}{\partial r} - \rho \frac{\partial \Phi}{\partial r} - \alpha \rho_* u^2$$

$$\frac{\partial (\rho E)}{\partial t} + \nabla \cdot (\rho E u) = - P \nabla \cdot u + \alpha \rho_* [0.5 \, u^2 + E_g]$$

where α is the specific mass loss rate of the stars and E_g is the thermal energy of new gas. Notice the new additions to the usual fluid equations. There is a "mass creation" term in the continuity equation due to mass entering the system from stellar mass loss. Because new gas entering the system initially has zero radial velocity, it exerts a drag on the surrounding fluid (the final term in the equation of motion). Associated with this gas is drag heating plus the energy content of the gas itself, which modifies the energy equation (the final terms).

When the temperature in the center of the galaxy is above the escape temperature, all of the gas entering the system will participate in a wind, provided that radiative losses occur on a timescale longer then the flow time of a wind, which is expected to be the case. The local mass loss rate is proportional to the stellar density, while the heating of the gas (thermalization and supernovae) is assumed to occur locally (before significant radial motion). The pressure gradient is therefore proportional to the star density, and for temperatures above the thermalization temperature, the force created by the pressure gradient will be greater than that of gravity. For the temperature meeting the above criteria, there is significant outward acceleration at all locations and the gas is driven out as a galactic wind.

The stellar density distribution of a galaxy is usually given by models in which the stellar density is slowly varying within a core radius. Within this region, the gas density in a wind is also constant and the velocity increases linearly with radius. The gas changes from subsonic to supersonic motion beyond the core radius ($>1.4 r_{core}$ for a modified King potential) and the location of the sonic point increases with decreasing temperature. At large radii, when mass loss from stars is negligible and the gravitational force $\propto r^{-2}$, the gas velocity is constant (the coasting solution) and the Mach number continues to rise as adiabatic losses decrease the gas temperature. If there is a surrounding medium with a non-zero pressure, the wind passes through a shock, is decelerated and merges with the ambient gas. The flow time of the gas (sometimes called the flushing time) is about 10^7 yr, so the mass of gas in the galaxy at any moment is typically 10^6-10^7 $M_\odot$. This small amount of hot gas would never have been detected with the Einstein observatory, which is why there was such a surprise when many elliptical galaxies were detected.

When the gas is above the escape temperature in only the outer part of the galaxy, a partial wind may occur. In this case, there is a stagnation radius beyond which a normal galactic wind occurs and within which the gas accumulates and eventually flows inward (a small cooling flow; see below). For a mathematically convenient but otherwise unrealistic set of conditions, Holzer and Axford (1970) found a steady-state solution with a wind and infall. White and Chevalier (1984) were unable to find such steady solutions for realistic galaxy parameters, but this does not prove they do not exist. In time dependent numerical hydrodynamic calculations (Bregman, in progress), partial wind solutions do occur, but it is not yet clear whether they persist for a Hubble time. Partial winds will have a signature that can be searched for: the X-ray surface brightness will decrease rapidly beyond the stagnation radius. Such behavior has been seen in NGC 4649, and more detailed from future missions will be able to test this hypothesis more directly.

b) Cooling Flows in Ellipticals

Hot gas that remains bound to an elliptical galaxy initially has a temperature no smaller than the thermalization temperature and therefore occupies the same volume as the stars. Insight into the eventual fate of the gas can be gained by examining the various timescales of interest. Based on the X-ray data, the cooling time of the gas in ellipticals is considerably less than the Hubble time. Within the inner few kpc where the density is 0.1-0.01 cm^{-3}, the cooling time is 10^7-10^8 yr, while at tens of kpc, density is 10^{-3} cm^{-3} and the cooling time is 10^9 yr. The sound crossing time across 1kpc (T=1×10^7 K) is 3×10^6 yr and across the entire X-ray emitting region (tens of kpc), it is 10^8 yr. The cooling time is about an order of magnitude greater than the sound crossing time (except at very small radii), so one might expect that the gas has time to find a steady state configuration, which would naturally be in hydrostatic equilibrium.

Gas in hydrostatic equilibrium is a useful tool for measuring the mass of the galaxy. The equation that describes hydrostatic equilibrium is simply

$$\frac{1}{\rho} \frac{\partial P}{\partial r} = \frac{-G\, M_g(r)}{r^2}$$

where $M_g(r)$ is the total mass of the galaxy (stars plus unseen matter) within radius r. This equation has three quantities that are functions of radius: M_g, P, and ρ. In principle, X-ray observations can determine the radial dependence of P and ρ (alternatively, T and ρ), and one can solve for $M_g(r)$. In practice, the existing data determine $\rho(r)$, but not T(r). Currently, if one is to determine a mass with this approach, an model for the temperature gradient must be adopted. Common models employed assume that the temperature is constant (isothermal) or that the temperature gradient is at the adiabatic limit (the non-convective model; Fabian et al. 1986). We can look forward to future X-ray telescopes that will measure T(r) directly.

Although gas bound to an elliptical is approximately in hydrostatic equilibrium, there is some motion of the gas caused by cooling processes. The density of gas is greatest in the center of the galaxy, so the cooling time is shortest there. Because the cooling time is long compared to the sound crossing time, cooling occurs at constant pressure. Consequently, as the gas cools, it is compressed in order to maintain constant pressure. As gas in the inner regions occupy a smaller volume, gas from larger radii flows inward.

The process is actually a bit more complicated because as gas flows inward it undergoes adiabatic heating (compression), which is the conversion of potential to thermal energy. Because the potential of a galaxy is approximately isothermal, the energy gained in falling some distance d is nearly independent of where the gas began its fall. That leads to compressional heating that is nearly independent of radius and would lead to an isothermal temperature profile in the absence of radiative losses. Instead, compressional heating is always a bit less than radiative cooling, so the temperature of the gas is expected to decrease into the center of the galaxy (as the radiative loss rate increases) and the gas flows inward.

To review the basic scenario for cooling flow ellipticals in the current epoch, the first stage is when stellar mass loss is thermalized but supernovae are not frequent enough to drive most of the gas from the galaxy. All of the hot gas in an elliptical galaxy may once have been in stars. The only other important heat source is compressional heating as the cooling gas flows inward; inflow velocities are small, typically 10 km/s. Eventually, one might expect that as the gas flows inward and its density continues to increase, that the cooling time becomes exceedingly short and the gas cools on the spot. Models that examine this possibility indicate that they would lead to a X-ray profile that rises in the inner several kpc much more sharply than is observed. This suggests that mass is not being conserved during the inward flow of gas but is being removed from the system over a broad range of radii. Even before it was known that the data required mass depletion, it was suggested that thermal instabilities could lead to non-uniform cooling and that cool gas would form in the flow (e.g. Mathews and Bregman 1978).

Thermal instabilities probably play an important role in the theory of cooling flows, and this is still an active area of interest. In a seminal paper, Field (1965) showed that even if heating was balanced by cooling, a perturbation in a static fluid would grow if the radiative cooling function had a certain temperature and density behavior. He examined traveling waves, and a zero frequency mode, sometimes referred to as the condensation mode. I will not discuss the traveling wave modes because it is difficult to find an astronomical object that will produce a infinite train of waves. The condensation mode is a simple situation in which pressure is everywhere the same but the density and temperature vary inversely (this gives rise to entropy perturbations). For cooling flows, the isobaric (constant pressure) condensation mode is the most important. This type of perturbation will grow if the following condition is satisfied:

$$2 - \frac{\partial \ln \Lambda}{\partial \ln T} > \frac{\partial \ln H}{\partial \ln n} - \frac{\partial \ln H}{\partial \ln T}$$

where Λ is the radiative cooling function and H is the heating rate per unit volume. We see that without heating, thermal instabilities grow for $d\ln\Lambda/d\ln T < 2$, which occurs when the temperature is above 10^5 K. Small perturbations grow algebraically rather than exponentially in time, with the characteristic timescale being the cooling time.

To make this process more appropriate for the environment of cooling flows, stability calculations were performed for an inflowing gas with a radial pressure gradient. Growth in a contracting background (due to gas flowing inward) leads to a more rapid rate of growth (Mathews and Bregman 1978; White 1986; Balbus 1986; Chevalier 1988) provided that the flow time is comparable to or shorter than the cooling time. Cowie, Fabian, and Nulsen (1977) were the first to point out that the loss of buoyancy in a cooling perturbation and the subsequent infall of that gas would act to stabilize the fluid against thermal instabilities. Their heuristic discussion has been expanded upon more recently by Malagoli, Rosner, and Bodo (1987) and Balbus (1988). These authors find that buoyancy will likely prevent the effective growth of radial perturbations. However, certain non-radial perturbations will grow, although possibly at a reduced rate. Based upon the growth rates and the inflow rates of the background flow, a small perturbation (<10%) would be carried close to the center of the flow before it cools. Since gas is estimated to cool at a large range of radii, at least some of the perturbations must be large (Thomas, Fabian, and Nulsen 1987) and their growth would occur in a nonlinear fashion.

David and Bregman (1988a,b) used a hydrodynamic code to study the nonlinear evolution of an entropy perturbation. They found that the growth rate became substantially more rapid than that derived from linear analysis when the magnitude of the density perturbation rose above 10%. A shock would develop in as the perturbation cools which might provide observable ultraviolet lines, but is unable to account for the optical emission lines seen in some systems. The spectrum of initial perturbations needed to yield the inferred Mdot(r) has been discussed by Nulsen (1986) and Thomas, Fabian, and Nulsen (1987) under the assumption that the cooling parcels of gas are locked into the background fluid by drag forces. They find that the initial range of densities at one radial location must be comparable to the background density itself.

This now raises the issue of how these large perturbations originate. Very little detailed work has been done on this topic, so one if free to speculate. It might be that not all mass loss from stars becomes thermalized and that such gas is cooler than the ambient medium and therefore provides the necessary perturbations. Future work in this area should illuminate this issue.

To summarize this digression about thermal instabilities, density perturbations of amplitude 10-100% will grow in the background fluid and become cooled gas distributed throughout the galaxy. The eventual fate of the cooled gas is probably stars, although it is difficult to test this prediction observationally because the small rate of the production of young stars would be difficult to see against the background light of the galaxy. If the cooled gas were in the form of HI it would have an unusual line profile, and HI with the predicted properties has been detected in the giant elliptical galaxy NGC 4406 (Bregman, Roberts, and Giovanelli 1988).

In my opinion, the fundamentals of cooling flow ellipticals are sound. The gas must be bound to the galaxy, because unbound gas (galactic winds) would not be detected. The cooling time in the galaxy is much less than a Hubble time, so it is likely that cooled gas is being produced, leading to the slow inward flow of gas in these systems. The basic theory leads to the observed relationship between the X-ray and optical luminosity (Sarazin 1985) and can reproduce the observed X-ray profile and the inferred cooling rate (although a perturbation spectrum must be assumed).

It has been argued that influence of supernovae at the current epoch is probably not too important because it would lead to the wrong X-ray to optical luminosity relationship and to the wrong X-ray profile (Sarazin 1985; Sarazin and White 1987). During the early epochs of a galaxy, supernovae are likely to be considerably more common and therefore more important. Vader (1986) has shown that early galactic winds driven by supernovae can explain certain metalicity properties of elliptical and dwarf elliptical galaxies. It is certainly possible, if not likely, that winds were common at early times, but as the supernova rate decreased below some critical value, stellar ejecta was retained and a cooling flow ensued. The situation may be slightly different in giant elliptical galaxies (Mathews and Loewenstein 1986) where a heavy halo has a deep enough potential well to stop a wind as the galaxy evolves.

In closing the discussion about gas in elliptical galaxies, it is necessary to return to the possibility of stripping as galaxies pass through dense cluster gas (Takeda, Fabian, and Nulsen 1985; Gaetz, Salpeter, and Shaviv 1986). This is likely to occur in rich clusters such as Coma, and theoretical calculations indicate that for orbits passing through the cluster center, elliptical galaxies would retain too little gas to be detectable; this is consistent with the data. However, in poor clusters such as Virgo and A 1367, ellipticals manage to retain their gas.

3. HOT GAS IN SPIRAL GALAXIES

a) Observational Features in the Galaxy

It is valuable to supplement the discussion of hot gas in external spirals (Trinchieri, this book) with an examination of various kinds of gas in the halo of our galaxy. The first suggestion that enriched neutral gas exists above the disk was that of Munch and Zirin (1961), who showed that the strength of absorption lines toward stars in the halo was greater than anticipated if neutral gas existed only in a thin disk. For neutral gas, Albert (1984) performed a definitive study in which she measured absorption lines of Ti, Na, and Ca in pairs of O stars close together on the sky but at significantly different distances. The closer of the two stars was at a height above the disk of about 0.4 kpc while the more distant star was 1-2 kpc above the plane. For each pair of stars, additional absorption lines were found against the more distant star. This absorbing gas has a scale height of about 1 kpc and the velocities are generally more negative than the velocities expected if the gas were corotating with the disk; the velocities extend from +40 km/s to -70 km/s. Danly (1985) examined ultraviolet absorption lines of low ionization species toward several halo stars. She found absorbing gas covering a slightly broader range of velocities than Albert, and there is clearer evidence for infall in the northern hemisphere.

There are other lines of evidence that support the presence of considerable neutral or warm gas within 1 kpc of the disk. Lockman (1984) and Lockman, Hobbs, and Shull (1986) found a thick disk component of HI with a scale height of about 1 kpc. In addition, Lockman (1985) finds that within 45° of the north and south galactic poles there is more gas with velocities less than -50 km/s than above +50 km/s. There seems to be a component of neutral gas participating in infall, and this may be the same gas observed by Albert and Danly.

In summary, clouds of neutral gas, which have a fraction of the total column density of the thin HI disk, appear to cover the sky, have a scale height of approximately 1 kpc, and on average have a net downward velocity (also see Jenkins 1985).

Warm ionized gas (10^4 K) is also present in the halo, but with a larger scale height. Based on the rate of increase in the column densities for C IV and Si IV with distance to background continuum sources (Savage and de Boer 1981; Pettini and West 1982; Savage and Massa 1986; Savage 1985), a scale height of about 5 kpc is deduced. The velocities of the absorbing gas extend to ±160 km/s and attributed to galactic rotation plus a random velocity component (Kaelble, de Boer, and Grewing 1985). Absorption by these higher ionization species seems to be less pervasive, suggesting that gas clouds in the upper halo are less common than the neutral gas at lower altitudes.

One of the most prominent HI features are the high velocity clouds, which cover about 25% of the sky and have velocities (by definition) 70 km/s < v (Hulsbosch 1984; van Woerden 1984; Giovanelli 1985). The distances to these clouds are not yet determined, although preliminary results by Cowie and Songalia (1987) suggest that one of these clouds is 1-2 kpc distant. While there likely to be a component of galactic rotation reflected in the velocities, many if not most of the clouds have velocities unallowed by galactic rotation. In addition, a net infall is suggested with a mass flux of a few $M_\odot$/yr, although this is highly uncertain because the distances are unknown. Without distance information, it is difficult to determine whether this gas is to be associated with the ionized gas found in the upper halo, or with the neutral gas in the lower halo.

The gaseous component that would be expected to exist in the halo is million degree gas. A problem with observing this gas is that the disk is filled with X-ray emitting gas and the Sun probably lies in a hot bubble (e.g., Cox and Reynolds 1987; Nousek et al. 1982). Consequently, it is difficult to separate the contribution of the disk from that of the halo. While it is possible to place limits on the amount of million degree gas above the disk, the constraints are not particularly severe.

An enhancement of soft X-ray emission around the galactic bulge been interpreted as hot gas that is probably bound to the bulge and with a temperature of 3×10^6 K (Garmire and Nugent 1981). This would be a scaled down version of the cooling flows seen in elliptical galaxies. However, this brightening toward the galactic bulge has been interpreted as a reheated supernova shell (Loop I; Iwan 1980), so the analysis of this feature is unclear.

b) Theoretical Models for Gas in the Corona

Perhaps at this point we can appeal to theory and ask at what heights above the disk one would expect gas to exist. Gas in hydrostatic equilibrium with the galactic potential in the neighborhood of the sun has an isothermal scale height of about 5 ($T/10^6$ K) kpc. Neutral gas supported by its own thermal pressure would have a scale height of less than 100 pc while warm ionized gas would have a scale height of about 100 pc. Yet we have already seen that there is a component of neutral gas with a scale height of about 1 kpc and the warm ionized gas has a scale height of several kpc. This argument suggests that either some agent is supporting the gas against gravity, or that the clouds were ejected from the disk or created in the halo and are falling under the force of gravity onto the disk. This reasoning has led to several models, such as galactic fountains and cosmic ray supported halos.

It has long been noted that the energy density in cosmic rays in the disk is comparable to that of random gas motions and magnetic fields. In addition, cosmic rays are expected to diffuse out of the disk into the halo and scatter off irregularities in the magnetic fields, which are frozen into the gas. This provides the coupling between the cosmic rays and the gas and permits the cosmic rays to exert a force on the gas. The density of gas that can be supported by cosmic rays is proportional to the cosmic ray flux divided by the product of the diffusion coefficient and the gravitational force of the galaxy. Unfortunately it is difficult to estimate the amount of gas that can be supported by this process because there is a great deal of uncertainty in the value of the diffusion coefficient (Cesarsky 1980) and in the flux of GeV protons out of the plane. The approach that most authors have taken is to assume interesting but reasonable values for the diffusion coefficient and the cosmic ray flux and solve the hydrostatic equations (Pikelner and Shklovskii 1957; Badhwar and Stevens 1977; Chevalier and Fransson 1984; Hartquist, Pettini, and Tallant 1984; Fransson and Chevalier 1985). In general, the thermal, magnetic, and cosmic ray pressures may be considered when solving the hydrostatic equations. The best developed model of cosmic ray supported gas is that of Chevalier and Fransson, who show that low density gas will exist at several kpc and will be ionized by the ambient quasar background. This model can reproduce the scale heights for C IV and Si IV and also predicts denser neutral gas within the first kpc of the disk.

A different class of models suggests that hot gas will rise above the halo and eventually give rise to the cooler gas that is detected. Supporters of such models point out that the energy content of hot gas produced by supernovae is about ten times greater than that of the cosmic rays. Furthermore, hot gas is extremely buoyant and should rise out of the disk. If more than 10% of this gas escapes from the disk, it should be a more important process than that involving cosmic rays.

Analogous to the situation with elliptical galaxies, if gas leaving the disk has a temperature above the escape temperature (about 5×10^6 K from the solar circle), it leaves as a galactic wind. Galaxies without cold gas in the disk are able to drive galactic winds throughout the entire galaxy in much the same way that elliptical galaxies are, provided that the supernova rate is sufficient and that cold gas has been removed from the system (Bregman 1978). For normal gas rich galaxies, it may still be possible to have some hot gas escape. A plausible scenario is that hot gas rises from the disk (probably with $T<T_{esc}$) and is heated by halo supernovae (e.g., McKee and Ostriker 1977; Heiles 1986). The emission properties of a wind of this kind has not been discussed in detail, although it may be rarefied and with a low enough emission measure to provide an explanation for why edge-on spiral galaxies show no evidence for hot halo gas.

If hot gas from the disk is bound to the galaxy, it will envelope the system (recall the scale height is about 5 kpc for T = 10^6 K). However, unlike the conditions in elliptical galaxies, the radiative cooling time is more nearly equal to the sound crossing time if the pressure in the halo is like that of the disk (which would be the case if hot gas in the disk were free to flow into the halo). The ratio of the cooling to the sound crossing time is:

$$t_c/t_{eq} \approx (T/10^6 \text{ K})^{2.1} (nT/10^3 \text{ Kcm}^{-3})^{-1}$$

which is correct in the temperature range 2×10^5-10^7 K; the pressure of the ISM in the disk is typically nT = 3×10^3 Kcm^{-3}. For low temperature flows, when this ratio is much less than unity (e.g., T = 10^5), gas that escapes the disk begins to flow upward, but soon cools radiatively into dense neutral or warm ionized clouds. Formation of clouds occurs with positive velocity, although the clouds eventually fall to the disk. In this case, there would be nearly the same amount of cooled gas rising as falling. The height to which the gas could rise is approximately given by the scale height, 0.5 kpc for T = 10^5 K (this scales linearly with T). This cycle of hot gas rising, cooling, and returning to the disk has been labeled as the "galactic fountain" model and has been investigated by a number of workers (Spitzer 1956; Shapiro and Field 1976; Chevalier and Oegerle 1979; Bregman 1980; Ikeuchi 1981; Salpeter 1985; Corbelli and Salpeter 1988).

Two features change at higher temperatures when the cooling time becomes comparable to the sound crossing time. The radial scale length of the disk is about 4 kpc, so gas that remains within a few kpc of the disk sees a uniform gravitational force. At temperatures of 10^6 K and greater, the scale height of the gas is greater than this characteristic length of the galaxy. As the gas rises, it finds a decreasing radial gravitational field and probably a radial pressure gradient. These effects lead to radial motion in addition to the normal vertical motion of the gas (Corbelli and Salpeter 1988 and Salpeter, in this volume, discuss another aspect of radial motion for galactic fountains).

The other change in the behavior of the gas is the location and velocity of cold clouds. Unlike the situation for short cooling times, where the clouds cool early in the ascent of the gas, at 10^6 K clouds are likely to cool either near the top of their ascent or during decent. This will lead to more negative velocity clouds than positive velocity clouds. Compared to lower temperature fountains, the velocities will be greater as will the distances from the disk.

An important consideration, and one that is poorly understood, is the flow of hot gas from the disk into the halo. The McKee-Ostriker (1977; also Cox and Smith 1974, Cox 1981) model would suggest that most of the volume of the ISM is occupied by hot gas, in which case hot gas flows easily into the halo. However, regions of hot dilute gas may be confined to superbubbles that are relatively isolated from each other. Then, hot gas would enter the halo only when superbubbles break through the disk, which is expected to occur (McCray and Kafatos 1987). Until a detailed description for the flow of gas from the disk to the halo is well understood, we are free to speculate on this matter. With increased supernova activity in spiral arms (Maza and van den Bergh 1976) it is likely that the flow of gas from spiral arms and from interarm regions are different. Suppose that gas from the arm region is the hotter of the two. If it were $1-2 \times 10^6$ K, a large fountain would form in which the formation of clouds might be identified with C IV and Si IV absorption as well as some of the high velocity clouds of HI. A lower temperature interarm fountain ($2-5 \times 10^5$ K) would populate the low halo with clouds and might be responsible for the neutral clouds with a scale height of 1 kpc and possibly some of the high and intermediate velocity clouds of HI.

Finally, one must consider the influence upon galactic fountains by planetary nebulae stars and Type Ia supernovae, both of which are expected to have scale heights greater than the thin disk of HI. Mass loss by planetary nebulae away from the plane may provide the seed material around which larger clouds will eventually cool (Salpeter 1985). More detailed work needs to be carried out to determine if this mechanism dictates the cloud condensation process. Supernovae would be very destructive to galactic fountains, because if enough energy were dispersed throughout the halo, a wind rather than a fountain would ensue. However, it is not viable to form a large number of clouds in a wind, so if cloud formation does occur in the halo, supernovae do not deposit a large amount of energy uniformly in the gas (Chevalier and Oegerle 1979).

To conclude this discussion of hot gas around spiral galaxies, it is probably naive to believe that there is only one process by which gas enters the halo. In addition to cosmic ray supported gas and galactic fountains, which may both be present, a lesser amount of gas may be cast into the halo by magnetic fields (Parker instability), and the Galaxy may be accreting gas. The Magellanic Stream is an example of gas tidally torn from a smaller galaxy that is now falling toward the Milky Way. Like other astronomical objects, spiral galaxies are likely to be sites of complicated and varied behavior.

4. REFERENCES

Albert, C.E. 1983, Ap.J., 272, 509.
Badhwar, G.D., and Stephens, S.A. 1977, Ap.J., 212, 494.
Balbus, S.A. 1986, Ap.J. (Letters), 303, L79.
Balbus, S.A. 1988, Ap.J., in press.
Bregman, J.N. 1978, Ap.J., 224, 768.
Bregman, J.N. 1980, Ap. J., 236, 577.
Bregman, J.N., Roberts, M.S, and Giovanelli, R. 1988, Ap.J., in press.
Cesarsky, C.J. 1980, Ann. Rev. Astr. Ap., 18, 289.
Chevalier, R.A. 1988, Ap.J., in press.
Chevalier, R.A., and Clegg, A.W. 1985, Nature, 317, 44.
Chevalier, R.A., and Fransson, C. 1984, Ap.J. (Letters), 279, L43.
Chevalier, R.A., and Oegerle, W.R. 1979, Ap.J., 227, 398.
Corbelli, E., and Salpeter, E.E. 1988, Ap.J., in press.
Cowie, L.L, Fabian, A.C., and Nulsen, P.E.J 1980, M.N.R.A.S., 191, 399.
Cox, D.P. 1981, Ap.J., 245, 534.

Cox, D.P., and Reynolds, R.J. 1987, Ann. Rev. Astr. Ap., 25, 303.
Cox, D.P., and Smith, B.W. 1974, Ap.J. (Letters), 189, L105.
Danly, L. 1985, in "Gaseous Halos of Galaxies", ed. J.N. Bregman and
    F.J. Lockman, NRAO, p. 45.
David. L.P., and Bregman, J.N., 1988a,b, Ap.J., in press.
Fabian, A.C., Thomas, P.A., Fall, S.M., and White, R.E. 1986,
    M.N.R.A.S., 221, 1049.
Field, G.B., 1965, Ap.J., 142, 531.
Forman, W., Jones, C., and Tucker, W. 1985, Ap.J., 293, 102.
Fransson, C.A., and Chevalier, R.A. 1985, Ap.J., 296, 35.
Gaetz, T.J., Salpeter, E.E., and Shaviv, G. 1987, Ap.J., 316, 530.
Garmire, G.P., and Nugent, J.J. 1981, B.A.A.S., 13, 787.
Giovanelli, R. 1985, in "Gaseous Halos of Galaxies", ed. J.N. Bregman
    and F.J. Lockman, NRAO, p. 99.
Gisler, G.R. 1976, Astr. Ap., 51, 137.
Hartquist, T.W., Pettini, M, and Tallant, A. 1984, Ap.J., 276, 519.
Heiles, C. 1987, Ap.J., 315, 555.
Hulsbosch, A.N.M. 1984, in IAU Symp. 106, "The Milky Way Galaxy", ed.
    by van Woerden, H., Allen, R.J. and Butler Burton, W., (D.Reidel:
    Dordrecht), p. 409.
Iwan, D.C. 1980, Ap.J., 239, 316.
Ikeuchi, S. 1981, Publ. Astr. Soc. Japan, 33, 211.
Jenkins, W.B., in "Gaseous Halos of Galaxies", ed. J.N. Bregman and
    F.J. Lockman, NRAO, p.1.
Kaelble, A., de Boer, K.S. and Grewing, M. 1985, Astr. Ap., 143, 408.
Lockman, F.J. 1984, Ap.J., 283, 90.
Lockman, F.J. 1985, in "Gaseous Halos of Galaxies", ed. J.N. Bregman
    and F.J. Lockman, NRAO, p. 63.
Lockman, F.J., Hobbs, L.M., and Shull, J.M. 1986, Ap.J., 301, 380.
MacDonald, J., and Bailey, M.E. 1981, M.N.R.A.S., 197, 995.
Malagoli, A., Rosner, R., and Bodo, G. 1987, Ap.J., 319, 632.
Mathews, W.G., and Baker, J. 1971, Ap.J., 170, 241.
Mathews, W.G., and Bregman, J.N. 1978, Ap.J., 224, 308.
Mathews, W.G., and Loewenstein, M. 1986, Ap.J. (Letters), 306, L7.
Maza, J., and van den Bergh, S. 1976, Ap.J., 204, 519.
McCray, R., and Kafatos, M. 1987, Ap.J., 317, 190.
McKee, C.F., and Ostriker, J.P. 1977, Ap.J., 218, 148.
Munch, G. and Zirin, H. 1961, Ap.J., 133, 11.
Nousek, J.A., Fried, P.M., Sanders, W.T., and Kraushaar, W.L. 1982,
    Ap.J., 258, 83.
Nulsen, P.E.J. 1986, M.N.R.A.S., 221, 377.
Pettini, M., and West, K.A., 1982, Ap.J., 260, 561.
Pikelner, S.B., and Shlovskii, I.S. 1957, Astr. Zh., 34, 145 (English
    translation in 1957, Soviet Astr. -- AJ, 1, 149).
Salpeter, E.E., 1985, Mitteilungen der Astronomischen Gesellschaft; 63.
Sarazin, C.L. 1985, in "Gaseous Halos of Galaxies", ed. J.N. Bregman
    and F.J. Lockman, NRAO, p. 223.
Sarazin, C.L., and White, R.E. 1987, Ap.J., 320, 32.
Savage, B.D., in "Gaseous Halos of Galaxies", ed. J.N. Bregman and
    F.J. Lockman, NRAO, p. 17.
Savage, B.D., and de Boer, K. 1979, Ap.J., (Letters), 230, L77
Savage, B.D., and de Boer, K. 1981, Ap.J., 243, 460.

Savage, B.D., and Massa, D. 1985, Ap.J. (Letters), 295, L9.
Shapiro, P.R., and Field, G.B. 1976, Ap.J., 205, 762.
Spitzer, L., 1956, Ap.J., 124, 20.
Takeda, H., Nulsen, P.E.J., and Fabian, A.C. 1984, M.N.R.A.S., 208, 279.
Tammann, G.A., 1974 in "Supernovae and Supernovae Remnants", ed.
Cosmovici, C.B., Reidel, Dordrecht, Holland.
Tammann, G.A. 1986, in the Annual Report of the Carnegie Institute of Washington, p. 59.
Thomas, P.A. 1986, M.N.R.A.S., 220, 949.
Thomas, P.A., Fabian, A.C., Arnaud, K.A., Forman, W., and Jones, C. 1986, M.N.R.A.S., 222, 655.
Thomas, P.A., Fabian, A.C., and Nulsen, P.E.J. 1987, preprint.
Vader, J.P. 1986, Ap.J., 305, 669.
van den Bergh, S. 1987, reported at the ESO Workshop on SN 1987a, in press.
van Woerden, H., Schwarz, U.J., and Hulsbosch, A.N.M. 1984, in IAU Symp. 106, "The Milky Way Galaxy", ed. by van Woerden, H., Allen, R.J. and Butler Burton, W., (D.Reidel: Dordrecht), p. 387.
White, R.E., and Chevalier, R.A. 1983, Ap.J., 275, 69.
White, R.E., and Chevalier, R.A. 1984, Ap.J., 280, 561.
York, D.G. 1982, Ann. Rev. Astron. Ap., 20, 221.

RAM PRESSURE STRIPPING AND GALACTIC FOUNTAINS

E.E. Salpeter
Center for Radiophysics and Space Research
Cornell University
Ithaca, NY
USA

ABSTRACT. Ram pressure stripping can occur when the intracluster gas in a galaxy cluster impinges on the interstellar gas of a member galaxy. The effects have to be discussed separately for spiral galaxies (sporadic effects) and for ellipticals (more nearly continuous). Various kinds of "galactic fountains" tend to replenish the hot corona in a galaxy halo. An extended HI disk in an isolated spiral galaxy can decrease the coronal density by initiating a cooling flow, which is very sensitive to the coronal temperature.

1. INTRODUCTION

I want to discuss two different topics, one involving the outer regions of an isolated spiral galaxy (galactic fountain), the other the interaction of a cluster galaxy (most likely an elliptical galaxy) with the intracluster gas (ram pressure stripping). In both problems hot "coronal" gas plays an important role, although in one case the "active medium" is coronal gas at temperatures well below 10^7 K in the halo of a spiral galaxy, in the other case it is the intracluster gas at temperatures above 10^7 K (the ICP in Cavaliere's talks). The two topics also have a great uncertainty in common, namely what the values of the transport coefficients are in the hot plasma.

For a fully ionized plasma where complete thermal equilibrium is violated only by an infinitesimally small temperature gradient (and where the magnetic field is identically zero) there is no controversy about the analytic expressions for the transport coefficients. The viscosity coefficient and the heat conduction coefficient then have their "gas kinetic" values, which can be found in the classic book by Spitzer (1967). The electron's contribution to the heat conduction is much larger than that of the positive ions, because the electron's thermal speed is larger than the sound speed by a factor of order $(M_p/m_e)^{1/2}$. Largely because of this factor, heat conduction tends to be more important than viscosity in many astrophysical problems (if the gas kinetic values apply) and the importance of both increases greatly as temperature increases (especially if one considers approximate pressure equilibrium between regions at different temperatures). Possible deviations from the gas kinetic values could thus be particularly important for processes involving the intracluster gas (see, eg., Livio, Regev and Shaviv 1978).

For some high temperature astrophysical problems the gas kinetic transport coefficients are so large that the corresponding microphysics implies collision mean free paths larger than distances over which the temperature changes appreciably. In such cases one would have to worry about the hot and the cold electrons constituting two distinct but interpenetrating fluids and the complications of "saturated heat conduction" [Cowie and McKee (1977); Balbus (1985]. However, I want to emphasize the opposite worry which Chiuderi has already mentioned, namely that magnetic fields and plasma oscillations can decrease collision mean free paths enormously. Such enhanced scattering (or at least path bending) will decrease the gas kinetic heat conduction coefficient, say, by some large factor η. In the rest of this talk I will pretend that η is a single number - possibly to be determined from analyzing observations - but in reality there are many complexities. First of all, the complications are likely to affect electrons more strongly than ions and η is likely to be more severe for heat conduction than for viscosity. Second, the irregular magnetic field depends on the geometry of the flow and η will be different for an expanding flow and a compressional flow. Third, the plasma instabilities which act as scattering centers may themselves be amplified by the "heat engine" implied by the temperature differential, which results in a non-linear problem. The coefficient η, which multiplies the temperature gradient in the heat-flow equation, itself depends on the temperature gradient. Values between about 10^{-1} and 10^{-6} have been suggested for η in various circumstances (see, eg. Friacas 1986).

A general overview is given in Section 2 for ram pressure stripping. In Section 3 I discuss "sudden stripping", which mainly applies to spiral galaxies, and "continuous stripping" for elliptical cluster galaxies in Section 4. A general overview for different types of "galactic fountains" is given in Section 5, the consequences for the coronal gas of one particular model in Section 6.

2. RAM PRESSURE STRIPPING OVERVIEW

The interaction of a cluster galaxy with the intracluster gas has to be viewed quite differently for spiral (plus irregular) galaxies and for elliptical (plus S0) galaxies. First of all, spirals are rare in rich, relaxed clusters which tend to have a smooth distribution of intracluster gas (see the review by Sarazin 1986). Most data so far on spirals therefore concentrates on unrelaxed clusters, especially on the Virgo cluster. Furthermore, the spirals tend to have different orbits from the early type galaxies in a cluster like Virgo - spirals in the cluster core have a larger dispersion of systemic velocities and are likely to be "just passing through the core" on a highly elliptical orbit and some are even likely to be "merely falling in for the first time" [Hoffman, Olson and Salpeter (1980); Tully and Shaya (1984)].

In addition to the kinematic suggestion of a spiral galaxy moving into a cluster core only rarely, a spiral also has a lot of primordeal gas in its HI disk. Theoretical calculations for spirals should therefore concentrate on the "sudden impact model", i.e. a galaxy suddenly "runs into" a region containing intracluster gas [Gunn and Gott (1972); Lea and DeYoung (1976)]. In such a calculation one can restrict oneself to the primordeal gas disk of the spiral galaxy, even though replenishment might also be of some importance (Gisler 1979). In Section 3 I give the dimensional analysis relevant to "sudden stripping" and summarize the observations.

Elliptical galaxies, on the other hand, populate the relaxed clusters, which tend to have a smooth distribution of intracluster gas, in orbits which are not extremely eccentric. For ellipticals it is then reasonable to consider "continuous models" where the galaxy moves with constant speed relative to intracluster gas, assumed to have constant density. Furthermore, for ellipticals we need not consider any primordeal gas but only the rate at which star deaths produce interstellar gas continuously. This leads to "steady state" models, discussed in Section 4 (for earlier work see Gisler, 1976).

3. SUDDEN STRIPPING

I want to summarize the simplest form of dimensional analysis for "sudden stripping", following the spirit of Gunn and Gott (1972) but carrying it out more crudely. Let M_{gas} be the total mass of pre-existing gas in a galaxy, M_{tot} the total mass which provides the gravitational field for the galaxy and R "the radius" of the galaxy. Without specifying the definition for "radius", nor the distribution of gas and total mass, consider the following order of magnitude relations and definitions of "typical" quantities:

$$\rho_{gas} \sim \frac{M_{gas}}{R^3}, \quad \sigma_{gas} \sim \frac{M_{gas}}{R^2}, \quad \rho_{tot} \sim \frac{M_{tot}}{R^2}, \quad \sigma_{tot} \sim \frac{M_{tot}}{R^2},$$

$$v_{rot} \sim \frac{1}{2} v_{esc} \sim (GM_{tot}/R)^{1/2}, \quad g \sim v_{rot}^2/R \sim GM_{tot}/R^2.$$

In these equations ρ and σ are the orders of magnitude of the average over the galaxy of the (three-dimensional) volume density and the (two-dimensional) surface density, respectively. v_{rot} and v_{esc} are typical values for rotational velocity inside the galaxy (or the internal velocity dispersion) and escape velocity from some point in the galaxy, respectively. Let ρ_∞ be the constant density of intracluster gas (or of a large extragalactic gas cloud) which approaches the galaxy at constant speed V_∞, which is assumed to be supersonic.

If one assumes that the important quantity is pressure, force and momentum transfer, rather than transfer of energy, the external quantity is simply the ram pressure, $\rho_\infty V_\infty^2$. From dimensional analysis the internal quantity to be compared with this is then $\rho_{gas} v_{esc}^2$. Using equation (1), the order of magnitude of this quantity can be rewritten in different ways, such as

$$\rho_{gas} v_{rot}^2 \sim gM_{gas}/R^2 \sim G\sigma_{gas}\sigma_{tot} . \qquad (2)$$

The effectiveness of ram pressure stripping is then positively correlated with the dimensionless ratio of $\rho_\infty V_\infty^2$ to any of the forms in equation (2). If, on the other hand, the ejection of the interstellar gas from the galaxy is achieved by the transfer of energy, ie. the interstellar medium is first heated and then escapes thermally, the scaling is different. The external quantity is then not the ram pressure but the rate of energy transport $\rho_\infty V_\infty^3$ and the equivalent internal quantity contains an extra power of v_{esc} or v_{rot} than the expression in equation (2). Detailed numerical modelling can, in principle, yield the theoretical power of (V_∞/v_{esc}) which multiplies ρ_∞ in the efficiency, but such are available so far only for the continuous stripping case (see Section 4).

In a spiral galaxy, the surface mass density $\sigma_{gas}(r)$ of the gaseous HI disk varies greatly with radial distance r from the center and the local volume density $\rho_{gas}(r,z)$ of the gas varies greatly with height z above the galactic plane, as well as with r. If the important effect is pressure, i.e. momentum transfer and not heating followed by evaporation, one can carry out a more "local" dimensional analysis for cylinders with their axis parallel to the vector velocity V_∞. In such an analysis the distribution of the interstellar gas along a cylinder does not matter, but only $\sigma_{gas}(r)$, the column density of interstellar gas for this whole cylinder. The dimensionless ratio which determines the stripping efficiency for the galactic disk at radial distance r is then of form

$$\rho_\infty V_\infty^2/\sigma_{gas}(r) \, g(r), \qquad (3)$$

where $g(r)$ is a more slowly - varying function of r than $\sigma_{gas}(r)$. A very characteristic prediction of ram pressure theory is then that (unless $\rho_\infty V_\infty^2$ is either very large or very small) the outermost regions of the gas disk (small σ_{gas}) should be stripped and the innermost (large σ_{gas}) should remain intact.

At least for spiral galaxies in the Virgo Cluster, it is clear that (a) there is a tendency for galaxies in the cluster core to be deficient in HI and (b) that deficiency stems from the gas disk having a smaller radius, not from a decrease in the central surface mass density of gas [Giovanardi et al. (1983); Haynes, Giovanelli and Chincarini (1984)]. This observational fact corroborates the theoretical prediction that ram pressure stripping mainly affects the outer regions of a spiral's galactic disk. However, observations so far have not corroborated any clearcut dependence of stripping efficiency on (i) V_∞ for different spirals in a cluster (Magri et al., 1987) nor on (ii) v_{esc} in a comparison of large spirals with dwarf irregular galaxies (Hoffman, Helou and Salpeter, 1988). This may be due to clumpiness of intracluster gas in Virgo, the complicated orbits of spirals and the history dependence of stripping (see, eg. Takeda, Nulsen and Fabian, 1984), rather than to a basic failure of the theory.

4. CONTINUOUS STRIPPING

Primordeal interstellar gas in an elliptical galaxy is likely to be unimportant even if the galaxy is relatively isolated, but star deaths continuously produce new interstellar gas at some constant rate $\dot{M}_{gas}$. As Fabian and others have described at this meeting, cooling flows and eventual star formation are likely to be important in various environments, including an isolated galaxy (see, eg., Nulsen, Stewart and Fabian (1984); Sarazin and White (1987)]. A steady state flow is then possible, where the mass of interstellar gas remains constant at some value $M_{gas,eq.} = \dot{M}_{gas} t_{accum}$, where t_{accum} is some timescale of order R/v_{esc}. If the galaxy is exposed to intracluster gas, the released interstellar gas will have a more complicated flow, but will also accumulate for a time of order t_{accum} before it either undergoes star-formation or merges with the intracluster gas. In dimensional analysis for ram pressure stripping from a whole galaxy, one need only replace M_{gas} in Section 3 by the expression $\dot{M}_{gas} R/v_{esc}$.

If one assumes that pressure (momentum transfer) is the important criterion, the dimensionless ratio which determines stripping efficiency is then

$$\frac{\rho_\infty V_\infty^2 R}{M_{gas,eq} v_{esc}^2} \sim \frac{\rho_\infty V_\infty^2 R}{\dot{M}_{gas} v_{esc}} . \qquad (4)$$

Again, if energy transfer were the important criterion, the expression in equation (4) would be multiplied by one additional power of

V_∞/v_{esc}). Numerical modelling, using a two-dimensional hydrocode but with heat-conduction and viscosity omitted ($\eta = 0$), has now been carried out for various values of V_∞ and of v_{esc} (Gaetz, Salpeter and Shaviv, 1987). The empirical scaling-result is that a ratio proportional to the expression in equation (4) multiplied by $(V_\infty/v_{esc})^{0.7}$ determines the stripping efficiency, intermediate between the momentum and the energy criteria. These calculations also show a tendency for the outer regions of an elliptical galaxy to be stripped somewhat more than the inner regions but the distinction is not very sharp.

Ellipticals in different regions in a cluster encounter different values for the intracluster gas density ρ_∞ and the predicted degree of stripping will vary. Nevertheless, the calculations by Gaetz et al. (1987) predict that most ellipticals in a rich X-ray cluster should have their gas mostly stripped. For the larger elliptical galaxies in more isolated regions the amount of X-ray emitting hot coronal gas is now known fairly reliably (see Canizares, Fabbiano and Trinchieri, 1987). Future X-ray observations for internal gas in elliptical galaxies will be most rewarding for galaxies in "marginally relaxed" clusters, ie. clusters with intracluster gas more smoothly distributed than in the Virgo cluster but with density ρ_∞ appreciably less than in a "typical X-ray" cluster. Observationally, the galactic emission should not be masked by cluster emission as much as in a richer cluster and the theoretical predictions are interesting: If heat conduction is negligible, as assumed by Gaetz et al., the larger ellipticals should not be stripped very severely; however, if the heat conduction has its full gas kinetic value the stripping should be quite severe (as indicated by preliminary calculations). Thus, future X-ray observations will almost be able to measure the parameter η described in the Introduction!

5. GALACTIC FOUNTAIN OVERVIEW

If an isolated spiral galaxy has hot coronal gas in its halo, it must have been put there by supernova explosions (and O-star winds) from the galactic disk. The supernova ejecta themselves are hot enough and fast enough that they would easily escape from the galaxy if they did not mix with ambient interstellar gas. On the other hand, the gaseous interstellar disk is massive enough that it could easily contain and cool the ejecta if supernovae went off singly and near the galactic plane. Hence, it is the "packaging" of supernova explosions which determines how much of the total supernova energy from the disk "breaks through" the disk into the corona and how much mass from the ambient gaseous disk is injected into the corona together with the energy. Even if the injection were a steady process there would be two parameters to be specified, the rate $\dot{M}$ of mass injection and the injection temperature T_{in}. In reality the process can be sporadic, which introduces further complications.

We have at least indirect evidence that there are large "super-bubbles" of coronal gas, both in our Galaxy (Heiles 1979) and in M31 (Brinks and Bajaja 1986) which are almost breaking through the galactic disk. These are likely to have been formed from the concerted actions of many supernovae and O-star stellar winds; other superbubbles which manage to break through the disk are likely to be a major injection mechanism for the coronal gas in the halo. The temperature T_{in} of the superbubble at injection depends on the ratio of total energy input to total mass picked up, but the bubble will also be launched with some vector bulk velocity V_{in}, which depends on the surroundings as well as on the heat content of the bubble. If a hot bubble were created just above a uniform cold gas disk, the bubble would "simply rise upward", but this verbal description leaves out an important ingredient. A spherical bubble in vacuum would merely expand, not rise up - in the actual situation the rising is due to the "rocket effect", ie. the bubble imparts momentum to the disk below and the recoil accelerates the bubble in the direction of the disk's density gradient. A real galactic disk contains spiral arms which contain giant molecular cloud complexes and an escaping superbubble is lanuched by recoiling from the complex, not just vertically but also with some horizontal component. The ratio of the speed of this bulk motion to the thermal speed appropriate for T_{in} will be roughly of order unity, but with a numerical factor which can vary appreciably.

Even in a single galaxy, the injection conditions will vary greatly from place to place and from time to time. There may be some episodes where the injection stays "violent", with T_{in} well above $10^7 K$, long enough to cause a temporary "galactic wind" (Mathews and Baker 1971); others with T_{in} well below $5 \times 10^5 K$ which merely cause a thickening of the disk rather than coronal flow (Salpeter 1985). I will discuss only intermediate cases which lead to some kind of a "galactic fountain" which rises into the halo corona but does not escape from the galaxy. For $T_{in} \sim (0.5 \text{ to } 2) \times 10^6 K$ one has the "normal" fountain (Bregman 1980) where cool droplets form in the corona and "rain down" onto the galactic disk not very far from where the supernova breakthrough occurred. In this case one has a fully developed thermal instability with hot fluid flowing up and cool fluid flowing down. The downward flow may not be in random individual droplets but in a hollow cylinder surrounding the upward columnar flow (Ikeuchi, private commmunication). In the next section I discuss an "outer fountain" with $T_{in} \sim (2 \text{ to } 7) \times 10^6 K$, violent enough to carry some material far from the injection region but not so violent as to cause escape from the whole galaxy.

6. A STEADY OUTER FOUNTAIN

For injection into the corona with $T_{in} \sim (2 \text{ to } 7) \times 10^6 K$ and an appreciable horizontal component of the bulk speed, some material (bubble with decreased angular momentum) will end up at smaller radial distances, but other material (with increased specific angular momentum) will move outward. Some of this material, which we call the "outer galactic fountain", will return towards the galactic plane at large enough radial distances (~ 25 kpc, say) so there are no young stars or supernovae there and therefore no hot upward flow. In this case we do not have the two-stream flow of the normal galactic fountain, but a single-flow pattern: upwards from the disk regions where star formation and supernova explosions are most common, then outward (or inwards) and then down again in a cooling flow.

If the coronal temperature is sufficiently large, $T \gtrsim 2 \times 10^6 K$, one finds that small deviations from a steady cooling flow are oscillatory without a thermal instability (Corbelli and Salpeter, 1988). This dichotomy of behavior stems from the strong temperature dependence of the radiative cooling rate; above some critical temperature of about $1 \times 10^6 K$ the radiative cooling time is longer than the gravitational fall time. In those cases one finds highly subsonic, steady cooling flows with oscillatory behavior (although it is not clear if the oscillations are overstable or not; see Malagoli, Rosner and Bodo 1987).

In essentially all spiral galaxies, the regular HI disk extends out to at least twice the radius of the "active disk" which has appreciable star formation in it. Corbelli's calculations on steady outer ("fountain") flow were motivated mainly by the effect such flow has on the HI distribution (Corbelli and Salpeter 1988). Here I only want to mention the effect an extended, inert HI disk has on the hot coronal gas in the halo. This disk essentially acts like a cold-plate trap in a laboratory vacuum system, i.e. it will induce a cooling flow from the hot gas onto the "cold plate" with the cooled gas accumulating near the galactic plane. For a given pressure in the corona ($nT \sim 500$ cm^{-3} K, say) the total rate of the cooling flow depends strongly on the coronal temperature T_c (roughly as $T_c^{-4.3}$) and is proportional to the area of the HI disk, but does not depend on the temperature and column density of the HI disk (compared to the corona it has negligible temperature and scaleheight anyway). For $T_c \sim 3 \times 10^6 K$ the cooling flow onto the "cold trap" of the extended disk could just about balance the mass-flow-rate outwards from the inner disk if ~ 0.5% of all supernova energy goes into such outward motion.

We can summarize crudely the effect of the HI extended disk on the hot corona as follows. If the coronal temperature T_c is below $3 \times 10^6 K$, then the outer disk acts as a "good vacuum cleaner" and one should observe little coronal X-ray emission far outside the region where supernovae are prevalent. If T_c is well above $3 \times 10^6 K$, on the other hand, there is little cooling flow onto the outer disk and the coronal gas should have a large extent because of its high temperature. There should thus be thermal X-ray emission from the outer regions of an isolated spiral galaxy in this regime, unless T_c is so large as to give a galactic wind. The wind proviso is more severe for dwarf irregular galaxies, because of their low escape velocity, so we can make the following prediction for thermal X-ray emission from the outer regions of late-type galaxies. It should be largest for those giant spirals where the HI disk does not extend very far beyond the stellar disk, smallest for dwarf irregular galaxies with an extended HI disk.

This work was supported in part by U.S. National Science Foundation Grant AST 84-15162.

REFERENCES

S.A. Balbus, Ap. J. 291, 518 (1985).
J.N. Bregman, Ap. J. 236, 577 (1980).
E. Brinks and E. Bajaja, Astr. Ap. 169, 14 (1986).
C. Canizares, G. Fabbiano and G. Trinchieri, Ap. J. 312, (1987).
E. Corbelli and E.E. Salpeter, Ap. J. in press, March 1988.
L.L. Cowie and C.F. McKee, Ap. J. 211, 135 (1977).
A.C. Friacas, Astr. Ap. 164, 6 (1986).
T.J. Gaetz, E.E. Salpeter and G. Shaviv, Ap. J. 316, 530 (1987).
G.R. Gisler, Astr. Ap. 51, 137 (1976).
G.R. Gisler, Ap. J. 228, 385 (1979).
C. Giovanardi, G. Helou, E.E. Salpeter and N. Krumm, Ap. J. 267, 35 (1983).
J.E. Gunn and J.R. Gott, Ap. J. 176, 1 (1972).
M. Haynes, R. Giovanelli and G. Chincarini, Ann. Rev. Astr. Ap. 22, 445 (1984).
C. Heiles, Ap. J. 229, 533 (1979).
G.L. Hoffman, D.W. Olson and E.E. Salpeter 242, 861 (1980).
S.M. Lea and D.S. DeYoung, Ap. J. 210, 647 (1976).
M. Livio, O. Regev and G. Shaviv, Astr. Ap. 70, L1 (1978).
C. Magri, M. Haynes, W. Forman, C. Jones and R. Giovanelli, NAIC Report 234, Cornell University (1987).
A. Malagoli, R. Rosner and G. Bodo, Ap. J. 319, 632 (1987).
W.G. Mathews and J. Baker, Ap. J. 170, 241 (1971).
P. Nulsen, G. Stewart and A. Fabian, MNRAS 208, 185 (1984).
E.E. Salpeter, Mitteilungen der Astronomischen Gesellschaft 63, 11 (1985).
C.L. Sarazin, Rev. Mod. Phys. 58, 1 (1986).
C.L. Sarazin and R.E. White, Ap. J. 320, 32 (1987).
L. Spitzer, Physics of Fully Ionized Gases, Wiley, New York (1967).
H. Takeda, P. Nulsen and A. Fabian, MNRAS 208, 261 (1984).
R.B. Tully and E.J. Shaya, Ap. J. 281, 31 (1984).

7. CLUSTERS OF GALAXIES

X-ray Observations of Clusters of Galaxies

Richard Mushotzky
Laboratory for High Energy Astrophysics
Goddard Space Flight Center
Greenbelt, MD

Introduction

Because Sarazin (1986) has written a very extensive review of the x-ray emission from clusters of galaxies I have allowed myself considerable leeway in writing this review. In particular it has allowed me to be incomplete in my references and to omit certain topics. I have chosen some topics because they have been somewhat ignored in the past literature and others for which both the observational and theoretical situation is unclear. This paper is therefore quite different from the oral talk I gave at Cargese. This review is somewhat different from the usual ones in that I hope that the reader finds the observational situation to be somewhat untidy and not well explained, and that the need for future data and theory is evident.

It is not clear how the optical and x-ray properties of clusters are related to each other. There is not a strong overlap in the data bases of clusters well studied in both the x-ray and optical which makes correlation analysis difficult. As I hope to show in this paper, it is not even clear that analysis of x-ray and optical data give the same values for many of the fundamental cluster properties, such as depth of the potential or core radius.

I. The X-ray Data

A. The Data Base- Detections and Identifications

There have been three surveys of the sky in x-rays performed by the Uhuru (Forman et al. 1977), Ariel-V (McHardy et al. 1981) and HEAO-1 (the A-2 experiment, Piccinotti et al. 1982 and the A-1 experiment, Wood et al. 1984) satellites. The detectors on these satellites were non-imaging mechanically collimated proportional counters. The Uhuru and Ariel-V detectors were sensitive in the 2-10 kev band while the two different HEAO-1 experiments covered a somewhat broader energy range (2-30 kev for A-2 and 0.5-12 kev for A-1). While the sensitivity over the whole sky was somewhat variable, both the A-1 and A-2 surveys reached limiting all sky fluxes of $\sim 2\times 10^{-11}$ ergs/cm^2-sec in the 2-10 kev band and the Uhuru and Ariel-V experiments had a somewhat higher threshold.

The ability to determine the optical counterparts of the x-ray sources discovered by these surveys depended, to a large part, on the size of the error boxes. At the limit of $\sim 2\times 10^{-11}$ ergs/cm^2-sec essentially all the high galactic latitude (b>20°) x-ray sources have been identified from error boxes of size less than 1/2 square degree. Among the 86 such objects in the 8.2 steradians of sky covered by the A-2 survey (Piccinotti et al. 1982) there are 31 clusters of galaxies. At this relatively high flux threshold, several more clusters have been identified at lower galactic latitudes This is the only complete x-ray sample available. The A-1 experiment (Wood et al. 1984 and Kowalski et al. 1984) has detected many fainter clusters in selected regions of the sky. However the error boxes from this survey are too large to allow complete identification.

The *Einstein* and EXOSAT missions were imaging x-ray telescopes that observed many selected clusters. The *Einstein* mission also detected reasonable samples of clusters of galaxies in a serendipitous mode (Maccacaro et al. 1982). The imaging experiments on these observatories worked primarily in the 0.5-3 kev band and had an angular resolution <1'. The imaging capability enabled much weaker objects to be studied, with some clusters being detected at fluxes of $\sim 2 \times 10^{-13}$ ergs/cm^2-sec in the 0.5-3.0 kev band. However, detailed, high quality imaging results typically required long exposures on clusters 10-50 times brighter than this threshold.

Because clusters of galaxies are quite large objects with <50% of the flux being contained in a metric circle of radius ~0.5 Mpc (~10' at z=0.03 for H_0=50/km/sec/Mpc) the *Einstein* Medium Survey (MSS, Maccacaro et al. 1982) was heavily biased in detecting clusters whose x-ray surface brightness is strongly centrally concentrated. In addition because of the method of analysis in that survey, which calculated the flux in a box with 2' sides, the fluxes of clusters are underestimated with respect to the "large beam" all sky survey fluxes for clusters closer than redshifts of ~0.3 .

Sealed proportional counters are not very sensitive to x-rays of energy less than ~2 kev, and therefore the all sky surveys were much less sensitive to clusters whose effective temperature was less than 2 kev. As we shall see (sec II.A.) this restricts these surveys to clusters more luminous than $\sim 10^{43}$ ergs/sec[†]. However, this energy restriction is not a limit in finding the more luminous clusters and the observed upper bound of $\sim 3 \times 10^{45}$ ergs/sec seems to be a true limit, at least for low, z<0.3, redshifts. The limited sensitivity of the all sky surveys effectively restricted the sample to low, z<0.2, redshift objects. As indicated in the introduction it is not altogether clear how the x-ray and optical properties of clusters are related. It is thus not surprising that the distribution of cluster types found in x-ray surveys is somewhat different from that found in optical surveys (such as the classic one by Abell 1958). However, what is unusual is that whenever an x-ray source has the characteristic properties of a cluster of galaxies, viz. extended smooth x-ray emission and/or thermal bremmstrahlung emission with a Fe line, inspection of the the Palomar sky survey so far has always found an associated cluster of galaxies, even though it might be a quite poor one. The x-ray surveys thus indicate that there are few (essentially none) collections of hot baryons in a potential well where there are no galaxies. This has important implications for theories of dark matter and biased galaxy formation.

So far there has been no reliable detection of x-ray emission from superclusters or other large scale optical structures. However, the upper limits have not been extremely restrictive; the HEAO-1 surveys did not have sufficient angular resolution to separate out putative supercluster emission from that due to the embedded clusters and the *Einstein* and EXOSAT missions were not sufficiently sensitive to surface brightness to see the expected extremely low surface brightness (but high integrated luminosity) supercluster emission.

[†] The ROSAT all sky survey will not be subject to this bias and thus should find large numbers of low luminosity clusters. It is anticipated that the survey will find luminous clusters to very high (z>0.7) redshifts.

B. X-ray Luminosity Functions

The all sky and other surveys mentioned in the previous section have allowed determination of the x-ray luminosity function of galaxies and comparison with the optical luminosity function (Bahcall 1979a,b). Fitting the cluster luminosity function in the range from 5×10^{43}-2×10^{45} ergs/sec, several groups have found that the best fitting model is a power law of form

$$f(L_x) = A_p (h_{50})^5 \{L_x/10^{44} h_{50}^{-2} \text{ ergs/sec}\}^{-p} \times (10^{44} \text{ ergs/sec})^{-1} \text{Mpc}^{-3}$$

Various authors have found consistent results with $p\sim 2.2$ and $A_p\sim 4\times 10^{-7}$. Extrapolation of the power law to lower luminosities does not provide a good fit to the A-1 survey data (Kowalski et al. 1984), but this could be a selection effect due to the bandpass of the A-1 instrument and its relatively lesser sensitivity to "cool" clusters. There are two other indications that the power law fit does not hold over a wider luminosity band: First of all, extrapolation of the power law implies that at $L_x \sim 10^{43}$ ergs/sec the total number of x-ray emitting clusters would exceed the number of Abell optical clusters (and of course diverges asymptotically at low L_x). However it is clear that many low luminosity, poor clusters are not "Abell" clusters. Thus the fact that the power law fit would exceed the number of Abell clusters at low x-ray luminosity is not a strong constraint. Secondly, the *Einstein* results of Abramopoulos and Ku (1983) based on a random but not complete sampling of Abell clusters also indicates a flattening of the luminosity function at low L_x in qualitative agreement with the A-1 results. These x-ray surveys are also in fairly good agreement with Bahcall's prediction of the x-ray luminosity function from the optical luminosity function data which predicts a flattening at low luminosities of the x-ray luminosity function.

The available data are consistent with virtually every optically selected cluster being an x-ray emitter and vice versa. At luminosities of less than 10^{42} ergs/sec it becomes quite difficult to distinguish cluster emission from that due to a superposition of individual galaxies. However, because individual galaxies have $\sim 1/30$ the x-ray luminosity at a given x-ray temperature of clusters there must be a considerable number of galaxies radiating (≥ 10) to produce the luminosity of a 10^{42} ergs/sec object. If the cluster were close enough imaging observations would show the x-ray emission being resolved into indivudal galactic components. Also, at this low a luminosity, the predicted x-ray temperatures are <1 kev, consistent with that expected from the potential wells of galaxies. Thus there is a natural lower bound to the x-ray luminosity of clusters and groups. The lowest upper limits from the A-1 and *Einstein* surveys are in the range of $\sim 10^{42}$ ergs/sec with a few exceptions. A typical *Einstein* upper limit of $\sim 2\times 10^{-13}$ ergs/cm^2-sec, for a "point-like" source corresponds to $L_x<10^{42}$ ergs/sec at a distance of ~ 200 Mpc. There are less than 100 Abell clusters closer than this. Thus the available x-ray data do not strongly constrain the form of the luminosity function at $L_x<10^{42}$ ergs/sec and it is possible that there do indeed exist optical clusters with little or no x-ray emission: that is all x-ray clusters are optical clusters, but not necessarily vice-versa.

At present there are no strong indications of evolution with redshift of the x-ray luminosity function of clusters (Kowalski et al. 1984, Henry and Lavery 1984) in either normalization or form. There is no information on evolution of cluster sizes, temperatures or structure. However, the data are not very good. The best constraints on the evolution of the slope of the luminosity function come from the *Einstein* results of Henry and co-workers, but these depend essentially on "how random" their "random" sample of high redshift clusters is. Given the large changes in galaxy colors seen in clusters at redshifts >0.3 it seems likely, *a priori*, that there will be x-ray cluster evolution. The limited results published so far from *Einstein* Medium Survey (MSS) indicate that the observed logN-log S of clusters is consistent with no evolution. However, the observed luminosity distribution of the MSS clusters (Schmidt and Green 1986) is inconsistent with the predictions from the above luminosity functions in that the MSS detects many more low luminosity objects than predicted and, given the general agreement on the number of sources, many fewer high luminosity objects. It is not clear what the origin of this discrepancy is. Part of the discrepancy may arise if there is a relation of cluster size and luminosity with more luminous clusters being larger. If this is true than the fixed beam size used by the MSS survey will systematically underestimate the luminosity of high L_x clusters.

The ROSAT all sky survey should measure any evolution in the x-ray luminosity function and in cluster parameters (such as core radius, density and structure) out to redshifts of ~0.5. It should not suffer from the same types of selection effects as the MSS. Similar results but of somewhat lower statistical accuracy, should also be obtained from analysis of the extended MSS data set. Because of its all sky coverage, great sensitivity to low surface brightness and large effective solid angle the ROSAT survey may also provide strong limits on supercluster emission. Selected Ginga observations of luminous high redshift clusters may provide information on any temperature or abundance evolution.

C. X-ray Spectra

1) Continuum

There have been several large samples of x-ray spectra of clusters of galaxies obtained from the OSO-8 (Mushotzky et al. 1978), Ariel-V (Mitchell et al. 1979), HEAO-1 (Mushotzky 1984) and *Einstein* (Mushotzky 1984) satellites. The first three experiments used proportional counters sensitive in the 2-40 kev range and the last was a solid state detector sensitive in the 0.6-4 kev band. At this meeting the results of the EXOSAT survey (Edge and Stewart) will be presented.

The main conclusion of these surveys, with ~35 objects in them, is that the total x-ray spectrum of almost every cluster can be moderately well described by isothermal bremmstrahlung emission (see section III below) with most of the detected clusters having temperatures in the 2-9 kev range.(figure 1)

As previously discussed the lower bound on the temperature is a selection effect caused by the bandpass of the sky survey instruments used to find the objects. However, the "cutoff" at ~10 kev effective temperature (that is the temperature determined from an isothermal fit to the spectrum) is a real one and is an upper bound to the depth of the potential well in clusters if the gas is isothermal. Given the relationship between temperature and luminosity, $L_x/L(0)_x \sim T^{5/2}$ (sec II.A), this observed upper cutoff in temperature also indicates a upper cutoff in luminosity which agrees with the observed lack of objects more luminous than $L_x \sim 3\times 10^{45}$ ergs/sec. This last statement is not a tautology because there are many more clusters with measured luminositites than temperatures.

For most clusters, the x-ray luminosity is dominated by emission from the hot gas. However, there appear to be a few cases where more than 10% of the x-ray luminosity in the 0.5-40 kev band arises from an active nucleus (AGN) located in the cluster. The best documented cases are NGC 1275 in the Perseus cluster (Abell 426, Rothschild et al. 1981) and M87 in the Virgo cluster (Lea, Holt and Mushotzky 1982). It is also predicted theoretically (Rephaeli 1980) that at high energies (E>20 kev) some

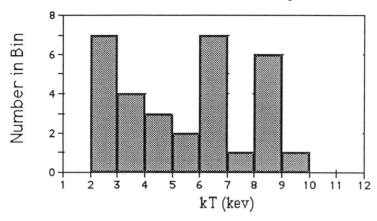

Fig 1 Distribution of isothermal x-ray temperatures of clusters of galaxies from the HEAO-1 A-2 data set.

clusters may be dominated by microwave background photons Compton scattered off relativistic electrons that appear to permeate the intercluster medium in a few clusters. Despite several claims there do not appear to have been any definitive detections of such radiation (Lea et al. 1981). These upper limits on hard x-ray emission put lower bounds on the magnetic fields in the intracluster medium which seem consistent with the equipartition values.

There appear to be a few clusters, mostly of low luminosity, which have "complex" spectra which may be described as a sum of thermal components (e.g. the Centaurus cluster, Mitchell and Mushotzky 1980). In the absence of spatially resolved spectra the origin of this type of behavior it is not clear but it may be due to "cooling flows" (see A. Fabian these proceedings), to radiation from individual galaxies (see Trinchieri these proceedings), to the co-existence in the IGM of both hot and cold phases, or to temperature gradients. Spatially resolved x-ray spectroscopy is needed to resolve this issue.

2) Line Emission

One of the major discoveries of the first moderate quality x-ray spectra of clusters was the existence of a strong Fe line radiation (Mitchell et al. 1976, Serlemitsos et al. 1977). The presence of Fe in the intracluster gas showed that the gas did not have primordial abundances and that at least some of it has been processed in stars. At the temperatures observed from most clusters, $kT > 2$ kev (cf fig 1), the dominant emission lines from a gas of roughly solar composition are expected to be those due to highly ionized Fe. There are several such lines but the strongest ones are the "6.7 kev" complex (actually a sum of the 6.67 kev triplet and the 6.98 kev singlet lines of Fe due to transitions from n=2 to n=1 in Fe^{24} and Fe^{25}) and the $K\beta$ lines at ~7.9 kev (due to transitions in the same ions between n=3 and n=1). There is now much information on the strength of these lines in different clusters (Mushotzky 1984, Rothenflug and Arnaud 1985, Henriksen and Mushotzky 1987, A. Edge this symposium). If most of the observed radiation is from isothermal or polytropic gas then the observed equivalent widths can be directly converted into abundances. While there appears to be some dispersion in the distribution function, most of the observed abundances tightly cluster around 1/2 solar.(Figure 6) (where solar is Fe/H= 3.2×10^{-5} by number).

From clusters whose mean ICM (intracluster medium) temperature is less than ~3 kev or which have a cooling flow (see III below), one expects to see radiation from the K lines of the lower atomic number abundant elements viz: O, N, Ne, Mg, S, Si, Ar and Ca and from the L lines of Fe. These lines appear in the 0.5-4.5 kev band. Most of the data on such line radiation were obtained with the spectroscopic instruments (the Solid State Spectrometer (SSS) and FPCS) on the *Einstein* observatory. The available data are somewhat difficult to interpret due to the strong temperature gradients in the cooling flows but detailed modeling indicates that these lower atomic weight elements have abundances ~1/2-1 solar. In the best determined case, that of M87, the fit to the cooling flow models indicate ~1.5 solar abundances, which is consistent with the optical data for this galaxy. At present there is no strong evidence for deviations from solar ratios, with the exception of a high ratio of oxygen to Fe seen in M87 and perhaps Abell 426 (Canizares, Markert, and Donahue 1987)

There are only two direct observation of the spatial distribution of the iron abundance in a cluster (A426, Ulmer et al. 1987; Virgo, Edge this symposium). Comparison of the Fe abundance determined in the core of ~ 10 clusters from the SSS observations of the Fe L complex with the abundance of Fe determined by the large solid angle proportional counters observations of the K lines indicates that there are no strong abundance gradients of Fe within ~2 core radii.

The measurement of the distribution of elemental abundances in the gas will only be possible with a imaging spectrometer system such as will flown on AXAF and XMM (see papers by Tanabaum and Peacock this symposium). Such observations may determine the origin of the metals e.g. ram pressure stripping of galaxies, expulsion of the gas out of the galactic potential by supernova etc. In addition these data will measure the evolution of the metals in the ICM with cosmic time. The mass of metals in the ICM is very large, ~$10^{11} M_0$ of Fe alone for a luminous cluster such as Perseus. Using normal supernova to produce this mass involves an energy of ~10^{62} ergs which is ~10% of the total binding energy of the cluster. Thus the creation of the metals in the ICM of clusters is not a small perturbation. Explanation of the regularities of the abundances from cluster to cluster and how such a vast amount of material has been expelled and mixed into the ICM are major problems in the theory of clusters.

D. Spatial Distribution

The vast majority of the detailed information on the spatial distribution of the hot gas in clusters has come from observations obtained with the imaging instruments on the *Einstein* observatory (see Forman and Jones 1982 for a recent review). I shall briefly try to summarize some of this large body of work.

1) X-ray centers

In a large fraction of bright clusters the x-ray surface brightness distribution is centered on a large and bright galaxy (almost always a D or cD galaxy). Very frequently this galaxy is also at the geometric and dynamic center of the total cluster but sometimes it is only at the center of a local density maximum. As pointed out by Beers and Tonry (1986) the x-ray centroid and the position of the D or cD galaxy near the cluster center differ, in the mean, by only $20h^{-1}$ kpc while the mean offset between the optically determined and x-ray centroids is $120\ h^{-1}$ kpc. Thus in some sense the x-ray emission is more sensitive to the local distribution of galaxies rather than to the global distribution. What is quite amazing is that the projected galaxy distribution is sensitive to the choice of the center. If the "optical" cluster center is chosen, the galaxy distribution resembles a "King" model with a well developed core. If the x-ray center is chosen the distribution looks more like a power law. Since the x-ray emission defines the center of the "local" gravitational potential (due to the relatively short relaxation time of the hot gas) while the center of the galaxy distribution is sensitive to the dynamical history of the cluster it is not clear what, exactly, the optical center represents. Extension of these results to more clusters and modeling of the evolution of the optical and x-ray images of clusters (Cavalieri this symposium) is vital to understanding this result.

2) Surface brightness profiles:
As shown by Forman and Jones (1984) almost all clusters which are sufficiently azimuthally symmetric to permit construction of a radial surface brightness profile, and which do not have cooling flows can be well fit in their central regions by a model of the form :

$$S(r)=S(0)[1+(r/a_c)^2]^{-3\beta+1/2} \qquad (1)$$

where $S(0)$ is the central surface brightness, a_c the "core radius" and β a measure of the asymptotic slope of the surface brightness law, are fitted parameters. The distributions of a_c and β are relatively narrow with $<\beta> \sim 2/3$ and $<a_c> \sim 0.3\ h_{50}^{-1}$Mpc (see figure 2).

The existence of a core to the x-ray surface brightness distribution seems well established from the well determined values of a_c in the fits to the x-ray data. As indicated above, it is not clear if the optical galaxy count distribution requires a core. The existence of such simple regularities in the x-ray profile is a rather surprising result but there are recent indications (West, Dekel and Oemler 1987) that there may be similar regularities in the optical surface brightness profile. Such simple patterns may be the "natural" result of virialized systems (White 1979). To my knowledge there have not been any other simple analytical forms (such as a deVacouleurs model) which have been fit to a large cluster sample. Thus it is not clear if the β model is unique or even the best fit simple physical and/or mathematical model. At distances larger than ~15' from the cluster centers there are very few x-ray images (basically only a few x-ray montages were obtained with the *Einstein* IPC which had an effective size of 34x34'). In addition the x-ray surface brightness at several effective core radii was below the ability of the IPC to detect for all but the brightest (largest S(0)clusters)). Thus the form of S(r) at large metric radii has not been well determined by direct imaging.

If the gas is isothermal, then equation 1 results in a density distribution of the form: :

$$n(r)=n(0)[1+(r/a_c)^2]^{-3\beta/2} \qquad (2)$$

As indicated by equations 1 and 2 there is a strong correlation between gas density and surface brightness, as expected from the fact that the Einstein IPC is relatively insensitive to temperature effects. There are no "typical" values of $n(0)$, but x-ray luminous clusters often have $n(0) > 2 \times 10^{-3}$ cm^{-3}. There appears to be an upper limit to $n(0)$ which may be set by the onset of a cooling flow, figure 8(see sec III). For typical values of β eq (2) shows that the total mass of the gas is diverging linearly with radial distance. This rather surprising result indicates that if the optical data are well fit by a King profile, the ratio of gas mass to galaxy mass is a constantly increasing function of radius (Cowie, Henriksen and Mushotzky 1987) and given the nominal densities of the gas, that the gas is the dominant baryonic component in the cluster.

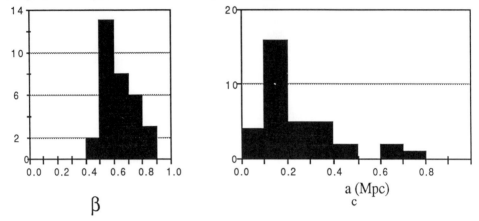

Figure 2 The vertical axis is the number of clusters in the Forman Jones (1984) survey with the given value of β or core radius. Only clusters with well measured values of a_c or β are plotted.

It is anticipated that the all sky ROSAT survey, which will be able to follow clusters out to large angular scales, will directly measure the surface brightness distributions of a large sample of clusters out to large metric radii.

Prior to the ROSAT satellite (and with the singular exception of the Coma cluster, Hughes 1987) we must rely on non-imaging detectors to provide surface brightness information at large angular distances from the cluster center. A comparison of the fluxes in the two HEAO-1 A-2 fields of view (3x3 and 3x1.5 degrees) led Nulsen et al.(1979) to conclude that most clusters do not have detectable x-ray emission beyond 1.5 degrees with only the Virgo and Perseus clusters having significantly larger extent (~3 Mpc at the distance of Perseus). However, this measurement did not have sufficient angular resolution to see similar physical extents in more distant clusters. In principle if the form of S(r) is specified, one can take the fluxes measured by various non-imaging but large field of view instruments and compare them to the the predictions of equation 1. In figure 3 the ratios of the fluxes measured by the HEAO-1 A-2 (3x1.5 degrees field of view) detectors to the fluxes seen by the EXOSAT ME (45x45' FOV) and *Einstein* MPC (45x45' FOV) detectors versus redshift are compared. We see that this ratio is a monotonic function of z . That is, as the distances from us increase the smaller beam detectors (MPC and ME) see more and more of the total flux (as measured by HEAO-1 larger FOV detectors). These ratios are in good agreement with the β model predictions at large distances from the center, or in other words the values of β determined by the ratios of the HEAO-1, EXOSAT ME, and*Einstein* MPC detectors (and the total flux seen in the IPC, Henriksen and Mushotzky 1985) agree with the values of β determined by the imaging IPC data in the central regions of the cluster (TABLE 1).

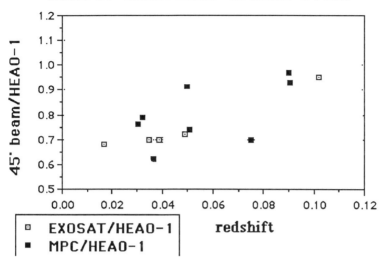

Figure 3 Ratio of small FOV(EXOSAT or MPC) detector fluxes to HEAO-1 fluxes for clusters. The EXOSAT data are taken from Henriksen and Mushotzky 1988 and the MPC data from Mushotzky 1988 .The errors in the ratios are not statistical and are due to relative calibration systematic errors

TABLE 1
Comparison of Values of β

NAME	IPC (image)	EXOSAT / IPC	HEAO-1 / EXOSAT
A262	0.6-0.6	0.42-0.52	0.45-0.55
A576	0.47-0.52	0.40-0.65	0.45-0.55

Thus there is evidence that the form of S(r) is can be well described by equation 1 out to radii of ~45' in at least a few clusters. In addition, the *Einstein* montage of the Coma cluster (Hughes 1987) shows a similar result. Thus the x-ray emission from the hot gas has a physical size of at least 2.5 Mpc (45' at z~0.03; if we look at figure 3 we note that the large and small beam fluxes do not seem to converge until z~0.08 which argues for measurable amounts of gas out to radii of 6 Mpc). This strengthens the conclusion drawn above that the hot x-ray emitting gas (baryons) has a mass which is a large fraction (>15%) of the total mass of the system. The Ginga satellite, with its greater sensitivity and large solid angle (~1x2 degrees), could provide a much larger sample of integral count rates of clusters to compare with the *Einstein* images to determine the large scale distribution of gas in many more clusters. Hopefully this larger data set will help to refine the largescale structure of clusters.

3) Morphology

Forman and Jones (1982) have recently reviewed this subject. They have categorized the *Einstein* images into two general classes : (1) large core radius clusters

(a_c~0.5 Mpc) and (2) small core radius clusters (a_c ~0.25 Mpc). At low x-ray luminosity clusters of type 1 have irregular x-ray structure, whereas at higher luminosities they appear smooth and symmetric. Clusters of type 2 always appear smooth and symmetric but at low x-ray luminosities the image seems to be dominated by emission from a central galaxy. Forman and Jones (1982) have connected many other properties of clusters together into an evolutionary scheme which naturally fits the morphological categorization. Unfortunately much of the data on which this classification has been based has not been published. It is also not clear how much of the categorization has depended on the relative insensitivity of the *Einstein* instruments to low surface brightness features which limited detailed analysis of many of the *Einstein* images. There may be other evolutionary schemes (Cavaliere this symposium) which can also reproduce the observed morphologies.

The x-ray images frequently show strong deviations from circularity which indicate that the potential well of the cluster is not spherically symmetric.In their original analysis of cluster images Jones et al. (1979) found that ~30% of the cluster images appear noticeably elliptical. Sometimes this ellipticity can be as interpreted (Fabricant et al. 1986) the superposition of two "clusters" along the line of sight. As the clusters separate on the plane of the sky one observes "double" clusters. It appears that a significant fraction of all clusters (perhaps as large as 1/4) have x-ray "double" images (Forman et al. 1981, Henry et al. 1981)

It is not clear if there is any systematic evolution of cluster morphology with redshift, supercluster environment or other extrinisic variable. However, as pointed out by Kaiser (1986) hierarchical clustering theories can make direct predictions about such evolution. We do not understand in detail what the x-ray images are telling us about cluster evolution. However, this is primarily due to a lack of detail predictions from the theory. The ROSAT all sky survey will be able to make statistical statements about the distribution of cluster morphology versus other parameters such as redshift (out to z~0.2) and luminosity, which will constrain theories of cluster formation and evolution.

4) Internal Structure

While many clusters appear to have a smooth x-ray profile, some clusters (in particular A1367 Bechtold et al. 1983) appear to have a clumpy, irregular distribution. This is most apparent in the low luminosity clusters. This may be partially understood as due to the influence of individual galaxies on the ICM and/or to x-ray emission from hot gas in the galaxies themselves. One that reason this is most visible in low luminosity clusters is the lower pressure in these objects. As shown below (figures 4 and 8) there is a correlation of central density and temperature with cluster luminosity; the pressure in the central regions goes as $L^{0.8}$. At a luminosity of less than 10^{44} ergs/sec, the mean central pressure in a cluster drops below the typical pressure in the ISM of a "normal" spiral galaxy, nT~10^4 cm^{-3}K. Thus one might naively expect that at lower luminosities and thus low central pressures that the effects of individual galaxies should become noticeable, as does seem to be the case. Of course in the outer regions of even dense, luminous clusters the pressures are also low. For even the most luminous clusters $L(x)$~3×10^{45} ergs/sec with central pressures of $nT>6\times10^5$ cm^{-3}K, at ~7 a_c the pressure has dropped below 10^4 cm^{-3}K and one might expect to see the effects of individual galaxies.

II. X-ray and Optical Correlations

A. X-Ray Temperature and Luminosity

The best correlation between x-ray properties of clusters is that between x-ray luminosity and x-ray temperature. The best fit powerlaw is of the form $L(x) \sim T^{5/2}$, figure 4, with the slope dependent on whether one considers 2-10 kev luminosity or bolometric luminosity. The best fit slope using the bolometric x-ray luminosity is $\sim 8/3$

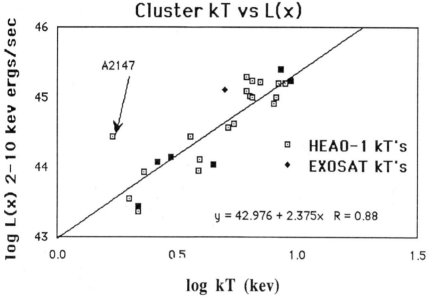

Figure 4 Cluster temperature versus 2-10 kev luminosity. The HEAO-1 temperature for A2147 is in error due to nearby sources in the beam. The MPC temperature is in good agreement with the correlation line. In this and following figures R is the correlation coefficient for the sample.

(A. Edge this symposium). This simple power law form holds over a factor of 10^3 in x-ray luminosity. The correlation has a variance of less than 30%. Because, at the present epoch, the correlation between luminosity and temperature is so tight it is a potential evolutionary test of clusters. That is, models of the evolution of cluster potential and x-ray emiting gas should be able to predict this relationship and how it should vary with epoch. I am not aware of any observational selection effects which can explain this relation.

Casting the x-ray bolometric luminosity in the form $L(x)_{bol} \sim <n^2 V> T^{1/2}$ and using the fact that the imaging data show that there is no systematic variation of core radius with x-ray luminosity, a little algebra shows that $n \sim T^{1.1}$. Thus denser clusters are hotter and more luminous. This is consistent with the "virial" relation, $T \sim GM_c/a_c$ (where M_c is the total mass inside a radius a_c) which "predicts" $T \sim n$, in the core of the cluster, if the x-ray gas density, n, is linearly related to the total, dark matter, density.

The direct correlation of total x-ray emitting gas mass, $\int n(r)dV = M_{gas}$, with x-ray temperature (figure 5) shows that $M_{gas} \sim T^{3/2}$ while the mass of gas in the core (data from Arnaud 1987) has $M_{gas} \sim T^{0.8}$. Of course the strong relation between M_{gas}

and T is an "artifact" of the correlation between L(x) and T. The core mass is almost independent of the total x-ray luminosity and thus is independent of that relation.

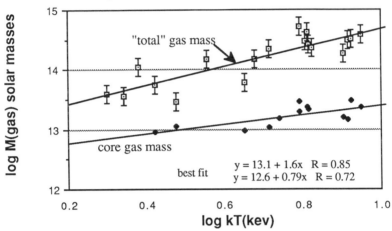

Figure 5 Mass of gas in the cluster versus x-ray temperature; the total gas masses are from Forman and Jones, while the core gas masses are from Arnaud(1987)

B. Fe abundance vs. Luminosity

Variations of the Fe abundance in clusters versus other parameters should help us understand the evolutionary origin of the Fe. For example, if massive clusters formed primarily by hierarchical mergers then they would be built up out of smaller systems. Given that no more gas is added to the system, more luminous clusters should have a least as large a metallicity as smaller systems. If, on the other hand, more massive systems are primarily formed at later times than low mass systems, out of larger scale fluctuations which take longer to "come over the horizon" (as predicted by cold dark mater theories), then the Fe abundance might show systematic differences between small and large clusters, since they are formed at very different times in the evolutionary history of their constituent galaxies.

The observational data (figure 6) show either no trend, Fe abundance is independent of luminosity (Rothenflug and Arnaud 1985, Edge 1987), or a slight correlation with less luminous systems having higher abundance (Henriksen and Mushotzky 1988). Unfortunately this correlation is sensitive to both the atomic physics used and to the detector systems. At low x-ray temperatures, T<2 kev, the atomic physics calculations appear to differ significantly from each other in the calculation of expected "6.7" kev Fe strength versus temperature (Raymond these proceedings, Rothenflug and Arnaud 1985), and part of the observed anti-correlation could arise from

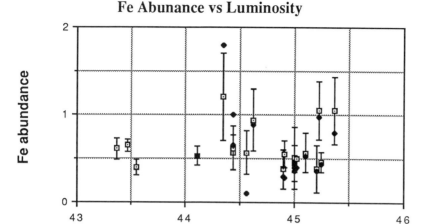

Figure 6 The black dots are the Raymond-Smith analysis of the HEAO-1 clusters while the boxes show the analysis of Rothenflug and Arnaud (with error bars)

use of the "wrong" atomic physics. Also, at these low temperatures, the "6.7" kev Fe line appears on a very steeply falling continuum and there are large, possibly systematic errors in the measurement of the line strength with low resolution proportional counters. Determination of the form of the Fe abundance versus L(x) will have to await better data.

C. Optical Richness and Density

It is now well established (Mushotzky 1984) that there is a strong correlation between the central density of galaxies (connoted N_0 by Bahcall 1977) and the x-ray luminosity and temperature. Because there has been no additional survey of N_0 after that of Bahcall, the latest summary of this subject appears in Mushotzky 1984. In that paper it was shown that N_0 correlates well with L_x and, of course, with kT_x. Because N_0 is quite easy to measure and is a good predictor of x-ray properties it should be determined for a much larger sample of clusters.

As originally suggested by by Jones and Forman (1978), there is a strong relationship (not correlation) between optical richness (optical richness as determined by Abell is only weakly correlated with N_0) and x-ray luminosity. However, it is difficult to quantify this relationship because of the way in which richness is determined. Kowalski et al. (1984) have shown that the mean x-ray luminosity is a monotonic function of optical richness but that the luminosity spread in each richness class is large. They point out that it is possible to find low x-ray luminosities in optically rich clusters and vice versa. Thus while the relationship is strong, it is not entirely clear what is being measured.

There is one additional strong correlation between optical galaxy counts and x-ray properties. Rothenflug et al. (1984) have shown a very strong correlation between the "total" optical luminosity, L_{opt}, and the mass of x-ray emitting gas M_{gas}, finding that $M_{gas} \sim L_{opt}^{3/2}$. Thus in more massive clusters the ratio of the mass of the x-ray emitting gas to the mass in stars is a monotonically larger fraction of the total mass of the system. For the most luminous systems the ratio of M_{gas}/L_{opt} can be as large as 50, about 1/2-1/3 of the M_{total}/L ratios calculated by Cowie, Henriksen and Mushotzky (1987) for several clusters. This correlation suggests that in massive clusters perhaps as much as 1/3 of the total mass of the system may be in baryons.

There are many other suggested relations between x-ray properties and optical galaxy counts, sizes, radio power, and other cluster properties. However, because of lack of space I have chosen not to discuss them in this review; see Kowalski et al. (1984), Abramopoulos and Ku (1983) and Sarazin (1986) for recent summaries of these relations.

D. Velocity Dispersion and X-ray Temperature

On general physical grounds one expects that the depth of the potential well of the cluster is measured by both the x-ray gas temperature and the velocity dispersion of the galaxies. The virial theorem relates the total mass to the velocity dispersion, $M_{tot} = 3R_G \sigma^2_r/G$, where R_G is the gravitational radius of the cluster and σ^2_r is the square of the line of sight velocity dispersion. As discussed earlier, if the gas is isothermal and the velocity distribution of the galaxies is isotropic, then

$$kT_{gas} = \mu m_p \sigma^2_r/\beta \qquad (3)$$

where β is a parameter to be determined (but if the physics is correct is the same as β in equations 1 and 2). If, as indicated by the x-ray surface brightness profiles many clusters have the same β value then $kT_{gas} \sim \sigma_r^2$. As shown below this is not correct. Struble and Rood (1982) point out that there is a bias in the determination of velocity dispersion for clusters at z>0.07 in the sense that only richer clusters are selected for determination of σ_r at these redshifts. However, this does not effect the actual value of σ_r nor the x-ray measurements. Thus this selection effect should not contribute to the scatter seen in figure 7.

We find that the relationship between kT_{gas} and σ_r is considerably flatter than predicted from equation 3 (figure 7). While our data do not seem to be well fit by a simple power law relationship the compilation by A. Edge and G. Stewart (this symposium) is compatible with a relation of the form $T_{gas} \sim \sigma_r^{1.5}$. Thus either β is a systematic function of kT_{gas} (or alternatively L_x), which does not seem to be indicated by the x-ray imaging data, or some of the assumptions of the model are incorrect. It is clear that the form of the divergence at high L_x (and thus high kT), with higher than predicted velocity dispersions at high kT is consistent with a radial distribution of galaxy velocities in these clusters or the effects of subclusters (e.g. the measured velocity dispersion is due to the effects of two or more subclusters of different velocities plus the velocity dispersion in each subcluster).

At low kT, the deviation is of the opposite kind, high kT for a given σ_r. Thus the most likely systematic errors are due to contamination in the x-ray spectrum such as an AGN lying in the cluster which would give a much flatter continuum, implying a high fit value for kT_{gas}. However, several of these low σ_r objects (such as A1060, Virgo and Centaurus) have very well determined x-ray spectra in which the effects of AGN have been detected and removed in the published temperatures.

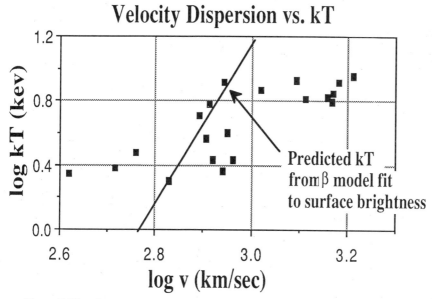

Figure 7 Plot of x-ray temperature versus optical velocity dispersion. The optical data have been taken from the compilations of Struble and Rood(1987) and Schmidt(1986).

It is not clear to this author why only clusters with velocities in the range from 700 to 900 km/sec should be well relaxed and with isotropic velocity distributions. However, there is a selection effect in that most of the clusters with measured surface brightness profiles have x-ray temperatures in the range from $3-7 \times 10^7$ degrees just where the β models give a good fit to the kT vs σ relation.

We also note that elliptical galaxies more or less "fit" on the cluster kT versus σ_r relation while they do not fit on the L(x) versus σ_r relationship.

E. Central Gas Density vs. Luminosity

Using the fits of Jones and Forman (1984) to the β model surface brightness distributions, we find that (see figure 8 below) that the the x-ray luminosity correlates with density, but there is large scatter. Note that this is not the actual density at the center of the cluster but the fitted density to the β model surface brightness law, and as such, it explicitly excludes the high density regions where the cooling flow occurs.

The slope of this regression, $n \sim L^{0.4}$, is in good agreement with the luminosity temperature relation if the core radius is not a strong function of L or T; that since $L \sim \langle n^2 \rangle V \cdot T^{1/2}$, using the above form predicts that $V \sim L^{0.2}$.

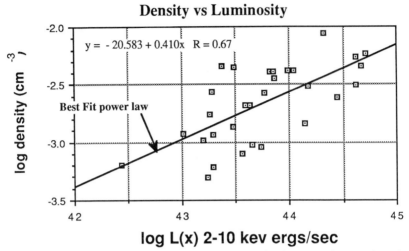

Figure 8 Density from β model fits versus x-ray luminosity. Note the upper bounds on density, n less than 0.01, and the large scatter in the relationship.

The last two correlations which I shall present here may be of great interest or may just be observational selection effects; it is unlikely, however that they are data analysis artifacts (C. Jones priv comm). In fits to the β model of surface brightness distribution for clusters, for each cluster there is a strong correlation of the core radius a_c and the slope of the surface brightness β as well as between central density n(0) and total luminosity L(x). I do not expect that this correlation to hold for the <u>sample</u> of clusters. However, we see in figure 9 that a_c and β are strongly related in the sense that large core radii clusters have smaller densities. Only steep surface brightness profiles clusters

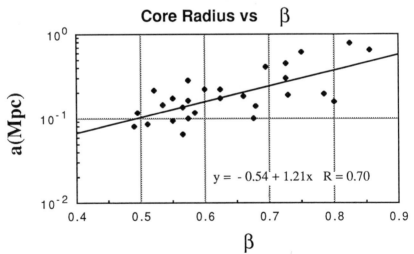

Figure 9: Plot of the mean cluster core radius a and the β parameter from the data of Jones and Forman (1984) I have selected only those clusters with well determined values of a and β.

have large core radii. If we adopt a semi-log form of the correlation (the solid line in figure 9) we find that the effective "size" (the "half power radius "within which 50% of the flux is contained) is remarkably constant from cluster to cluster at ~1- 2 Mpc. C. Jones and W. Forman (K. Arnaud priv comm) have remarked that most of the low β, small core radius clusters are cD systems, as is natural from their classification scheme.

The other "result" is the anti-correlation between central density and core radius. The sample of Jones and Forman (1984) shows that larger central densities are well correlated, figure 10, with smaller core radii . Such a correlation will tend to give a narrower spread in total luminosity vs β than naive application of the surface brightness model would otherwise predict. This may be a signature of the evolutionary status of a cluster. If we assume, as we have above, that the gas density is simply related to the total mass density then the virial relation,

$4\pi G \rho_0 a_c^2 = 9\sigma^2$ where ρ_0 is the central mass density, combined with the approximate relationship of $a_c \sim n^{-2/3}$ indicates that there is only a weak relationship between central mass density and temperaturewhich is inconsistent with the discussion on page 11 Clearly more thought is necessary.

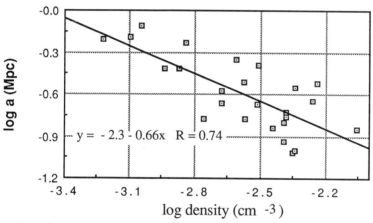

Figure 10 Core radius vs fitted "central" density (see definition on page 14), data from Forman and Jones 1984

III. Special Topics

There are two additional topics that I would like to cover in this somewhat idiosyncratic review of bulk cluster properties. These are spectral evidence for cooling flows and evidence for non-isothermal temperature distributions. However because of the detail available in the NATO ASI Workshop on Cooling Flows and the detailed discussion of the situation in the papers presented by J. Bregman and A. Fabian at this meeting (and due to a lack of space), I shall concentrate on the evidence for non-isothermal distributions.

A. Non-Isothermal Temperature Distributions in Clusters

As pointed out in the papers by Fabian and Bregman in this School, (cf Mushotzky 1987) the application of the equation of hydrostatic equilibrium to the hot gas in clusters allows determination of their mass and mass profile. The main uncertainty in the application of this equation to clusters has been the absence of spatially resolved x-ray temperature determinations. There are only 3 clusters large enough to have have spatially resolved spectra obtained by collimated proportional counters, A426 (Perseus) by Ulmer et al. 1987, A1656 (Coma) Hughes (1987) and Virgo (Edge et al 1987, this symposium). This makes determination of the temperature distribution in clusters very model dependent. In fact, even the existence of temperature gradient beyond the core (where the cooling flow can be) is somewhat controversial. The 3 cases with spatially resolved spectra available have shown: 1) negative temperature gradient from the center outward (Coma), 2) virtually no temperature gradient (Virgo) and 3) a negative temperature gradient in the central core, interpreted as being due to the cooling flow (Perseus).

Henriksen and Mushotzky (1986,1988) have shown that the high signal-to-noise HEAO-1 A-2 integral spectra of ~10 clusters cannot be well fit by an isothermal bremmstrahlung model. They have shown that these spectra in the 2-30 kev band are not very sensitive to the presence of a cooling flow and have therefore assumed that poor quality of the fit of an isothermal model to the integrated spectrum is due to a temperature gradient in the cluster. They have fit the surface brightness profile and the integrated spectrum with a polytropic model; that is

$$(n(r)/n(0))^{\gamma -1} = T(r)/T(0).$$

This type of model can be well constrained by the available data and is physically reasonable, but is by no means unique. They find that the temperature gradients are not very strong and many clusters can be well characterized by polytropic models with $\gamma \sim 5/4$. Cowie, Henriksen. and Mushotzky(1987) point out that this type of model can strongly constrain the mass profile in clusters and, if correct, the best fit integral masses for clusters are about 1/2 of those calculated from naive application of the virial theorem. If the polytropic models are correct the central regions of clusters are hotter (~14 kev in the center for Coma compared to the 8.2 kev for the isothermal model) than previously believed and thus the dark matter is centrally concentrated rather than lying preferentially at the outer regions of the cluster.

Future Ginga observations of more clusters will produce a much larger set of high quality integral spectra which will test whether the deviation from isothermality is a common feature in nearby clusters. It will, however, require an x-ray imaging spectrometer with a broad band pass to measure directly any temperature gradient. Such measurements, in addition to determining the total mass and mass distribution in clusters and thus "solving the dark matter problem" may also help determine how clusters form(ed) and how their potential well evolves.

IV. Conclusion

I hope I have shown that while the HEAO-1, EXOSAT and *Einstein* missions have produced a large body of data on clusters, much of it is currently not well understood . Both the ROSAT and Ginga missions have the potential for making a large contribution to our understanding of clusters. However, it will require spatially resolved high sensitivity spectroscopy to make a giant step forward in our understanding.

Acknowledgements: I would like to thank K. Arnaud for use of his data before publication, G. Madejski for help with MPC analysis and, A. Fabian and A. Edge for extensive discussions. The clarity of the paper has benefited greatly from discussions with D. Cioffi, K. Jahoda and K. Arnaud. I would also like to thank R. Pallavicini for the invitation to lecture at Cargese.

References

Abell,G. 1958 Ap .J. Suppl 3,211
Abramopoulos, F. and Ku, W. 1983 Ap.J. 271, 446
Arnaud, K. et al. 1987 preprint
Arnaud K. 1987 in preparation
Bahcall, N. 1977 Ap.J.(Letters) 217, L77
Bahcall, N. 1979a Ap.J. (Letters) 232, L33
Bahcall, N. 1979b Ap.J. 232, 689
Bechtold, J. Forman, W., Giaconni, R. Jones, C. Schwarz, J. Tucker, W. and Vanspeybroeck, L. 1983 Ap.J. 265, 26
Beers T. and Tonry, J. 1986 Ap.J. 300, 557
Canizares,C. Markert,T., and Donahue, M. 1987 in NATO Workshop " Cooling Flows in Clusters of Galaxies" edited by A. Fabian
Cowie, L., Henriksen, M. and Mushotzky, R. 1987 Ap. J. 317, 593
Fabricant, D., Beers, T., Geller, M. Gorenstein, P. Huchra, J. and Kurtz, M. 1986, Ap. J. 308, 530
Forman, W., et al 1977 Ap.J. Suppl 38, 357
Forman, W., Bechtold, J, Blair, W., Giaconni, R, Vanspeybroeck, L. and Jones, C. 1981 Ap.J. Letters 234, L133
Forman, W. and Jones C., 1982 Ann Rev. Astron and Astrophy 20, 547
Henriksen, M. and Mushotzky, R. 1986 Ap.J. 302, 287
Henriksen, M. and Mushotzky, R. 1988 Ap.J. submitted
Henry, J., Henriksen, M, Charles, P., and Thorstensen,J. 1981, Ap.J. Letters 243, L137
Henry, J. and Lavery, R. 1984 Ap.J. 280, 1
Hughes, J. 1987 preprint
Jones C, Mandel, E., Schwarz, J., Forman, W. and Murray, S. 1979, Ap.J. Letters 234, L21
Jones C. and Forman, W. 1978 Ap.J. 224, 1
Jones C. and Forman, W. 1984 Ap.J. 276, 38
Kaiser, N 1986 M.N.R.A.S. 222, 323
Kowalski M, Ulmer, M., Cruddace, R. and Wood, K. 1984 Ap.J. Suppl 56, 401
Lea, S., Reichert, G., Mushotzky, R., Baity, W., Gruber, D., Rothschild, R and Primini, F. 1981 Ap.J. 246, 369
Lea, S. Mushotzky, R. and Holt, S 1982 Ap.J. 262, 24
Maccacaro, T. et al. 1982 Ap.J. 253, 504
McHardy, I, Lawrence, A., Pye, J. and Pounds, K. 1981, M.N.R.A.S. 197, 893
Mitchell, R., Culhane, J., Davison, P. and Ives, J. 1976, M.N.R.A.S. 178, 75p
Mitchell, R., Dickens, R., Bell-Burnell, S and Culhane, J. 1979, M.N.R.A.S. 189, 329
Mitchell, R. and Mushotzky, R. 1980 Ap.J. 236, 730
Mushotzky, R. Serlemitsos, P, Smith, B. Boldt, E, and. Holt, S. 1978 Ap.J. 225, 21

Mushotzky, R 1984 Physica Scripta T7, 157
Mushotzky, R 1987, Astro. Lett and Communications 26, 43
Mushotzky, R. et al. 1988 in preparation
Nulsen P., Fabian, A. Mushotzky, R. Holt, S., Marshall, F. and Serlemitsos. 1979 M.N.R.A.S. 189, 183
Piccinotti, G., Mushotzky, R., Boldt, E., Holt, S., Serlemitsos, P., and Shafer, R. 1982, Ap.J. 253, 485
Rephaeli, Y. 1980 Ap.J. 241, 858
Rothenflug, R. Arnaud, M., Boulade, O and Vigroux, L. 1984 in X-ray Astronomy '84 pg 391 edited by M. Oda and R. Giaconni ISAS publication
Rothenflug, R. Arnaud M. 1985 A&A 147, 337
Rothschild, R., Baity, W., Marscher,A. and Wheaton, W. 1981, Ap.J. Letters 243, L9
Sarazin, C. 1986 Rev. Mod Phys 58, 1
Schmidt, K-H 1986 Astron Nach 307, 69
Schmidt, M. and Green, R. 1986 Ap.J. 305, 68
Serlemitsos, P., Smith, B., Boldt, E, Holt, S. and Swank, J. 1977 Ap.J. Letters 211, L 63
Struble, M. and Rood H. 1987 preprint
Ulmer, M., Cruddace,R, Fenimore,E., Fritz, G. and Snyder,W. 1987 Ap.J. 319, 118
West, M., Dekel, A. and Oemler, A. 1987 Ap.J. 316, 1
White, S. 1979, M.N.R.A.S. 189, 831
Wood, K. et al. 1984 Ap. J. Suppl, 56, 507

THEORY OF INTRACLUSTER GAS

A.C. Fabian
Institute of Astronomy
Madingley Road
Cambridge CB3 0HA
U.K.

ABSTRACT.

Clusters of galaxies typically contain about 10^{14} M$_\odot$ of metal-enriched gas. The distribution and properties of this gas are reviewed, with an emphasis on the role of radiative cooling of the densest gas.

1. INTRODUCTION

Clusters and groups of galaxies contain enormous quantities of diffuse gas. The sound speed of the gas is similar to the velocity dispersion of the cluster which is typically 500 to 1200 km s^{-1} and so the gas predominantly radiates X-rays. Diffuse X-radiation is the principal source of information on the intracluster medium (ICM). There is further indirect evidence for the gas in 'head-tail' radio sources and from theories of the propagation of double-lobe radio sources. A comprehensive review of the properties of the ICM is given by Sarazin (1986) and of observations by R.Mushotzky in these Proceedings.

Most of the observed intracluster gas has an electron density, n_e, in the range of $10^{-4} - 10^{-2}$ cm^{-3} and a temperature $T \sim 10^7 - 10^8$ K, and is contained within a radius of 1 to 2 Mpc. The total bremsstrahlung luminosity is $\sim 10^{42} - 10^{46}$ erg s^{-1}. The 6.7 keV iron emission line is observed in all clusters that are bright enough for a detection to be made (see e.g. Rothenflug & Arnaud 1986) and the gas has ~ 0.4 times solar abundance in iron. The work of Canizares et al. (1979; 1982) and Mushotzky et al. (1981) on cooling flow clusters shows O, Ne, Si and S also.

Some cluster properties show correlations which are fairly good if just X-ray quantities are used (e.g. X-ray luminosity L_X versus gas temperature T_X; see the contributions in these Proceedings by A. Edge and R.Mushotzky). The temperature correlates with optically-determined velocity dispersion (which is a notoriously uncertain quantity) and with the central galaxy number density. Generally, the deeper the potential well, the more gas and galaxies it contains and the more luminous it is in X-rays. The gas is at a temperature close to that given by the Virial Theorem, i.e.

$$kT_{gas} \approx \frac{GM_{cluster} m_p}{R}. \qquad (1)$$

This review is an expanded version of part of my Saas-Fee Lectures and deals first with the simplest ideas of the gas distribution. It shows how the mass distribution of clusters and groups will eventually be determined accurately. I then discuss the effects of cooling on the densest gas in the cluster core. These lead to cooling flows which are depositing hundreds of solar masses per year of cooled gas in some clusters. Consequently, clusters of galaxies appear to contain the largest regions of star formation at the current epoch. Most of the stars formed must be of low mass and so are a kind of dark matter. Detailed studies of cooling flows suggests that the intracluster gas is inhomogeneous on small scales. This should be remembered when the gross properties of the intracluster gas is discussed.

2. THE GAS DISTRIBUTION

The intracluster gas acts as a fluid on galactic scales. The electron-electron coupling time, $t_{e-e} \approx 2.10^5 T_8^{\frac{3}{2}} n_{-3}^{-1}$ yr, where $T = 10^8 T_8$ K and $n = 10^{-3} n_{-3}$ cm^{-3}, and the electron-ion coupling time is 1840 times larger. The gas is therefore expected to behave locally as a Maxwellian distribution. The crossing time of the cluster exceeds these 2-body timescales,

$$t_{cross} = \frac{R}{\langle v^2 \rangle^{1/2}}$$

$$\approx 10^9 R_{\text{Mpc}} v_8^{-1} \text{ yr}. \tag{2}$$

This timescale is also roughly equal to the free-fall timescale in the cluster potential. The radiative cooling time of the gas, due to bremsstrahlung, is

$$t_c \approx 7.10^{10} T_8^{\frac{1}{2}} n_{-3}^{-1} \text{ yr}. \tag{3}$$

Cooling may therefore be important ($t_c < t_a$, the cluster age) in high density regions and where the temperature is low. The mean-free-path of an electron in the intracluster gas,

$$\lambda_e \approx \lambda_i \approx 23 T_8^2 n_{-3}^{-1} \text{ kpc} \tag{4}$$

$$\sim R_{galaxy},$$

whereas the gyroradius,

$$r_g \approx 3.10^8 Z^{-1} T_8^{\frac{1}{2}} \left(\frac{m}{m_e}\right)^{\frac{1}{2}} B_{\mu G}^{-1} \text{ cm}. \tag{5}$$

Although $\lambda_e \sim R$ in the outskirts of a cluster, tangled weak magnetic fields will keep the gas behaviour fluid-like.

Cluster gas is generally optically thin since its Thomson depth is ~ 0.1 per cent. However, Gilfanov et al. (1987) have noted that resonance lines may be scattered. This is relevant to abundance gradient determinations if the gas is static, particularly if low-energy lines are used. Turbulent motions of $> 100 \text{ km s}^{-1}$ will cause this effect to be minimized. Cooled gas may provide a distributed source of absorption (see e.g. Crawford et al. 1987).

Non-hydrostatic pressure variations are eliminated on the sound crossing time of the gas,

$$t_s = \frac{R}{c_s} \approx t_{cross} = 10^9 R_{\text{Mpc}} T_8^{-\frac{1}{2}} \text{ yr}, \tag{6}$$

where the size of the region is R_{Mpc} Mpc. In the inner Mpc of the cluster, which is the region well-studied by X-ray detectors, t_s is much less than the age of the cluster. Any radial flows of the gas must be subsonic otherwise impossibly large mass flow rates are implied ($>$ (10^{14} M$_\odot$)/(10^{10} yr) = 10^4 M$_\odot$ yr^{-1}). Intracluster gas is then close to hydrostatic support so that

$$\frac{dP_{gas}}{dr} = -\rho_{gas}\frac{d\phi}{dr} = -\rho_{gas}g, \qquad (7)$$

where $g = GM(<r)/r^2$. This means that measurements of P_{gas} and ρ_{gas}, the gas pressure and density as a function of radius r (i.e. $n_e(r)$ and $T_e(r)$), allow $\phi(r)$ and so $M_{cluster}(r)$ to be determined (see later). The gas will arrange itself (convect) so that isobaric surfaces are on equipotentials which will be roughly spherical, even for quite a lumpy or flattened mass distribution. Analysis therefore assumes spherical equipotentials, or at most two sets of spherical equipotentials (e.g. A754; Fabricant et al. 1986). Hydrostatic equlibrium will break down somewhere at the edge of a cluster where $t_s \sim t_a$, i.e.

$$R > 10 T_8^{\frac{1}{2}}\left(\frac{t_a}{10^{10} \text{ yr}}\right) \text{ Mpc}. \qquad (8)$$

The relevant age for the cluster, t_a, is about 10^{10} yr, although as clusters are probably forming now by the accretion of subclusters (Geller 1984), a more appropriate time may be half that value. If the outer regions of the cluster gas have an adiabatic profile with temperature decreasing outwards then hydrostatic equilibrium may be lost at 3 to 5 Mpc.

If the gas is isothermal (which it probably is not) then hydrostatic equilibrium implies

$$kT\frac{dn_e}{dr} = -n_e\mu m\frac{GM(r)}{r^2}, \qquad (9)$$

and

$$\rho_{gas} \propto \exp\left(-\frac{\phi(r)}{(kT/\mu m)}\right). \qquad (10)$$

Also, if the galaxies have an isothermal velocity distribution (also unlikely) then

$$\rho_{gal} \propto \exp\left(-\frac{\phi(r)}{\sigma_{los}^2}\right). \qquad (11)$$

where σ_{los} is the line-of-sight velocity disperion of the cluster. Then

$$\rho_{gas} \propto (\rho_{gal})^\beta, \qquad (12)$$

where

$$\beta = \frac{\mu m \sigma_{los}^2}{kT} \qquad (13)$$

(introduced by Cavaliere & Fusco-Fermiano 1976 as τ). We might expect $\beta \sim 1$ so $c_s^2 \approx \sigma_{los}^2$. However, the application of the equations indicates otherwise. The above isothermal-isothermal model suggests the use of King's (1966) approximation of an isothermal distribution where

$$\rho_{gal}(r) = \rho_{gal}(0)(1 + \left(\frac{r}{a}\right)^2)^{-\frac{3}{2}}, \qquad (14)$$

so

$$\rho_{gas}(r) = \rho_{gas}(0)(1 + \left(\frac{r}{a}\right)^2)^{-\frac{3\beta}{2}}, \qquad (15)$$

where a is the core radius. This profile can then be fitted to X-ray images of clusters (emissivity$\propto \rho_{gas}^2$). Jones & Forman (1984) obtain

$$\langle \beta_{image} \rangle = 0.65, \tag{16}$$

so

$$\rho_{gas} \approx \rho_{gas}(0)(1 + \left(\frac{r}{a}\right)^2)^{-1}. \tag{17}$$

On the other hand, X-ray spectral and optical velocity-dispersion measurements of T and σ_{los} give

$$\langle \beta_{spec} \rangle \approx 1.2 \approx 2 \langle \beta_{image} \rangle. \tag{43}$$

The reasons for this discrepancy are not yet clear, although it is unlikely that the gas and galaxies are isothermal (see e.g. paper by R. Mushotzky). β_{spec} is therefore not defined for each cluster, although some mean for a cluster involving $\bar{T}$ and $\bar{\sigma}_{los}$ could be constructed. This need not correspond to β_{image} as they are weighted differently. There may also be considerable velocity anisotropy in the galaxy distribution. If clusters are lumpy, as seems to be the case, and some significant component of σ_{los} is due to individually bound subclumps, then the relevant σ_{los} for estimating the potential may be overestimated. There may be no discrepancy in the Coma cluster. The high quality data obtained with the Einstein Observatory allow more sophisticated fits to be made so that a single parameter like β loses its usefulness.

Earlier attempts to overcome a lack of knowledge of the equation of state of the cluster gas assumed that it is polytropic i.e $P \propto \rho^\gamma$. This does not necessarily mean that γ is the ratio of specific heats * and is little more than a mathematical expediency. Using it in the equation of hydrostatic support yields

$$\frac{\gamma}{(\gamma-1)} \frac{k}{\mu m} \frac{dT}{dr} = -\frac{d\phi}{dr}, \tag{19}$$

so that

$$T = T_c + \frac{(\gamma-1)}{\gamma} \frac{\mu m}{k} (\phi_c - \phi) \tag{20}$$

and

$$\frac{\rho}{\rho_c} = \left(1 + \frac{(\gamma-1)}{\gamma} \frac{(\phi_c - \phi)}{(kT_c/m)}\right)^{1/(\gamma-1)}. \tag{21}$$

The subscripts refer to values at the centre. The density equation limits to the exponential isothermal form as $\gamma \to 1$.

The polytropic approach is still in common use, especially when needing to extrapolate to large radii. There is no particular reason, however, why the polytropic γ has to have a single fixed value throughout a cluster. It just measures how the cluster was set up and the conditions when the core was formed may have been quite different from those when the outermost atmosphere arrived. Furthermore, the clumpiness of clusters (Geller

---

* The polytropic γ, γ_p, parametrizes the entropy profile in the cluster atmosphere which is presumably due to initial (or early) conditions. It does not mean that $P \propto \rho^\gamma$ if work is done on the gas. The ratio of specific heats γ, γ_{sh}, is relevant when work is done on the gas or it is displaced. γ_{sh} is probably always 5/3 for rapid changes. Note that gas with an isothermal profile, $\gamma_p = 1$, can have $\gamma_{sh} = 5/3$.

1984, Forman et al. 1981) means that the cluster potential is time-dependent on large-scales (see also work by Cavaliere and colleagues). The core radius is also poorly defined. Most estimates of the total mass of gas in a cluster rely on some large extrapolation assuming γ is constant. This is the main reason why there are conflicting results in the literature (see e.g. Cowie, Henriksen & Mushotzky 1987 and The & White 1987). Note that the expected breakdown of hydrostatic equilibrium discussed earlier means that the extrapolation should not be extended much beyond $\sim 3\,\mathrm{Mpc}$.

It is possible to obtain gas density and temperature profiles without assuming an equation of state (Fabian et al. 1981). The X-ray surface brightness profile can be deprojected, assuming some geometry (e.g. spherical) and a distance to the cluster, to yield count emissivities as a function of radius. The emissivity depends upon n_e and T_e as well as the detector response and the effects of intervening photoelectric absorption, the last two of which are assumed known. A further relationship between n_e and T_e is obtained from the pressure via the equation of hydrostatic equilibrium. The densities obtained in this way are usually determined to better than 10 per cent, whereas the temperatures are somewhat dependent upon the assumed value of g (usually estimated from σ_{los}) that is used in the hydrostatic equilibrium equation. One pressure, typically the outer pressure, is required to start the solution and this is usually adjusted so that most of the cluster gas has a temperature consistent with X-ray spectral measurements. These spectra have usually been obtained with wide field-of-view experiments. When spatially resolved spectra become available with future X-ray satellites such as ROSAT, AXAF and XMM (see talks by Truemper, Tananbaum and Peacock), then we shall be able to solve directly for n_e and T_e without requiring g (which can then be measured, of course).

X-ray measurements of the hot gas in elliptical galaxies and clusters will eventually provide a powerful means for determining their gravitational mass profiles (see e.g. Mushotzky 1987a). The equation of hydrostatic equilibrium may be written as

$$\frac{d\phi}{dr} = -\left(\frac{kT_{gas}}{\mu m}\right)\left(\frac{\mathrm{d}\ln\rho_{gas}}{dr} + \frac{\mathrm{d}\ln T_{gas}}{dr}\right). \tag{22}$$

$$\phi = \int \frac{GM}{r^2}dr \tag{23}$$

is thus obtained from the gas pressure and density profiles. Although the density profiles measured so far are robust with respect to potential changes, the temperatures carry large uncertainties. Some progress has been made using Einstein Observatory IPC spectra of the cluster emission around M87 (Fabricant & Gorenstein 1983) and Focal Plane Crystal Spectrometer spectra of the inner regions (Stewart et al. 1984). These measurements all indicate that M87 is surrounded by an extensive dark halo. Confirming evidence is obtained from velocity measurements of the globular clusters surrounding that galaxy (Mould, Oke & Nemec 1987; Huchra & Brodie 1987).

In the absence of any other spectral data, we can estimate a lower limit to the gravitational binding mass of an early-type galaxy from observing gas at radius r_o with a pressure P_o and temperature T_o (Fabian et al. 1986). The method is analogous to those used on stars to obtain limits on central conditions from surface properties, excepting this time we observe the centre and do not know how far away the 'outside' is. All that is assumed is that the gas is pressure-confined by an 'unseen' hydrostatic, convectively-stable atmosphere in which the pressure decreases outward to some external pressure P_∞. Remember that the central pressure in a cluster ($nT \approx 10^5 - 10^6$) is much higher than that in a spiral ($nT \approx 10^3 - 10^4$) and so P_o is likely to far exceed P_∞. If the gas is not confined then it will expand and the cluster mass loss arguments already mentioned rule that out. The key point about this outer atmosphere is that the total binding mass is minimized when its temperature gradient (which decreases outward) is steepest. This occurs for an atmosphere which is marginally convectively stable and so follows an adiabat. The total gravitational binding mass,

$$M_T \geq \frac{5}{2} \frac{kT_o r_o}{G\mu m} \frac{\left[1 - \left(\frac{P_\infty}{P_o}\right)^{\frac{2}{5}}\right]}{(1 - r_o/r_\infty)}. \tag{24}$$

This limit applies to all systems in which gas is in hydrostatic support (e.g. galaxies, groups and clusters). Dimensionally, it resembles a virial mass (as any such combination should). It is the numerical coefficient that is important.

3. THE SUNYAEV-ZELDOVICH EFFECT

X-ray emission is not the only means of observing the hot intracluster medium. The low energy photons of the cosmic microwave background can be Compton-scattered on passing through the gas and experience an energy shift, $\Delta\epsilon$, given by

$$\left\langle \frac{\Delta\epsilon}{\epsilon} \right\rangle \approx \frac{4kT_e}{m_e c^2}, \tag{25}$$

where ϵ is the initial photon energy. The proportion of photons scattered is given by the Thomson depth,

$$\tau_T = \sigma_T \int n_e dl. \tag{26}$$

This leads to a microwave dip in the direction of the cluster in the Rayleigh-Jeans part of the cosmic blackbody spectrum

$$\frac{\Delta T}{T} = -\frac{2k\sigma_T}{m_e c^2} \int n_e T_e dl. \tag{27}$$

This is the Sunyaev-Zeldovich effect and measures the integral of the pressure along the line of sight.

It has now been detected in several clusters (Birkinshaw, Gull & Hardebeck 1984). Since the X-ray luminosity is proportional to $\int n_e^2 T_e^{\frac{1}{2}} dl$, a combination of X-ray and microwave measurements can lead to an independent estimate of cluster distance. To understand this, consider a spherical region of gas of radius R at distance D, subtending an angle θ at the observer. The X-ray luminosity

$$L_X \propto n_e^2 T_e^{\frac{1}{2}} R^3 \tag{28}$$

and

$$\frac{\Delta T}{T} \propto n_e T_e R. \tag{29}$$

The observer measures the flux

$$F_X \propto \frac{L_X}{D^2} \propto n_e^2 T_e^{\frac{1}{2}} \theta^3 D, \tag{30}$$

so, substituting for n_e,

$$F_X \propto \left(\frac{\Delta T}{T}\right)^2 \frac{\theta}{T_e^{\frac{3}{2}} D}. \qquad (31)$$

Measurements of F_X, $\frac{\Delta T}{T}$, and T_e then lead to an estimate of D. If the redshift of the cluster is known, say from optical measurements, then the Hubble constant is obtained from one cluster and q_0 from two! Of course in practice, the cluster is not a uniform isothermal sphere of gas and a statistical approach has to be made. Van Speybroeck (1987) estimates that H_0 will be determined to an accuracy of 10 per cent with AXAF, a factor of ten improvement on the current uncertainty.

4. THE ORIGIN OF THE ICM AND GALAXY STRIPPING

The intracluster medium is chemically enriched and so is not primordial, or at least any primordial fraction is small. The mass of gas is comparable to that in stars (i.e. excluding dark matter) and so current stellar mass-loss is relatively unimportant ($\sim 100\,M_\odot\,yr^{-1}$ throughout the cluster). The gas originated from stars and galaxies, probably as winds and fountains (see the paper by J.Bregman) during an early phase and through ram-pressure stripping (see the paper by E. Salpeter) later on. The sound speed of any gas ejected from galaxies and then mixed would be comparable to the velocity dispersion of the galaxies. Some early large-scale heating could have taken place and have contributed to the $\langle \beta \rangle$ problem discussed in Section 2.

The existence of an enriched intracluster medium shows that large galaxies process $\sim 10^{11}\,M_\odot$ of gas early on to yield roughly solar abundance. Supernova heating can cause gas to be expelled from the weak potential wells of small galaxies but not from clusters and groups. A systematic study of the specific mass of intracluster gas per galaxy as a function of gas temperature (which is presumably related to the depth of the potential well) should allow the importance of heating to be estimated eventually.

Gas bound to an individual galaxy may be liberated in a galaxy - galaxy collision (Spitzer & Baade 1951) or by ram-pressure stripping in the ICM. This can be understood by considering a spherical mass of gas of density ρ_i and radius R in a galaxy of mass M and velocity dispersion σ falling through a medium of density ρ_0 at velocity v. The ram force due to the ICM

$$F_{ram} \approx \rho_0 v^2 \pi R^2 \qquad (32)$$

and the maximum gravitational restoring force is

$$F_{grav} \approx \rho_i \frac{4}{3}\pi R^3 \frac{GM}{R^2}. \qquad (33)$$

The gas is then pushed out if

$$\rho_0 v^2 > \frac{4\rho_i GM}{3R}, \qquad (34)$$

so for stripping, $\rho_0 v^2 \gtrsim 3\rho_i \sigma^2$. This criterion is for sudden stripping of gas that has previously accumulated in the galaxy. If there is continuous mass loss at a rate per unit volume ($\alpha \sim \rho_\star/\tau$, where $\rho_\star$ is the stellar density and τ is the mass loss time, $\sim 10^{12}$ yr) then the condition for continuous stripping is

$$\frac{1}{2}\rho_0 v^3 \geq \int \alpha |\phi| dz, \qquad (35)$$

where ϕ is the gravitational potential. This is derived from the kinetic energy flux into the gas (Takeda et al. 1984) and reduces to

$$\rho_0 v^3 \gtrsim 16 \Sigma \sigma^2 \tau, \tag{36}$$

where Σ is the column density of stars.

Gas pushed out of a galaxy will be shredded, dispersed and decelerated rapidly on a time

$$t \sim 3 \left(\frac{R}{v}\right) \left(\frac{\rho_i}{\rho_0}\right)^{\frac{1}{2}}. \tag{37}$$

When a galaxy moves subsonically through the ICM, its ISM may be stripped by viscous processes, turbulence and the effects of the Kelvin-Helmholtz instability (Nulsen 1982). The rate of mass loss is given by

$$\dot{M} \simeq \pi R^2 \rho v \left(\frac{12}{R_e}\right)^{\frac{1}{2}}, \tag{38}$$

where the Reynold's number $R_e = 3(R/\lambda)(v/c_s)$. If conduction is uninhibited then gas may be evaporated from a galaxy at a rate

$$\dot{M} \approx 8\rho c_s \pi R^2, \tag{39}$$

(Cowie & Songaila 1977).

An interesting example of stripping occurs in 'head-tail' radio sources. O'Dea et al. (1987) have used the direction of these tails to estimate the orbits of cluster galaxies.

5. COOLING FLOWS

The gas density in the core of many clusters is sufficiently high that the radiative cooling time is less than the age of the cluster (i.e $t_{cool} < H_0^{-1}$). The weight of the overlying gas then causes the gas density to rise and establishes a cooling flow. The inflow velocities are highly subsonic. X-ray observations of this phenomenon indicate that the mass flow rates can be substantial and in the range of hundreds of solar masses per year. It appears that we are observing the continued formation of central cluster galaxies.

The cooling time of the gas can be obtained from density and temperature profiles estimated by the method given in Section 2. Where $t_{cool} < H_0^{-1}$, the rate at which mass is deposited through cooling, $\dot{M}$, is given by

$$L_{cool} \approx \frac{5}{2} \frac{kT}{\mu m} \dot{M}, \tag{40}$$

where L_{cool} is the total luminosity within that region. The factor of 5/2 represents the enthalpy of the gas (i.e. the sum of thermal energy and PdV work done on cooling). Short cooling times are associated with high gas densities and bright regions and so X-ray images of cooling flows show a highly peaked surface brightness profile. This is common in 30 - 50 per cent of rich clusters (Stewart et al. 1984b), in some poor clusters (Schwartz, Schwarz & Tucker 1980; Canizares et al. 1983) and in elliptical galaxies (Nulsen, Stewart & Fabian 1984; Thomas et al. 1986). Recent unpublished work by K. Arnaud suggests that the fraction may be still higher. Most nearby clusters (Virgo, Centaurus, Hydra, Perseus

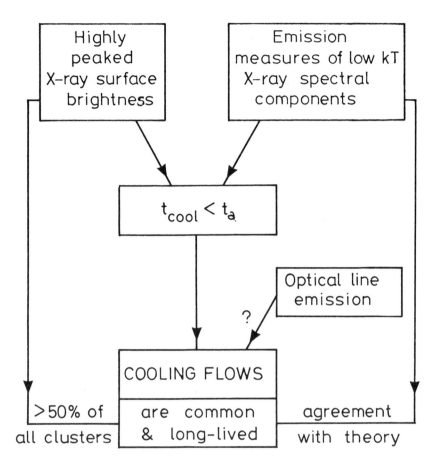

Figure 1. Evidence for cooling flows.

etc.) contain cooling flows and it is only those with two dominant central galaxies such as Coma that do not show a sharp X-ray peak. Constraints on more distant clusters are not very strong due to the relatively poor spatial resolution of the commonly-used imaging proportional counter (IPC) on the Einstein Observatory.

Unambiguous evidence for cooling is provided by the low X-ray temperature spectral components observed in high resolution spectra. Mushotzky et al. (1981, 1987b) and Canizares et al. (1979, 1982, 1987) find emission lines characteristic of gas in the 5.10^6 to 2.10^7 K range in the core of the Perseus cluster, where the mean gas temperature is nearer 8.10^7 K. Lines are also observed in M87 (Lea et al. 1982) and in A496 (Nulsen et al. 1983). The emission measures ($\int n_e^2 dV$) of gas at the lower temperatures agree well with a cooling interpretation (Mushotzky & Szymkowiak 1987b; Canizares et al. 1987). Ulmer et al. (1987) find evidence for a temperature decrease in the core of the Perseus cluster from a strip scan measurement. Much work has been carried out on the Perseus cluster as it is the brightest X-ray cluster. The Ophiuchus cluster, which is the second brightest cluster and coincidentally lies in a direction close to the Galactic Centre, also contains a cooling flow (Arnaud et al. 1987).

5a Mass deposition in Cooling Flows

The mass deposited as cooled gas can be very substantial:

$$\dot{M} \approx \begin{cases} 1\,M_\odot\,\text{yr}^{-1} & \text{(isolated ellipticals)} \\ 10-100\,M_\odot\,\text{yr}^{-1} & \text{(poor clusters)} \\ 10-1000\,M_\odot\,\text{yr}^{-1} & \text{(rich clusters)} \end{cases} \qquad (41)$$

(The lower estimates are a consequence of instrumental resolution and sensitivity and do not mean that there is some lower cutoff.) The total mass accumulated within a Hubble time can thus easily exceed $10^{12}\,M_\odot$ in the higher $\dot{M}$ cases. As the surface brightness profile gives $L(r)$ we can determine $\dot{M}(r)$, the mass deposition profile. The X-ray images are not as peaked as they could be if all the matter flowed to the centre. We generally find that

$$\dot{M}(r) \propto r. \qquad (42)$$

(Arnaud 1985). This means that $\dot{M}(r) \sim 1\,M_\odot\,\text{yr}^{-1}\,\text{kpc}^{-1}$ in the cores of many rich clusters of galaxies.

Distributed mass deposition is presumably due to a range of densities present at all radii. The gas is inhomogeneous. Nevertheless, it is instructive to consider the equations of a homogeneous flow, which roughly represent the mean conditions of an inhomogeneous flow. We have the equation of continuity;

$$\dot{M} = 4\pi r^2 \rho v, \qquad (43)$$

the pressure equation (ignoring highly subsonic flow terms)

$$\frac{dP}{dr} = -\rho \frac{d\phi}{dr}, \qquad (44)$$

and an energy equation,

$$\rho v \frac{d}{dr}\left(\frac{5}{2}\frac{kT}{\mu m} + \phi\right) = n^2 \Lambda, \qquad (45)$$

where Λ is the cooling function. If the cooling region (where $t_{cool} < H_0^{-1}$) is at constant pressure ($d\phi/dr = 0$), then $n \propto T^{-1}$ and

$$\rho v \frac{d}{dr}\left(\frac{5}{2}\frac{kT}{\mu m}\right) = n^2 \Lambda \qquad (46)$$

and if

$$\Lambda \propto T^\alpha, \qquad (47)$$

then
$$nv\frac{dn^{-1}}{dr} \propto n^{2-\alpha},\tag{48}$$

so
$$v\frac{dn}{dr} \propto n^{3-\alpha}.\tag{49}$$

From continuity, if $\dot{M}$ is constant,
$$v \propto n^{-1}r^{-2},\tag{50}$$

so
$$\int_\infty^n \frac{dn}{n^{4-\alpha}} \propto \int_0^R r^2 dr\tag{51}$$

and
$$n \propto R^{-3/(3-\alpha)}.\tag{52}$$

This is proportional to $R^{-\frac{6}{5}}$ for bremsstrahlung. The density rises inward as the temperature falls. Constant pressure is a fair approximation to the core region of a cluster. Gravity is not particularly important, except perhaps for focussing the flow, until the gas has cooled to about the virial temperature of the central galaxy. Then the gas heats up as it flows in further and the pressure rises (Fabian & Nulsen 1977). The flow velocity $v \simeq r/t_{cool}$, which is highly subsonic.

When the flow is inhomogeneous we can estimate $\dot{M}(r)$ by assuming that the gas is composed of a number of phases, the densest of which cools out of the flow at the radius under consideration (Thomas et al. 1987), or by model fitting (White & Sarazin 1987). In the first approach, the cooling region is divided into a number of concentric shells of size compatible with the instrumental resolution. The luminosity, δL_i of the ith shell can then be considered to be the sum of the cooling luminosity of the gas cooling out at that radius from the mean temperature T_i at rate $\delta \dot{M}$ and the luminosity of gas flowing across the shell experiencing temperature and potential changes ΔT_i and $\Delta \phi_i$;

$$L_i = \delta \dot{M}_i \frac{5}{2}\frac{kT_i}{\mu m} + \left(\frac{5}{2}\frac{k\Delta T_i}{\mu m} + \Delta\phi_i\right).\tag{53}$$

In our most detailed approach (Thomas, Fabian & Nulsen 1987), we have allowed for as many phases at a radius as there are shells within that radius and have integrated the cooling function and spectrum carefully. A typical mass deposition profile is shown in Fig. 2. It agrees fairly well with that obtained by assuming that the gas is homogeneous, principally because most of the energy is lost on cooling from the average cluster temperature $T_{cluster}$ at temperatures close to $T_{cluster}$. This new approach does allow us to measure the spread of densities in the gas at any radius. It is this which determines the manner in which mass is deposited (Nulsen 1986). We infer that the intracluster gas must contain a density spread of at least a factor of two (Figure 3). This may not be surprising when it is recalled that it has been enriched in metals which must have mixed different gases together.

The result that $\dot{M}(r) \propto r$ means that the deposited matter has $\rho \propto r^{-2}$ which is essentially an isothermal halo such as inferred for the dark matter around galaxies. It is assumed that whatever condenses out of the cooled gas orbits about, or through, the central galaxy such that its mean radius is similar to that where it was formed.

One last comment on our inference that $\dot{M} \propto r$ concerns time-dependence. The central cluster galaxy presumably has some finite velocity with respect to the cluster gas

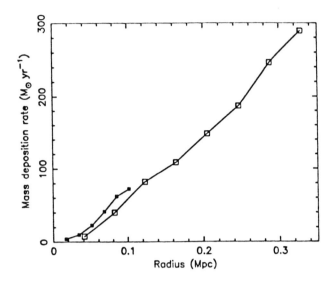

Figure 2. Mass deposited within radius r by the cooling flow in A2199, from Thomas et al. (1987).

induced by tidal forces from passing galaxies. Also the cluster gas is probably turbulent. Consequently a sharply-focussed flow is unlikely even in the case of an initially homogeneous gas. The relevant potential that gas at radius R is subject to is that averaged over its cooling and residence time at R to $R - \delta R$. This potential may be relatively flat over the central 100 kpc owing to motion of the central galaxy and to turbulence. The net result is likely to be a highly inhomogeneous flow within this region with only a small fraction cooling onto the central galaxy. Distributed mass deposition of the remaining gas must still occur.

In many cases, the rate of mass deposition is large, with $50-500\,M_\odot\,\mathrm{yr}^{-1}$ being common for a cluster of galaxies. An outstanding problem now presents itself, namely what does the cooled gas form? If stars, then the IMF must be non-standard or the central galaxies would be much bluer and brighter than is observed.

5b Alternatives to Cooling

The problem of what forms from a cooling flow has caused the assumptions behind the phenomena of cooling flows to be questioned. A heat source, for example, could offset the cooling so that little gas is actually deposited. Heat sources, or fluxes, that have been considered are cosmic rays (Tucker & Rosner 1982; Rephaeli 1987), conduction (Bertschinger & Meiksin 1986), galaxy motions (Miller 1986) and supernovae (Silk et al. 1986). None of these proposals successfully confronts the evidence from spectroscopic X-ray line measurements of low temperature components that are entirely consistent with a cooling interpretation. It

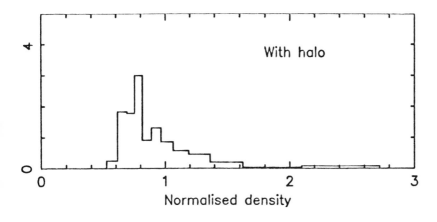

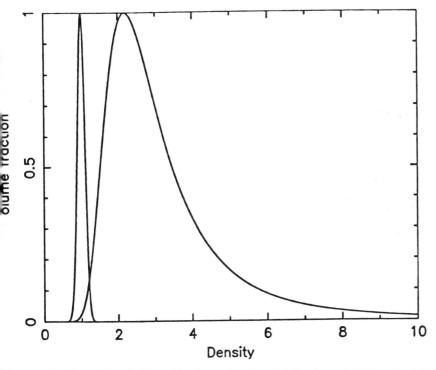

Figure 3. Upper Panel; Normalized gas density distribution at 100 kpc in A496. Lower Panel; Development of idealized gaussian density distribution (left) after 20 per cent of the gas has cooled out (right). (Both from Thomas et al. 1987).

would indeed be a strange heat source that allowed the gas to lose 90 per cent of its thermal energy in cooling from 80 million K down to 8 million K but then prevented it cooling further. There is, of course, no evidence for large stockpiles of gas at some intermediate temperature. The X-ray spectroscopic data also indicate that cooling flows are long lived since gases with a wide range of cooling times give the same mass cooling rate.

To consider heat sources in more detail, we note that it is generally difficult to keep the gas stable whilst heating it (Stewart et al. 1984). Most heat sources, such as cosmic rays, heat at a rate proportional to the gas density, whereas the cooling varies as the density squared. Heating may then cause the gas to become more unstable by tending to increase the temperature and pressure of the lowest density phases but allowing the denser gas to carry on cooling. If conduction occurs unimpeded then cooling of the core gas can set up a temperature gradient that is offset by a conductive heat flux. Whilst such a situation can occur, it is restricted to only a small part of parameter (initial density and temperature) space. The energy equation (at constant pressure) is

$$\frac{\rho v}{\mu m}\frac{5}{2}k\frac{dT}{dr} = n^2 \Lambda - \frac{1}{r^2}\frac{d}{dr}\left(r^2 \kappa \frac{dT}{dr}\right). \tag{54}$$

The first term on the r.h.s. is the cooling, which at constant pressure and for bremsstrahlung, varies as $T^{-\frac{3}{2}}$. The second term is the conductive heat flux which varies as $T^{\frac{7}{2}}$ (it does not depend on density). Their widely different temperature dependences makes it difficult to allow a balance where one term does not dominate. Conduction dominates where the temperature is high and cooling where it is low. Generally, cooling dominates at the centre of a flow and conduction can be important further out (Nulsen et al. 1982). If conduction is dominant, then it tends to make the gas almost isothermal, in disagreement with observations. The X-ray spectroscopic observations (Canizares et al. 1979, 1982, 1987; Mushotzky et al. 1981, 1987b) then demonstrate that conduction is inhibited, probably by tangled magnetic fields which greatly decrease the effective electron mean-free-path.

Supernova heating from stars formed from cooled gas (Canizares et al. 1982; Lea et al. 1982; Silk et al. 1986) can at most change the mass deposition estimates by a factor of two. This is because the energy from a supernova can only heat a mass of gas sufficient to form another supernova progenitor (and an IMF's worth of lower-mass stars) to 8.10^6 K. There is no evidence for supernovae around the central galaxies in cooling flow clusters (see e.g. Caldwell & Oemler 1981).

Finally on the topic of heating, it is worth remembering that cooling flows occur in a wide variety of clusters, both with and without strong radio sources (e.g. Cyg A vs. AWM4) and in deep and shallow potential wells (e.g. Perseus vs. Hydra). Any heat source necessary to counteract the radiative cooling would represent a major heat flow, of $\gtrsim 10^{62}$ erg per cluster. In my view, the current lack of understanding of star formation in our own Galaxy means that it is not a simple business to extrapolate to other situations. Star formation can proceed at hundreds of solar masses per year without necessarily being evident optically, provided that only low mass stars are formed. There is no problem with the total mass deposited, which is distributed out to 100 - 300 kpc from the central galaxy and does not pile up within its centre. Clusters are full of dark matter and there can be no problem with some (or even all) of it being baryonic.

5c The Evolution of the Cooling Gas and Star Formation

In the limit of zero viscosity, the gas in a cooling flow should be homogeneous and cool into a central singularity (Nulsen 1986; Malagoli et al. 1987; Balbus 1987). This is because

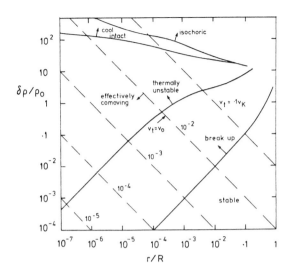

Figure 4. Expected behaviour of gas blobs of size r with excess density $\delta\rho$ in a background flow of density ρ at radius R, from Nulsen (1986).

gravity causes any denser gas to fall ahead and join gas with similar properties. Only near the centre of the flow, where $v \sim v_{ff}$ the free-fall velocity, would the flow become inhomogeneous. However a real flow does have viscosity, is turbulent and contains magnetic fields. Consequently, I do not believe that the linear perturbation analyses are particularly relevant to a real cooling flow. The non-linear behaviour of gas blobs in a flow has been explored by Nulsen (1986, see Fig. 4).

A large gas blob of size r and overdensity $\delta\rho$ will try to move ahead of the mean flow and reaches a terminal velocity

$$v_T \simeq v_{Kepler}\sqrt{\left(\frac{\delta\rho}{\rho_0}\frac{r}{R}\right)}. \tag{55}$$

This relative motion will then cause the blob to spread out and fragment (Nittman, Falle & Gaskell 1982). r/R is reduced and the relative velocity of the overdense gas, v_T, is reduced. Magnetic fields can help to pin the gas to the mean flow so that it comoves. The net result is that large, slightly overdense blobs at large radii from the centre of the flow are turned into an emulsion of smaller and very overdense blobs at smaller radii. The densest gas will cool out of the flow (i.e. $T \to 0\,\mathrm{K}$) at intermediate radii. The density distribution of gas at a given radius will tend to evolve a 'cooling tail' (volume filling fraction $f \propto \rho^{-(4-\alpha)}$, where α is the exponent of the cooling function; Nulsen 1986; Thomas, Fabian & Nulsen 1987, see Fig. 3 and below). This allows mass to be deposited in a distributed manner. If a spread of densities exists throughout the cluster, then gas may be deposited by cooling well beyond the radius where the mean cooling time is H_0^{-1}.

Density inhomogeneities will have been introduced early into the cluster gas by the production of metals, quasars and from the stripping of gassy galaxies and winds. Later it will be further mixed and inhomogenized by the infall of subclusters. A further spread of density will occur because of the motion of the central cluster galaxy and turbulence, as discussed in Section 5a. Convection, with magnetic fields and viscosity, will then create a limited range of densities throughout the gas.

The manner in which cooling gas is deposited can be seen in the following way, for a constant-pressure, spherically-symmetric flow. We shall consider a case in which $\dot{M}(r)$ is constant with time. This imposes a strict condition on f, which essentially must have the shape of the cooling tail. The mass of gas per unit volume with density between ρ to $\rho + \delta\rho$ is $dM = \rho f d\rho$ and this is constant in the steady, time-independent case. Thus $\rho f \dot{\rho}$ =constant, and since $\dot{\rho} \propto \rho^{3-\alpha}$ (equation 49),

$$f \propto \rho^{-(4-\alpha)}. \tag{56}$$

Again, $\dot{M}(r) = 4\pi r^2 \bar{\rho} v$ is time-independent and so at a given radius, since $\bar{\rho}$ behaves as ρ,

$$r^2 \dot{\rho} v = \text{const.} \tag{57}$$

i.e.

$$r^2 \rho^{3-\alpha} \frac{dr}{dt} = r^2 \rho^{3-\alpha} \frac{dr}{d\rho} \dot{\rho} = \text{const.} \tag{58}$$

So,

$$r^3 \propto \rho^{-(5-2\alpha)} \tag{59}$$

and

$$\dot{r} \propto \rho^{-\frac{(5-2\alpha)}{3}-1} \dot{\rho}. \tag{60}$$

Therefore,

$$v \propto \rho^{\frac{(1-\alpha)}{3}} \tag{61}$$

and finally,

$$\dot{M} \propto r^2 r^{-\frac{3}{(5-2\alpha)}} r^{-\frac{(1-\alpha)}{(5-2\alpha)}}, \tag{62}$$

so,

$$\dot{M} \propto r^{\frac{3(2-\alpha)}{(5-2\alpha)}}. \tag{63}$$

Then, if $-1/2 < \alpha < 1/2$, as expected for reasonable cooling functions, $\dot{M} \propto r^\eta$ with $5/4 > \eta > 9/8$. The observed value of $\eta \sim 1$ is consistent with this result. A cooling flow in which the central potential is relatively flat and $M(r)$ changes only slowly will give a similar value of η. What we do not know is what shape f has initially. Cooling will give it the $\rho^{-(4-\alpha)}$ shape over some fairly wide range of r anyway. Initial conditions (i.e. how mixing and convection leave the density distribution in the core) will determine the wings of the pre-cooling f-distribution and so the outermost and innermost behaviour of $\dot{M}(r)$. General results for different potentials and f-distributions are derived rigorously by Nulsen(1986).

Gas blobs can drop out of pressure equilibrium on cooling if $t_{cool} < t_{cross}$, the sound crossing time;

$$\frac{t_{cool}}{t_{cross}} \propto \frac{T^3}{R}. \tag{64}$$

using the simple approximation to the cooling function below 10^7 K. This just means that the gas is cooling faster than sound waves can maintain constant pressure. For typical cluster pressures ($nT \sim 10^5 - 10^6 \,\mathrm{cm^{-3}}$ K) then the cooling becomes isochoric (constant density rather than constant pressure) at temperatures $T \leq 3.10^6 r_{\mathrm{kpc}}^{0.32}$ K (Cowie, Fabian & Nulsen 1980), where $r = r_{\mathrm{kpc}}$ kpc. Even parsec size blobs will become isochoric around the peak of the cooling curve. The cooling is still initially rapid below 10^4 K and the gas may cool to 100 K before being repressurized by shocks.

Such shocks might be thought to lead to optical line radiation which could account for some of the optical and ultraviolet line emission observed at the centres of some cooling flows (see e.g. Kent & Sargent 1979; Heckman 1980; Hu, Cowie & Wang 1983; Johnstone, Fabian & Nulsen 1987). It is not enough, however, as the number of Balmer photons per hydrogen atom passing through the shock, H_{rec}, is only ~ 1 and the observed line luminosities require $H_{rec} \sim 100$ to 1000 (Johnstone, Fabian & Nulsen 1987).

$$L(\mathrm{H}\beta) \approx 10^{39} H_{rec} \left(\frac{\dot{M}}{100\,\mathrm{M_\odot\,yr^{-1}}} \right) \mathrm{erg\,s^{-1}} \tag{65}$$

is unobservable unless $H_{rec} \geq 10$ (depending upon the distance to the cluster and the area of the emission region).

There must be some extra source of ionization which could be photoionization by an active nucleus (Robinson et al. 1987), hot stars (Johnstone, Fabian & Nulsen 1987) or possibly cosmic rays. The precise source has not been identified at the moment. Johnstone & Fabian (1987) show that, in the case of NGC 1275, the ionizing source must be distributed.

The fate of the cooled gas blobs depends upon the competing roles of gravity, conduction and further cooling. The details are at present unknown. The colours and magnitudes of the central galaxies in cooling flows indicate that only a small fraction of the X-ray inferred mass deposition rate, $\dot{M}_X$, forms stars with a disk-galaxy IMF (see e.g. O'Connell 1987). Most disk galaxies such as our own are bright and blue through forming stars at $\leq 5\,\mathrm{M_\odot\,yr^{-1}}$, whereas a large cooling flow such as that around NGC 1275 in the Perseus cluster is depositing several hundred solar masses per year. It is presumed that most of this matter condenses into low mass stars (Fabian, Nulsen & Canizares 1982; Sarazin & O'Connell 1983). There are many differences between the conditions in a disk galaxy and those in a cooling flow. The pressure of the gas is 100 to 1000 times higher, which can reduce the Jeans mass by a factor of 10 to 30 (Jura 1977). Dust is likely to be absent in the hot gas, due to sputtering (Draine & Salpeter 1979), and unlikely to form before the gas has cooled and condensed. This may in turn inhibit the formation of GMC, which seem to be necessary for massive star formation in our Galaxy. The size of the cooled gas blobs may also be much smaller as well. Perhaps we are chauvinists with respect to star formation and only recognise star formation if it resembles that which is obvious in our Galaxy.

There is definitely a small amount of visible star formation in some central galaxies. A-stars are apparent in NGC 1275 (Rubin et al. 1977) to an extent such that the 'normal' star formation rate, $\dot{M}_\star, \sim 2$ per cent of $\dot{M}_X$ (Fabian, Nulsen & Arnaud 1984; Gear et al. 1985). The central galaxy in A1795 (O'Connell 1987), PKS 0745-191 (Fabian et al. 1985), NGC 6166 and M87 (Bertola et al. 1987) all show evidence for a few per cent of $\dot{M}_X$ passing into 'normal' stars. More widespread evidence for star formation is obtained from the 4000Å break, D, in the continuum spectra of the central galaxies. This break is characteristic of late-type stars and is filled in (weakened) by the presence of young stars. Johnstone, Fabian & Nulsen (1987) find that D correlates with $\dot{M}$. Up to a few per cent of $\dot{M}_X$ appears to pass into 'massive' stars (mostly A-stars of a few $\mathrm{M_\odot}$). Only in PKS

0745-191 (where $\dot{M}_X$ is 500 - 1000 M$_\odot$ yr^{-1}, Arnaud et al. 1987) is $\dot{M}_\star \sim 30-100$ M$_\odot$ yr^{-1}. D is only weakened in those objects with observed emission lines and so there is some correlation between massive star formation and optical line emission. Whether the stars give the ionizing flux that produces the emission lines, or the emission lines indicate large gas blobs within which the conditions are appropriate for the formation of some massive stars, is not yet known.

Roughly it can be stated that

$$\frac{\dot{M}(\text{normal IMF})}{\dot{M}(\text{low mass stars})} \approx \frac{M(\text{visible})}{M(\text{dark})} \tag{66}$$

for the underlying galaxy. This, together with the density profile of distributed gas, suggest that cooling flows may be related to galaxy formation.

Some other evidence for star formation in cooling flows is provided by the X-ray 'plume' of M86 (Forman et al. 1979) which is presumably the result of ram pressure stripping in the Virgo cluster and some distorted optical emission, presumed to be stars (Nulsen & Carter 1987). Finally, it is possible that the shells around E and S0 galaxies - and the arcs in some clusters (Soucail et al. 1987; Petrosian & Lynds 1986) - may be due to disturbances triggering star formation (Loewenstein, Fabian & Nulsen 1987).

5d Distant Cooling Flows and Galaxy Formation

We have already noted that cooling flows may have some relationship to galaxy formation. If a substantial fraction of the mass of a protogalactic cloud was virialized so that $H^{-1} > t_{cool} > t_{ff}$, then the gas can cool into the galaxy as a cooling flow. The densest (probably central) gas blobs may collapse without being virialized and form massive stars that enrich and mix the remaining gas. Of further interest in this mechanism is the possibility that much of the dark matter is baryonic, i.e low mass stars formed in the flow.

The obvious way to search for distant cooling flows is to make soft X-ray observations, but that is not possible until ROSAT, AXAF and XMM are launched. In the meantime we must make use of indirect methods. Extended optical emission lines are relatively common around radio galaxies and quasars and are observed out to redshifts of almost 2. The line emission from some 3CR radio galaxies is remarkably similar to that of nearby cooling flows such as that around NGC 1275. If there is a surrounding flow then the gas must be at high pressure and confined by hotter gas. Evidence for high pressure gas surrounding the quasar 3C48 is obtained from a photoionization calculation applied to the observed [OIII]/[OII] line ratio of the gas at 30 kpc from the nucleus (Fabian et al. 1987). The spectrum of the nucleus ($L \sim 10^{46}$ erg s^{-1}) at optical and ultraviolet wavelengths and in the X-ray band (Wilkes & Elvis 1987) indicates a power-law ionizing spectrum of energy index 1.3. The ionization parameter $\xi = L/nR^2$ can then be adjusted to give the observed line ratio at the observed gas radius (30 kpc). The density inferred there is then 30 cm^{-3} so that the pressure is 3.10^5 cm^{-3} K. This is comparable to the pressure that gas at 10^4 K would have if confined by the ~ 100 M$_\odot$ yr^{-1} cooling flow in the poor cluster MKW3s. There is evidence accumulating that many radio-loud quasars are in poor clusters (e.g. Yee & Green 1984). Hintzen & Romanishin (1987) also suggest that 3C275.1 lies in a cluster cooling flow from the appearance of its extended [OIII] gas. Crawford (1987) has applied the above method to spectra of a number of radio-loud quasars and has obtained similar results. If the gas in these quasars is not presssure-confined by extensive surrounding hot gas then 10 times more gas is required in order to explain such extensive emission.

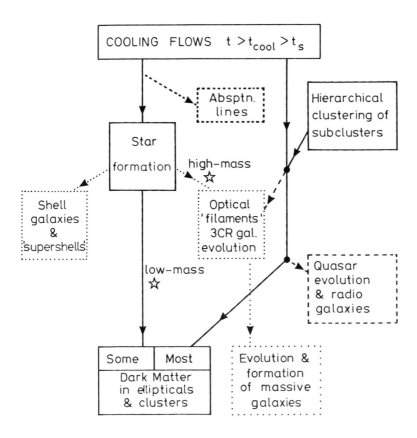

Figure 4. The 'Strong Cooling Flow Hypothesis'.

One picture for the evolution of cooling flows (Fabian et al. 1986) suggests that they were common in poor clusters (cf. MKW3s). Rich clusters are then assembled from the mergers of these subclusters and the merger process can then disrupt the individual flows. A large flow may survive relatively undisturbed if a small subcluster collides with a larger one, but the merger of two large subclusters in the formation of a Coma-like cluster could well spread the cooling gas throughout the core. Thus an unfocussed flow could persist even in Coma. The merging of subclusters will also make the intracluster medium inhomogeneous, as required. This picture could also mean that strong flows were once common around all large elliptical galaxies and have produced much of the galaxies and, in particular, much of their dark matter.

Only a small fraction of the cooling gas is needed at the centre of a galaxy to power even a quasar. Activity, and in particular radio emission (Valentijn & Bijleveld 1983; Jones & Forman 1984) is common in cooling flows. The decrease in the numbers of very powerful quasars at the present epoch can then be related to the assembly of clusters from quasar-active subclusters which are subsequently disrupted. Cooling-flow-deposited dark matter is spread throughout the cluster and so much cluster dark-matter may be baryonic.

The 'strong' cooling flow hypothesis, outlining some phenomena that may be related to, or connected with, cooling flows is shown in Figure 4.

6. Summary

Intracluster gas provides us with the largest bodies of optically-thin gas for study. Relatively simple models have had some success in modelling their properties, but it is beginning to be appreciated that the gas is complex and multiphase, in other words, just like most gas anywhere. Considerable progress remains to be made in understanding the large scale distribution of the gas, in the role of magnetic fields and in their evolution, particularly in terms of the interaction of subclusters (see also A. Cavaliere's talk).

In the cores of clusters, cooling flows appear to be common and represent large regions of star formation, particularly of low-mass stars. Star formation must then be environment-dependent. This means that at least some dark matter is baryonic and is forming now from X-ray emitting gas.

Acknowledgements

I thank Roderick Johnstone for much help in preparing this paper, Roberto Pallavicini for inviting me to Cargese and the Royal Society for supporting my work.

References

Arnaud, K.A., 1985. *PhD thesis*, University of Cambridge.
Arnaud, K.A., Johnstone, R.M., Fabian, A.C., Crawford, C.S., Nulsen, P.E.J., Shafer, R.A. & Mushotzky, R.F., 1987. *Mon. Not. R. astr. Soc.*, **227**, 241.
Balbus, S., 1987. Preprint.
Bertola, F., Gregg, M.D., Gunn, J.E. & Oemler, A., 1987. *Astrophys. J.*, **303**, 624.
Bertschinger, E. & Meiksin, A., 1986. *Astrophys.J*, **306**, L1.
Birkenshaw, M., Gull, S.F., & Hardebeck, H., 1984. *Nature*, **309**, 34.
Caldwell, C.N. & Oemler, A., 1981. *Astr. J.*, **86**, 1424.
Canizares, C.R., Clark, G.W., Markert, T.H., Berg, C., Smedira, M., Bardas, D., Schnopper, H. & Kalata, K., 1979, *Astrophys. J.*, **234**, L33.
Canizares, C.R., Clark, G.W., Markert, T.H., Berg, C., Smedira, M., Bardas, D., Schnopper, H. & Kalata, K., 1979. *Astrophys. J.*, **234**, L33.
Canizares, C.R., Clark, G.W., Jernigan, J,G. & Markert, T.H., 1982. *Astrophys. J.*, **262**, L33.
Canizares, C.R., Stewart, G.C. & Fabian A.C., 1983. *Astrophys. J.*, **272**, 449.
Canizares, C.R., Markert, T.H. & Donahue, M.E., 1987. In *Proceedings of NATO ARW Cooling Flows in Clusters and Galaxies*, ed. A.C.Fabian, Reidel, in press.
Cavaliere, A. & Fusco-Femiano, R., 1976., *Astr. Astrophys.*, **49**, 137.
Cowie, L.L. & Songaila, A., 1977. *Nature*, **266**, 501.

Cowie, L.L., Fabian, A.C. & Nulsen, P.E.J., 1980. *Mon. Not. R. astr. Soc.*, **191**, 399.
Cowie, L.L., Henriksen, M.J. & Mushotzky, R.F., 1987. *Astrophys. J.*, **312**, 593.
Crawford, C.S., 1987. In *Proceedings of NATO ARW Cooling Flows in Clusters and Galaxies*, ed. A.C.Fabian, Reidel.
Crawford, C.S., Crehan, D.A., Fabian, A.C. & Johnstone, R.M., 1987. *Mon. Not. R. astr. Soc.*, **224**, 1007.
Draine, B.T. & Salpeter, E., 1979. *Astrophys. J.*, **231**, 77.
Fabian, A.C. & Nulsen, P.E.J., 1977. *Mon. Not. R. astr. Soc.*, **180**, 479.
Fabian, A.C., Hu, E.M., Cowie, L.L & Grindlay, J.,1981. *Astrophys. J.*, **248**, 47.
Fabian, A.C., Nulsen, P.E.J. & Canizares, C.R., 1982. *Mon. Not. R. astr. Soc.*, **201**, 933.
Fabian, A.C., Nulsen, P.E.J. & Canizares, C.R., 1984. *Nature*, **311**, 733.
Fabian, A.C., Nulsen, P.E.J. & Arnaud, K.A., 1984. *Mon. Not. R. astr. Soc.*, **208**, 179.
Fabian, A.C., Arnaud, K.A., Nulsen, P.E.J. & Mushotzky, R.F., 1986. *Astrophys. J.*, **305**, 9.
Fabian, A.C., Thomas, P.A., Fall, S.M. & White, R.A., 1986. *Mon. Not. R. astr. Soc.*, **221**, 1049.
Fabian, A.C., Crawford, C.S., Johnstone, R.M. & Thomas, P.A., 1987. *Mon. Not. R. astr. Soc.*, **228**, 963.
Fabricant, D. & Gorenstein, P., 1983. **267**, 535.
Fabricant, D., Beers, T.C., Geller, M.J., Gorenstein, P., Huchra, J.P. & Kurtz, M.J., 1986. **308**, 580.
Forman, W., Schwarz, J., Jones, C., Liller, W. & Fabian, A.C., 1979. *Astrophys. J.*, **234**, L27.
Forman, W., Bechtold, J., Blair, W.,Giacconi, R., Van Speybroeck, L. & Jones, C., 1981, *Astrophys. J.*, **243**, L133.
Gear, W.K., Gee, G., Robson, E.I. & Nott, I.G., 1985. *Mon. Not. R. astr. Soc.*, **217**, 281.
Geller, M.J., 1984. *Comments on Astrophys. Space. Sci.*, **10**,47.
Gilfanov, M.R., Syunyaev, R.A. & Churazov, E.M., 1987. *Sov. Astr. Lett.*, **13**, 17.
Heckman, T.M., 1981. *Astrophys. J.*, **250**, L59.
Hintzen, P. & Romanishin, W., 1986. *Astrophys. J.*, **311**, L11.
Huchra, J & Brodie, J. 1987. Preprint.
Hu, E.M., Cowie, L.L. & Wang, Z., 1985. *Astrophys. J. Suppl.*, **59**, 447.
Johnstone, R.M., Fabian, A.C. & Nulsen, P.E.J., 1987. *Mon. Not. R. astr. Soc.*, **224**, 75.
Johnstone, R.M. & Fabian, A.C., 1987. *Mon. Not. R. astr. Soc.*, submitted.
Jones, C. & Forman, W., 1984. *Astrophys. J.*, **276**, 38.
Jura, M., 1977. *Astrophys. J.*, **129**, 268.
Kent, S.M. & Sargent, W.L.W., 1979. *Astrophys. J.*, **230**, 667.
King, I., 1966. *Astr. J.*, **71**, 64.
Lea, S.M., Mushotzky, R.F. & Holt, S.S., 1982. *Astrophys. J.*, **262**, 24.
Loewenstein, M., Fabian, A.C. & Nulsen, P.E.J., 1987. *Mon. Not. R. astr. Soc.*, **229**, 129.
Malagoli, A., Rosner, R. & Bodo, G., 1987. *Astrophys. J.*, **319**, 632.
Miller, L., 1986. *Mon. Not. R. astr. Soc.*, **220**, 713.
Mould, J.R., Oke, J.B. & Nemec, J.M., 1987. *Astr. J.*, **92**, 53.
Mushotzky, R.F., Holt, S.S, Smith, B.W., Boldt, E.A. & Serlemitsos, P.J., 1981. *Astrophys. J.*, **244**, L47.
Mushotzky, R.F., 1987a. *Astrophys. Lett.*, **26**, 43.
Mushotzky, R.F., 1987b. In In *Proceedings of NATO ARW Cooling Flows in Clusters and Galaxies*, ed. A.C.Fabian, Reidel.
Nittmann, J., Falle, S.A.E.G. & Gaskell, P.H., 1982. *Mon. Not. R. astr. Soc.*, **201**, 833.
Nulsen, P.E.J., 1982. *Mon. Not. R. astr. Soc.*, **198**, 1007.

Nulsen, P.E.J., Stewart, G.C., Fabian, A.C., Mushotzky, R.F., Holt, S.S., Ku, W.H.M. & Malin, D.F., 1982. *Mon. Not. R. astr. Soc.*, **199**, 1089.
Nulsen, P.E.J., Stewart, G.C. & Fabian, A.C., 1984. *Mon. Not. R. astr. Soc.*, **208**, 185.
Nulsen, P.E.J., 1986. *Mon. Not. R. astr. Soc.*, **221**, 377.
Nulsen, P.E.J. & Carter, D., 1987. *Mon. Not. R. astr. Soc.*, **225**, 939.
O'Connell, R.W., 1987. in *Proc. I.A.U. Symp. No. 127*, ed. De Zeeuw, T., 167, Reidel.
O'Dea, C.P., Sarazin, C.L. & Owen, F.N., 1987. *Astrophys. J.*, **316**, 113.
Petrosian, V & Lynds, R., 1986. *B.A.A.S.*, **18**, 1014.
Rephaeli, Y., 1987. *Mon. Not. R. astr. Soc.*, **225**, 851.
Robinson, A., Binette, L., Fosbury, R.A.E. & Tadhunter, C., 1987. *Mon. Not. R. astr. Soc.*, **227**, 97.
Rothenflug, R. & Arnaud, M., 1986. *Astr. Astrophys.*, **147**, 337.
Rubin, V.C., Ford, W.K., Peterson, C.J. & Oort, J.H., 1977. *Astrophys. J.*, **211**, 693.
Sarazin, C.L., 1986. *Rev. Mod. Phys.*, **58**, 1.
Sarazin, C.L. & O'Connell, R.W., 1983. *Astrophys. J.*, **258**, 552.
Schwartz, D.A., Schwarz, J. & Tucker, W.H., 1980. *Astrophys. J.*, **238**, L59.
Silk, J., Djorgovski, G., Wyse, R.F.G. & Bruzual, G.A., 1986. *Astrophys. J.*, **307**, 415.
Soucail, G., Fort, B., Mellier, Y. & Picat, J.P.,1987. *Astr. Astrophys.*, **172**, 44.
Spitzer, L. & Baade, W., 1951. *Astrophys. J.*, **113**,413.
Stewart, G.C., Canizares, C.R., Fabian, A.C. & Nulsen, P.E.J., 1984a. *Astrophys. J.*, **278**, 536.
Stewart, G.C., Fabian, A.C., Jones, C. & Forman, W., 1984b. *Astrophys. J.*, **285**, 1.
Takeda, H., Nulsen, P.E.J. & Fabian, A.C., 1984. *Mon. Not. R. astr. Soc.*, **208**, 261.
The, L.S. & White, S.D.M., 1987. Preprint.
Thomas, P.A., Fabian, A.C. & Nulsen, P.E.J., 1987. *Mon. Not. R. astr. Soc.*, **228**, 973.
Thomas, P.A., Fabian, AC., Arnaud, K.A., Forman, W. & Jones, C., 1986. *Mon. Not. R. astr. Soc.*, **222**, 655.
Tucker, W.H. & Rosner, R., 1982. *Astrophys. J.*, **267**, 547.
Ulmer, M., Cruddace, R.G., Fennimore, E.E., Fritz, G.G. & Snyder, W.A.,1987. *Astrophys. J.*, **319**, 118.
Valentijn, E.A. & Bijleveld, W., 1983. *Astr. Astrophys.*, **125**, 223.
Van Speybroeck, L., 1987. *Astrophys. Lett.*, **26**, 127.
White, R.E. & Sarazin, C.L., 1987. *Astrophys. J.*, **318**, 612, 621, 629.
Wilkes, B.J. & Elvis, M. 1987. *Astrophys. J.*, **323**, 243.
Yee, H.K.C. & Green, R.F., 1984. *Astrophys. J.*, **280**, 79.

INTERGALACTIC PLASMA IN CLUSTERS: EVOLUTION

A. Cavaliere
Astrofisica, Dip. di Fisica, II Università di Roma, Italy
S. Colafrancesco
Dip. di Astronomia, Università di Padova, Italy.

ABSTRACT

The statistical sampling of the X-ray sources associated with the Intra-Cluster Plasma in distant clusters and groups of galaxies will provide a direct view of the clustering process and of the history of the ICP in the making. We discuss the information that may be derived from the number counts and the redshift distribution of such sources, referring to the results from the present stage of the Medium Sensitivity Survey by HEAO 2 and from its continuing extensions. Additional implications of cosmogonic interest may be tested with future and deeper X-ray missions.

1. INTRODUCTION

From the earliest data gathered by the first X-ray observatory UHURU, the Clusters of Galaxies were recognized as one of the two classes of powerful extragalactic sources: the *extended* ones as opposed to the *compact* Active Galactic Nuclei (Cavaliere, Gursky and Tucker 1971).

In fact, in the systematic surveys to follow (Gursky et al. 1972, Giacconi et al. 1974) many Abell (1958) clusters of richness classes $R \geq 1$ and distance classes $D \leq 3$ (Bahcall and Bahcall 1975, Jones and Forman 1978) appeared as bright X-ray emitters with luminosities in the range $L \sim 5~10^{43} \div 10^{45}$ erg/s, a result confirmed by SAS-C (Markert et al. 1979) and by ARIEL V (McHardy et al. 1981).

The bulk of the cluster luminosity in X-rays is constituted by thin thermal bremsstrahlung emission from a hot and diffuse ICP (*Intra-Cluster Plasma*) at temperatures $T \sim 10^8$ K and densities $n \lesssim 10^{-3}$ cm^{-3}. The point, stressed by Solinger and Tucker (1972, and bibliography therein), is now proven by the extensive spectral evidence reviewed in these Proceedings by R. Mushotzky, concerning both the shape of the continuum and the emission of very high excitation lines consistent with such conditions.

So the integrated X-ray luminosity is given by

$$L \propto g^2(t)\, M\, \rho\, T^{\frac{1}{2}}, \qquad (1.1)$$

an expression involving two sets of parameters: i) those describing the dynamical state of the group or cluster as a whole, namely the total mass M and the correponding density

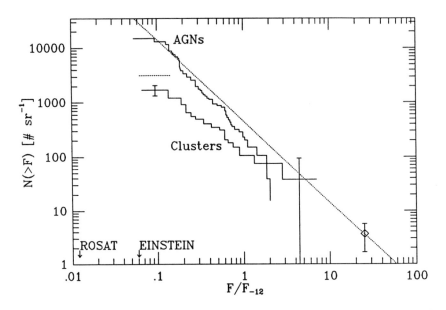

Fig. 1. The counts of Clusters and of Active Galactic Nuclei in the band $0.3 \div 3.5$ keV from the preliminary results of the MSS by Gioia et al. (1984). The high flux point is reduced by Maccacaro et al. (1982) from Piccinotti et al. (1982). The dotted segment indicates the faintest bin of the MSS extended to 68 clusters, Gioia et al. 1987. A line of constant slope $\beta = 1.5$ is given for reference.

ρ; ii) those concerning the ICP, namely its mass that we write as $M_{ICP} \equiv Mg(t)$ and its temperature $T \lesssim T_{vir} \sim GM/r_{vir}$, with the approximate equality holding near a relaxed state. The spectral distribution is approximated by $F_E \propto E^{-0.4} \exp(-E/kT)$.

The advent of the X-ray telescopes (HEAO 2, Giacconi et al. 1979) brought two major developments. The mapping capability unveiled rich morphologies in the brightness distribution of these sources (Jones and Forman 1984). Resolution and sensitivity permitted to pinpoint and to image low luminosity ($L \simeq 5 \; 10^{42} \div 5 \; 10^{43}$ erg/s) poor clusters and groups (Biermann, Kronberg and Madore 1982, Abramopoulos and Ku 1983, Kriss et al. 1983), thus nailing down the general results from the identifications of many such sources begun with HEAO 1 by Schwartz et al. 1980, Ulmer at al. 1981, and carried out by Johnson et al. 1983, Kowalsky et al. 1984. Preliminary contributions from EXOSAT have been reported at this Meeting.

It is now apparent that soft X-ray surveys will constitute the right channel not only to map the variances and the remnants (in the form of clumpiness) of the formation process in the relatively local clusters (Cavaliere et al. 1986, Oemler, West and Dekel 1987), but also to build up statistical samples of clusters out to substantial redshifts.

These samples will trace back into the cosmologically significant past the two kinds of processes contributing to L, namely: i) the dynamics of the clustering; ii) the history of the mass and the energy of the ICP, including the astration (and the preceding star formation) that must have produced a considerable component of the ICP.

2. THE STATISTICAL DATA BASE

At present, we have substantial observational information concerning the local luminosity function $N(L, z \lesssim 0.1)$ of clusters and groups. The data in the energy range $2 \div 6$ keV (cf. Schwartz et al. 1980, Henry et al. 1982, Johnson et al. 1983, Kowalsky et al. 1984) indicate a power law shape $N(L) \propto L^{-\gamma}$ with $\gamma \simeq 1.7$ in the range $L \simeq 10^{43} \div 5 \, 10^{44}$ erg/s followed by a gradual cut off (see also Soltan and Henry 1983). This is close to the overall slope $\gamma \simeq 2$ of the very local luminosity function derived by Piccinotti et al. (1982).

For $z > 0.15$, we cannot resolve yet the z-dependent luminosity function $N(L, z)$, but we have already some information concerning an integral of it out to $z \simeq 0.5$, namely the counts-flux relationship $N(> F)$ (for technical details, see Appendix A). The analysis of a sub-sample of the Medium Sensitivity Survey out of archived material from HEAO 2 (MSS, in the band $0.3 \div 3.5$ keV, Gioia et al. 1984) yielded counts with a slope $\beta \simeq 1.1 \pm 0.2$: the best value is considerably flatter than the "Euclidean" value ($\beta = 1.5$) and quite flatter than that found for AGNs in the same survey (see fig. 1). A recent expansion of the subsample from 19 to 68 clusters plus 66 candidates (Gioia et al. 1987) confirms the sharp flattening of the counts down to $F \simeq 10^{-13}$ erg/cm^2 s (cf. their fig. 6).

To explain such flat counts the cosmological bending of $N(> F)$ alone is not enough, and the minimal requirement is a luminosity function not only constant but also very broad (see Appendix A). But recently Schmidt and Green remarked that, if the local luminosity function of Piccinotti et al. (1982) were to be extrapolated uniformly in cosmic time out to large z, then: i) the MSS should have detected twice as many clusters as were actually identified; ii) the mean luminosity thus predicted would be ~ 10 times larger than the value derived from the observed luminosity distribution.

It may be that recalibrations, or incompleteness for faint extended sources at the present stage of the MSS will account for these discrepancies. Waiting for more statistics, we take hint from them to discuss more generally the implications of the counts.

The starting point is that to preserve for all $z \gtrsim 0.1$ a luminosity function broad and constant is not easy. On the contrary, straight scaling laws based on a simple clustering scenario and on dominance of primordial ICP would rather indicate an $N(L, z)$ steep and strongly evolving (cf. Kaiser 1986). But the result is sensitive to the ICP behaviour.

3. SCALING $N(L,z)$ FROM $z \sim 0$

We will retain here the hierachical clustering scenario (HCS, see Peebles 1980, summarized in Appendix B), one that is most definite and comparatively successful (cf. Dekel 1987). This envisages the non-linear gravitational structures to form bottom up from galaxies aggregating into groups, these into poor, and eventually into rich clusters: with progressing cosmic epoch, the same mass gets reshuffled (on average) into ever larger self-gravitating units. Conversely, looking back in time one expects the average condensations in previous stages of the hierarchy to be smaller and more numerous, intrinsically denser and cooler: eqs. B.1 give the scaling laws obeyed by characteristic quantities such as the mass M_c, the size r_c, the density ρ_c and the temperature T_c (Peebles 1974, White 1982).

As noted by Kaiser (1986), the luminosity functions $N(L, z)$ (number/ unit comoving volume and unit luminosity) will scale with z (i.e., in look-back time) as to normalization and argument following the general law

$$N(L, z) \propto \frac{1}{M_c L_c} N_o(\frac{L}{L_c}) , \qquad (3.1)$$

where $M_c(z), L_c(z)$ are the characteristic masses and luminosities at the redshift z.

The *density* component

$$\frac{1}{M_c} \propto (1+z)^{6/(n+3)} , \qquad (3.2)$$

takes care of the steep increase in normalization ("density evolution") corresponding to the crowding (even in comoving coordinates) of ever smaller systems.

The other, *luminosity* component

$$\frac{1}{L_c} N_o(\frac{L}{L_c}) \propto L^{-\gamma}(L_c)^{\gamma-1} , \qquad (3.3)$$

arises when, at set number of objects, the luminosities shift in time; its form follows from $N(L,z) \, dL = N_o(L_o) \, dL_o \propto L_o^{-\gamma} \, dL_o$ and $L \propto L_o L_c(z)$. It may be termed a "luminosity anti-evolution" when the average luminosities decrease as z increase, the case here in point. In fact, the scaling of $L_c(z)$ may be divided into two steps:

a) If the ICP mass fraction $g(t)$ were constant, then after eqs. 1.1 and B.2 the scaling

$$L_c(z) \propto (1+z)^{(5+7n)/2(n+3)} \qquad (3.4)$$

would apply. For our pivotal values of the index $n \simeq -1$ (see Appendix B), the decrease of $M_c(z)T^{1/2}(z)$ tends to offset the increase of $\rho_c(z)$ so that after eq. 1.1 $L_c \propto (1+z)^{-1/2}$ decreases moderately (cf. Kaiser 1986). This factor raised to the power $\gamma - 1 \simeq 0.7$ affects but little the evolutionary behaviour of $N(z)$.

b) But the *history* of the ICP mass fraction g(t) may matter considerably, because as long as a condensation escapes merging into a higher level of the hierarchy, its luminosity increases $\propto g^2(t)$, and even when it merges it returns the accrued mass fraction of ICP to the new ambient, where the accruing may continue. We shall recall briefly the relevant information, referring to Sarazin (1986) for a comprehensive review and discussion.

The first point to note is that clusters must contain a *mass* in ICP comparable to the mass of luminous material in galaxies, to explain the X-rays observed and to insure reasonably continuous dispositions in the clusters' potential wells (cf. Cavaliere 1980).

As to the *origin*, not all ICP can be primordial since spectroscopy (cf. Holt and McCray 1982, Bleeker et al. 1984, Mushotzky 1984, Ulmer et al. 1987) indicates current abundances of Fe and other metals $\simeq 0.5$ the Solar value over a wide range of L, implying advanced astration of the *measurable* material. But here two main options open: infall, or ejection from galaxies.

Considering that several mixing processes (convection, galaxy motions, subcluster mergings and associated large scale hydrodynamics, reviewed by Sarazin 1986) tend to homogenize the composition, it may be argued that the measured abundances are representative of the bulk of the ICP: based on this and on conventional stellar mass loss rates, middle-of-the-way estimates have about half the ICP produced by the stars and half infallen from truly intergalactic medium, but the variance of the evaluations is considerable on either side. To begin with, the homogeneity argument is less convincing if infallen material actually dominates the ICP. On the infall side, primordial gas has to dominate if the efficiency for star formation out of barionic material was $< 50\%$. Another line of argument, following recent works by Arimoto and Yoshii (1987a and b) and Matteucci and Tornambè (1987) infers that astration could have produced a high Fe/H ratio but a low

H mass, from consideration of the full distribution of cluster galaxies including the dwarfs (F. Matteucci and Vettolani, private communication); if so, dilution by a considerable primordial component is required. On the other side, it is not granted that intergalactic medium is available with amount and energy appropriate for massive infall (see Gunn and Gott 1972, Cowie and Perrenod 1978). Furthermore, it is possible that up to 90% of barionic mass was condensed into stars before the protogalaxies really gathered into single entities: 10 % diffuse barions in protogalaxies still would provide enough dissipation for the generation of high density structure on a galactic scale at relatively late epochs (cf. Silk 1985, Baron and White 1987). Unfortunately to some extent, the *temperature* of the ICP is not a problem but neither constitutes a strong discriminant, because in either case the particle kinetic energy tends to match the potential one to within a factor ~ 1.

The *timing* of the ICP production or infall is more uncertain yet, but it is here that the statistics of X-ray clusters will provide useful constraints. Considering first stellar origin, models for $g(t)$ ought to include production of raw gas over stellar time scales as pioneered by Perrenod 1980. But recent ideas about star formation (see, e.g., Scalo and Struck-Marcell 1986) lead us to consider such production not only as a short aftermath of an initial burst of star formation, but also as a continuing process from a more even (on gross average) star formation rate.

Subsequent injection of such gas into the ICP was conceivably effected over rather short time scales by a number of physical mechanisms (see the review by Sarazin 1986): ablation both static (Gunn and Gott 1972) and dynamical (Gisler 1979, Takeda et al. 1984), thermal evaporation (Cowie and McKee 1977, Cowie and Songaila 1977), viscosity and turbulent mixing (Livio et al. 1980, Nulsen 1982). All these mechanisms have in common the existence of some threshold density for them to set in rather abruptly. Other processes like galactic winds (Yahil and Ostriker 1973) and galactic collisions (Norman and Silk 1979, Sarazin 1979), have been proposed to first form and then to strip out early gaseous galactic halos thus building up the first ICP: these would give slow overall ICP injection rates and may rather provide the threshold density that triggers the other, faster mechanisms. When (at $t = t_m$ say, typically during the groups era) the threshold conditions were met, each of the above mechanisms could provide a mass build up rate depending in detail on the physical state of the ICP (density, temperature) and on galactic parameters (cross sections, $< v^2 >$, internal density or rate of gas production); but all these scales were conceivably rather fast, ~ 1 Gyr, in cluster or group cores (cf. Sarazin 1986). Subsequently, the ICP mass fraction $g(t)$ kept growing larger throughout the clustering process if the gas production from stars kept going: cases like M86 and perhaps A 1367 (see Forman and Jones 1984) show directly that strippings still occur now.

On average, later forming, richer clusters had a larger initial fraction of ICP, less dense because the overall density decreased from step to step of the clustering hierarchy. The time condition $t \geq t_m$ translates into a condition involving characteristic masses (Appendix B) and so for $M < M_m$ these structures did not emit appreciably, but for $M \geq M_m$ they shone as bright X-ray sources and their X-ray luminosities increased in time following $L \propto g^2(t)$ as long as they survived as distinct entities.

A similar behaviour of $g(t)$ obtains also in a case of (differential) infall: e.g., more mass infalls as diffuse barions, because star formation was conceivably more efficient inside protogalaxies than outside (e.g., "failed" galaxies in biased galaxy formation schemes cf. Rees 1986, Dekel and Silk 1986). The analogy extends to the presence of an effective M_m

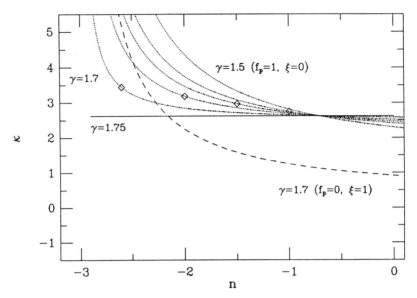

Fig. 2. Vs. the effective power index n it is plotted the exponent κ in the scaling law $N \propto (1+z)^\kappa$. Diamonds mark the asymptotic relationship $\gamma = (11-n)/8$.

if the diffuse barionic component to be accreted had random energies in the range 0.1 to 1 keV.

Summing up, we shall parametrize the uncertainties about the overall time scale of the ICP history representing the *time-integrated mass fraction* by the analytical form $g(t) \propto t^\xi$ ($\xi \geq 0$). For an initial burst of star formation followed by decaying ICP production $\xi < 1$ applies; for example, $\xi = \frac{2}{5}$ corresponded to thermal evaporation and $\xi = \frac{2}{3}$ to dynamical ablation in the evaluations made by Perrenod in 1980. For a more continuous star formation rate, constant on gross average, the injection rate can easily keep pace, and the limit $\xi \to 1$ will apply. In the infall case ξ is not so limited.

Thus, on time scales longer than t_m eq. 3.4 is to be completed as

$$L_c(z) \propto (1+z)^{(5+7n)/2(n+3)} * (1+z)^{-3\xi} . \qquad (3.5)$$

Hence $L_c(z)$ may either increase with z or more easily decrease, the latter condition prevailing when $\xi > (5+7n)/6(n+3)$, hence always for $n < -5/7$ since ξ is positive or zero. For example, when $n = 0$, $L_c \propto (1+z)^{5/6} * (1+z)^{-3\xi}$ decreases for $\xi > 5/18$; when $n = -1$, then $L_c \propto (1+z)^{-1/2} * (1+z)^{-3\xi}$ decreases for any ξ. The *luminosity* component, eq. 3.5 in eq. 3.3, combines with the *density* component of eq. 3.2 to yield an overall $N \propto (1+z)^\kappa$ with $\kappa = [7 + 5\gamma + 7n(\gamma - 1)]/2(n+3) - 3\xi(\gamma - 1)$, see fig. 2. Note that $H_o^{-1}(dN/Ndt)_o = -\kappa$, and for a heuristic estimate of the deviation of the slope from 1.5 compare such value for κ with $2(1+\alpha) \simeq 2.8$ ($\alpha \simeq 0.4$ for the bremsstrahlung spectrum) within the square bracket in eq. A.9.

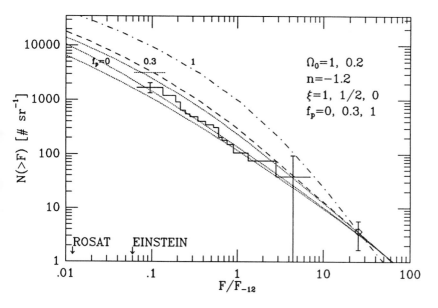

Fig. 3. Counts for various amounts of early ICP ($\Omega_o = 1$ unless otherwise specified). Dotted lines: $f_p = 0$, with in ascending order $\xi = 1$, $\Omega_o = 0.2$; $\xi = 1$; $\xi = 1/2$. Dashed line: $f_p = 0.3$, with $\xi = 0, 1$ for the early and the late component, respectively. Dot-dashed line: $f_p = 1$ ($\xi = 0$).

In general, we may envisage a two-component ICP – *early* and *late* – and introduce the compound mass fraction $G(t) = [f_p + (1 - f_p)g(t)]$ where f_p is the early component, i.e., that present in the condensations before the groups era. When $f_p \to 1$ (and hence $\xi = 0$), then κ ranges from $\simeq 2.7$ to 2.9 as n ranges from -1 to -2, having set $\gamma = 1.7$ (actually, we shall see that $\gamma \to (11-n)/8$): the indication is towards Euclidean (or slightly super-Euclidean) counts, in qualitative agreement with the numerical results illustrated in fig. 3. When instead $f_p \to 0$ applies, then an additional $\Delta\kappa = -3(\gamma - 1)\xi$ intervenes as an anti-evolutionary effect to flatten the counts sharply.

In these heuristic evaluations we focussed onto the case $\Omega_o = 1$ and neglected details like corrections due to the observational window or to the redshift of the spectrum, as well as effects of the detailed shape of the luminosity functions. But the heuristic argument may be refined to consider also the initially steepening effect of the spectral integration over the observational window, as well as the flattening effect of the spread of T with L to be discussed in next Sect. It can be shown that another, positive addition thus results, $\Delta'\kappa = (1 - \alpha)(7/4 - 3\xi/2)(\gamma - 1)$: for $f_p = 1$ ($\xi = 0$) a larger $\kappa \simeq 3.5$ obtains, pressing – as it were – for counts still steeper at the bright end. There is a limit, however, best understood by interpreting the (differential) counts as the envelope of the luminosity functions weighted at each z with the appropriate cosmological volumes, see eq. A.2 and fig. 8; as such, the counts can be only as steep as permitted by the slope (which varies with L near the cut off) of the dominant luminosity functions, namely those at large z

for $f_p \to 1$. The net result is that the actual counts-flux relationship tends to follow the shape of the dominant, distant luminosity functions, as can be seen from figs. 3 and 8.

Considering now the realistic two-components ICP, we may determine f_p from the existing data. At this point, however, one has to resort to numerical computations, that will also include all the complications like the observational energy window, the effects of the detailed shape of the luminosity functions and of the spread $T(L)$, and also FRW cosmologies with $\Omega_o \leq 1$.

4. COMPUTING $N(L,z)$

To go beyond the heuristic arguments, Cavaliere and Colafrancesco (Ap.J. in press) use the formalism of the continuity equation (Cavaliere, Morrison and Wood 1971)

$$\frac{\partial N}{\partial t} = S - \frac{\partial}{\partial L}(\dot{L}N) \quad < \text{ or } > \quad 0, \qquad (4.1)$$

that represents in full the apparent decrease or increase in cosmic time at given L, "evolution" or "anti-evolution", of $N(L,t)$; this is due to the object birth or death rate S [that changes the normalization] and/or to object dimming or brightening, $\partial(\dot{L}N)/\partial L$ with $\dot{L} < 0$ or > 0 [which by itself changes N after $N(L,z)dL = N_o(L_o)dL_o$].

Here the expression of $\dot{L}$ is to be derived from eq. 1.1. To recapitulate, under the HCS two opposite trends drive the luminosity change over cosmic time: a) an average dimming at each step of the clustering process, due to changes of $\rho(t_i) \propto \rho_u(t_i)$ of the surrounding universe when smaller systems are reshuffled into larger and more diffuse ones; a) a brightenig due to the increase (for $n \lesssim -1$, see B.1) of $M_c T_c^{1/2}$ and, in addition, to the effect $\propto g^2(t)$ of the continued increase of the ICP mass fraction (important unless early gas truly dominates).

The initial luminosity of a typical X-ray cluster at formation follows $L_i \propto G^2(t_i)\rho(t_i) MT_i^{1/2}$, whence $L_i \propto G^2(t_i)M^{4/3}\rho^{7/6}(t_i)$ obtains since in equilibrium $T(t) \propto M^{2/3}\rho^{1/3}(t_i)$ holds. Note that the latter behaviour implies both a considerable *spectral evolution* on average, $T_c = T_c(z)$ consistent with eq. B.2; in addition, at any given z a considerable *spread* $T \propto L^{1/2}$ is implied (see Cavaliere, Danese and De Zotti 1978 for a prediction of such a spread and Mushotzky, these Proceedings, for corresponding observations in the local clusters). Both effects are included in our computations.

Again under the HCS, the expression for S ought to describe production of new bound systems beyond the current upper mass cutoff at the expenses of many smaller ones. A model is given in Appendix B.

Thus the luminosity function $N(L,t)$ may be computed numerically from eq. 4.1 cast in the integrated form

$$N(L,t) = \int_{t_m}^{t} dt_i S(L_i,t_i) \frac{dL_i(t_i)}{dL} + N_m(L,t_m) \qquad (4.2)$$

where $L = L_i g^2(t_i)/g^2(t)$. The term $N_m(L,t_m) = N(M,t_m)(dM/dL)_m$ initializes when necessary ($f_p > 0$) the computation of the distribution of the observable luminosities $L \geq L_m$. The counts $N(>F)$ and the z-distributions are then derived using standard relationships recalled in Appendix A.

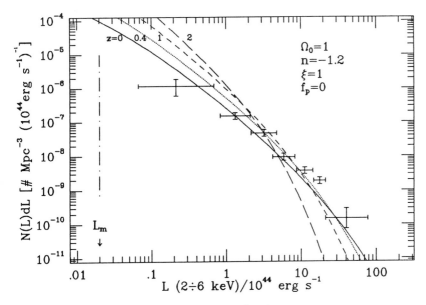

Fig. 4. The X-ray luminosity function $N(L,z)$ for clusters in the band $2 \div 6$ keV at four different redshifts $z = 0, 0.4, 1, 2$. Parameters of the evolutionary model are given on the right. The temperature corresponding to $L = 10^{45}$ erg/s is $T_{oc} = 9$ keV for the richest objects (corresponding to $T \sim 4$ keV for $R = 1$). ICP production is assumed to set in at $L_m \simeq 2\,10^{42}$ erg/s.

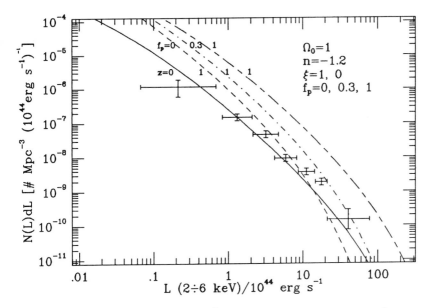

Fig. 5 $N(L, z = 0)$ and $N(L, z = 1)$ for various amounts of early ICP: $f_p = 0, 0.3$ ($\xi = 0, 1$ for the early and late components respectively), 1 ($\xi = 0$).

A conservative setting for the relevant parameters is given here and is discussed in next Sect. We pivot on the fiducial value $L_c(t_0) \sim 10^{45}$ erg/s for the cut off of the local luminosity function (cf. Henry et al. 1982, Kowalsky et al. 1984). The threshold for the ICP to begin its build up (when the early component $f_p \to 0$) is taken to be $L_m \lesssim 5 \; 10^{42}$ erg/s ($2 \div 6$ keV), the minimum luminosity observed for groups. If the total mass corresponding to $L_c(t_o)$ is taken $\sim 10^{15} M_\odot$ as suggested by the above Authors, then – sensibly – L_m corresponds to $\sim 10^{13} M_\odot$ and to an effective cut off at $z_m \simeq 2.5$ after Eqs. (3.1) and (3.2). As for the source function $S(L,t)$, values $n \simeq -1.2$ for the power spectral index of the linear perturbations in the mass range $10^{13} \div 10^{15} M_\odot$ are suggested by the cold dark matter model (Blumenthal et al. 1984) and inserted in eq. B.2. As for the increase of the ICP mass fraction, the exponent of $g(t) \propto t^\xi$ will be given two fiducial values, namely 1/2 (representative of prompt build up, e.g. an initial burst dynamically ablated) and 1 (continuous star formation rate with fast removal of the gas produced, or constant infall). But our crucial parameter will be the early component f_p, for which we shall discuss the extremes 0 and 1 and will assess the value indicated by the current data. A Hubble constant $H_0 = 50$ km/s Mpc is used throughout. FRW cosmologies with $\Omega_0 = 0.2$ and 1 will be considered.

Fig. 3 compares the counts expected for various values of f_p. Note: First, for any f_p the shape of the luminosity functions in the following figs. explains the relatively small curvature of the counts. Second, the counts are ultimately cut off when redshift and decreasing $T_c(z)$ shift the spectrum below the observational energy band. Third, for $f_p = 1$ the integration is cut off at $z_M = 3$; a proper account of the extensive cooling flows that in this case would set in at larger z when H^{-1} exceeds the cooling time (Fabian and Rees 1978), would steepen the counts even more.

Figs. 4 and 5 show in their rest frame the luminosity functions $N(L,z)$ in the band $2 \div 6$ keV corresponding to the counts in fig. 3. Consider $f_p \to 0$. For $L \gtrsim L_c(t)$ the *positive* part of the source function S is dominant, resulting in an anti-evolutionary effect on a time scale $\tau_S < H^{-1}$ (for very rich clusters, $N(L, z \simeq 0)$ cuts off at the upper end following the cut off of the local source function). In addition, the brightening tends to compensate for the average dimming from step to step of the hierarchy, thus increasing $L_c(t)$ on a time scale $\tau_{\dot{L}} < H^{-1}$. Overall, a *strong anti-evolutionary* effect (flat counts) is hence expected at the bright end. For $L < L_c(t)$, the luminosity function is affected by the *negative* part of the source function that tends to impose the self-similar shape $N(L) \to L^{-(11-n)/8}$ given by eq. B.2; the brightening $\dot{L} \propto L$ from eq. 1.1 of the survivor groups preserves the shape generated by the source function, while at any given L it compensates somewhat for the loss of objects. All these figs. are normalized to the same local luminosity function, normalized in turn to the data from the HEAO 1 survey (Johnson et al. 1983) and consistent with the upper limits by Kowalsky et al. 1984.

Fig. 6 gives the redshift distributions computed for the flux ranges typical of HEAO 1 and of HEAO 2, and extended into the depth that will be probed by ROSAT, cf. Trümper 1984, with the advantage of a lower energy range $0.1 \div 2$ keV. Note that for $f_p = 1$ large redshifts tend to dominate even for bright fluxes, a circumstance that we encounterd already in the heuristic discussion of the counts.

Fig. 7 shows what differences are expected for the evolution of $N(L,z)$ in a low density universe. Note that the effective cut off scale, cf. Appendix B, is somewhat larger in this case.

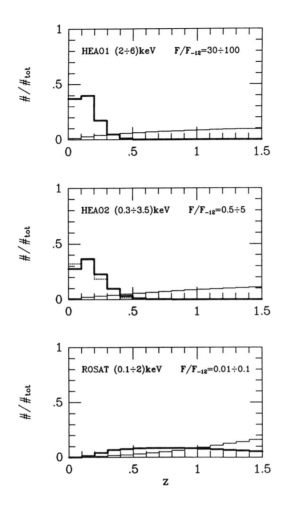

Fig. 6. Redshift distributions in the indicated flux ranges, normalized to their area. Thick lines correspond to the luminosity functions $N(L,z)$ for $f_p = 0$ ($\xi = 1$), illustrated in fig. 4. Thin lines correspond to those for $f_p = 1$ ($\xi = 0$) illustrated in fig. 5. Dotted lines in the middle panel mark the results from the MSS (Gioia et al. 1987), that excludes clusters with $D \leq 3$.

5. CONCLUSIONS AND DISCUSSION

To conclude, the shape of the counts constrains the timing for the ICP to build up. *If* counts with slope as *steep* as, or steeper than the "Euclidean" value will be found, then most of the ICP must have been within the condensations since the groups era: both the subsequent infall, or the production and injection from the stars must constitute minor additions.

If instead the prelimininary evidence of *flat* counts will be confirmed, a value $f_p \gtrsim 1/2$ for the early component will be ruled out; indeed, a value $f_p \lesssim 0.3$ is indicated even if the build up of the rest of the ICP is slow ($\xi = 1$). On the other hand, a value $f_p > 0.1$ is required in the HCS by dissipational collapse of galaxies, to provide for galaxies high contrasts and right amounts of rotation (cf. Fall and Efstathiou 1980). Thus a strong constraint is set to the ratio of the mass fraction in gas within groups since the epoch of their formation relative to that produced (or infallen) subsequently.

It is obviously tempting but still premature to relate the timing here discussed to the other indication that $z \sim 0.5$ may be an important epoch in cluster life and in fact in the ICP formation, i.e. the Butcher and Oemler (1978) effect in its conventional interpretation.

Low values of f_p required by flat counts are robust against varying other parameters away from the values set at the beginning of Sect. 4, since reasonable variations tend rather to *steepen* the counts. When the early component $f_p = 0$, values of $\xi < 1$ steepen the counts near their bright end and approach the behaviour for large f_p, see fig. 1, as they describe ICP effectively produced at early z. When $f_p = 1$, cutting at $z_M > 3$ would steepen the faint end because earlier, more numerous sources would be counted. A similar effect for similar reasons would be caused by rising z_m or lowering L_m, when $f_p \to 0$. Values $n < -1.2$ would smooth somewhat the bright end of the luminosity functions and that of the counts, but they would steepen more sharply the faint ends. In fact, in a cold dark matter scenario such lower values of n imply smaller masses, down to galactic values, consistent with lower L_m and higher z_m than here considered: this would reinforce the steepening at the faint ends as far as $T_c(z)$ permits full detection. Finally, a decrease of the cut off $L_c(t_o)$ below our generous 10^{45} erg/s would again steepen the bright end of the luminosity functions and hence that of the counts; in addition, at constant $T_c(t_o)$ more sources at high z would be retained inside the observational energy band and thus the steepening would be enhanced.

Little early ICP implies also the "receding" behaviour of the $N(L, z)$ seen in fig. 4. This behaviour would explain the evaluations of Schmidt and Green (1986) of an observed mean luminosity quite smaller than expected under the uniformity assumption.

In the case of ICP production from stars, the upper limit to f_p implies very efficient star formation from "cold" primordial barions (< 10 keV, see Ikeuchi and Ostriker 1986), and a slow rate of production or injection before the groups era. In the case of infall, the upper limit means continuing, differential infall conceivably because star formation favours the inside rather than the outside of the condensations to become groups or clusters. If differential infall were associated to barionic dissipation vs. collisionless dark matter, then cooling flows ought to be developing into the recent epochs (a similar outcome should be expected if the brightening in cosmic time were due to processes confined to the cores, the cluster volumes undoubtedly best sampled by the observations).

Telling safely apart infall from stellar production will require information concerning the strength of the Fe lines at substantial z. For example, the overall Fe/H ought to vary much more under dilution by constant infall of an initial burst of stellar production and injection, than in the case of dominant and continuous stellar production. This constitutes a program suited to AXAF and to XMM (cf. H. Tanabaum and H. Schnopper, these Proceedings).

A word is here in point about the classical goal of determining the value of Ω_o. The dependence of the luminosity functions on Ω_o is not strong, and because the data are still

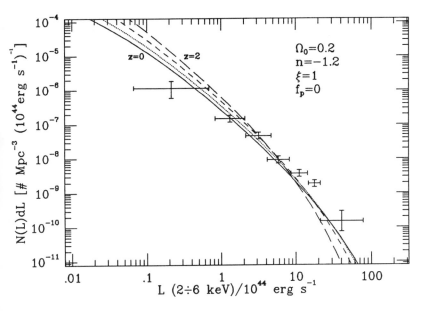

Fig. 7. The luminosity functions for $z = 0, 0.4, 1, 2$ in an $\Omega_o = 0.2$ FRW cosmology. Other parameters and data as in fig. 4.

nearly local consistency obtains both with $\Omega_o = 1$ and $\Omega_o = 0.2$. But a comparison of figs. 7 and 4 visualizes how deeper data from AXAF (Giacconi et al. 1980, H. Tananbaum these Proceedings) may test the issue.

Finally, a point over which we resonated with H. Tanabaum at this Meeting. It is already clear that the cluster counts will remain different and flatter than those of the AGNs, see fig. 1. By itself, this result already rules out an otherwise interesting possibility raised, e.g., by Segal and Nicoll 1986 and by Wampler 1987, that the evolution of the source populations might be only apparent and would only mirror general cosmological effects. The observations confirm instead that population evolutions differ intrinsically following the different astrophysics of the sources.

Aknowledgements. We thank E. Giallongo and F. Vagnetti for help in our numerical analysis, and G. De Zotti, I. Gioia., F. Lucchin, S. Matarrese, F. Matteucci and especially N. Vittorio for informative discussions and comments. Work performed under grants by MPI and CNR (GNA and PSN).

APPENDIX A: THE NUMBER COUNTS

The counts-flux relationship (cf. Weinberg 1972, Zeldovich and Novikov 1975) in terms of the comoving luminosity function reads

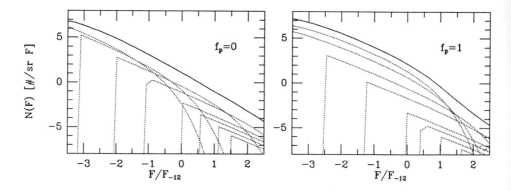

Fig. 8. Differential counts for $f_p = 0, 1$ as envelopes of the luminosity functions, displaced ($F \propto L/D^2*$ observational window) and volume-weighted after eq. A.2.

$$N(>F) = \int dL \int_{z_F(L)}^{0} dz \frac{dV(z)}{dz} N(L,z)(1+z)^3 = 4\pi c \int dL \int_{t(F)}^{t_o} dt \frac{r^2(t) R_o^3}{R(t)} N(L,t) \quad (A.1)$$

with $t(F)$ or equivalently $z(F)$ defined implicitly by the expression involving the luminosity distance: $D(z) = (1+z)^{\frac{1-\alpha}{2}} \sqrt{L/4\pi F}$ (α = spectral index, set to 1 for bolometric luminosities). The relationship $t(z)$ is provided by specific FRW cosmological models or, for nearly local values, by general series developments, cf. Weinberg (1972). Equivalently,

$$N(>F)/4\pi = 4\pi \frac{c}{H_o} \int_F^\infty dF \int_0^{z_F(L)} dz \frac{N[L(F,z),z] D_L^4(z)}{(1+z)^{4-\alpha}\sqrt{1+\Omega_o z}} \quad (A.2)$$

The inner integrand constitutes the z-distribution at given F. The inner integral gives the differential counts $N(F)$, represented in fig. 8.

On the other hand, in the limit of low redshifts $z \ll 1$ or short look-back times $\delta t \equiv t_o - t \ll t_o$, eq. A.1 may be developed into

$$N(>F) \propto \int dL \int_{t(F)}^{t_o} dt\, \delta t^2\, N(L, t_o) \times$$
$$\times \{1 - [1 + H_o^{-1}(-3H_o + \frac{\partial \ln N}{\partial t})_o] H_o\, \delta t\} \quad (A.3)$$

with the corresponding approximation for $t(F)$ (see Weinberg, 1972).

We replace $\partial N/\partial t$ with its expression from the continuity equation eq. 4.1 in the text

$$\frac{\partial N}{N \partial t} = \frac{S}{N} - \frac{\partial}{N \partial L}(\dot{L}N) =$$

$$\frac{1}{N}\frac{\partial N}{\partial t} = \frac{1}{\tau_S} + \frac{1}{\tau_L}(\gamma - 1) . \qquad (A.4)$$

where we have used for the *time scales* associated with the object generation (source term S) and with the luminosity shifts [term $\partial(\dot{L}N)/\partial L$] the defining expressions

$$\tau_S = \frac{N}{S}, \quad \tau_L = \frac{L}{\dot{L}}, \qquad (A.5)$$

taken L-independent for the sake of simplicity. Likewise we define the slope of the local luminosity function

$$\gamma = -\frac{\partial \ln N(L, t_o)}{\partial \ln L} . \qquad (A.6)$$

Note the appearence of the effective time-scale $\tau_L/(\gamma - 1)$ for the statistical effect of the luminosity shifts.

Considering also that $L_\nu \propto \nu^{-\alpha}$, eq. A.3 integrates to

$$N(> F) \propto F^{-3/2} \int dL\, N(L, t_o) L^{3/2} \times$$

$$\times \{1 - \frac{3}{4}[2(1+\alpha) + \frac{1}{H_o}(\pm\frac{1}{\tau_S} \pm \frac{\gamma - 1}{\tau_L})]\frac{H_o}{c}(\frac{L}{4\pi F})^{1/2}\} , \qquad (A.7)$$

where the double signs $-$ and $+$ correspond to evolutive (number decrease and/or dimming in cosmic t) and to anti-evolutive (number increase and/or brightening) behaviour, respectively.

Finally, the low-z approximation for the integral counts may be cast in the form

$$N(> F) \propto F^{-3/2}[1 - B(F_o/F)^{1/2} + O(F^{-1})] ; \qquad (A.8)$$

B marks the deviation from the "Euclidean" shape $N(> F) \propto F^{-3/2}$ (corresponding to a uniform population in a flat space with no cosmological effects) and we find it to read

$$B = \frac{3}{4}[2(1+\alpha) + \frac{1}{H_o}(\pm\frac{1}{\tau_S} \pm \frac{\gamma - 1}{\tau_L})]\frac{D_o}{R_H}\frac{\langle \ell^2 \rangle}{\langle \ell^{3/2} \rangle} . \qquad (A.9)$$

We have introduced the adimensional moments $\mu = 2, 3/2$ of the luminosity function

$$\langle \ell^\mu \rangle \equiv \langle (\frac{L}{L_o})^\mu \rangle = \int dL (L/L_o)^\mu N(L, t_o) \qquad (A.10)$$

and have used the Hubble radius $R_H = c/H_o$ and the distance $D_o = (L_o/4\pi F_o)^{1/2} \simeq 500$ Mpc ($L_o \sim 10^{45}$ erg/s, $F_o \sim 10^{-11}$ erg/cm^2 s).

The expression stresses the role of the time scales for the two basic kinds of evolution outside the local neighborhood of the comoving luminosity function $N(L,t)$: the scale τ_S for breakdown of object number conservation, and the effective scale $\tau_{\dot L}/(\gamma-1)$ for average shifts in luminosity. The uniform case corresponds to the limits $\tau_{S,\dot L} \to \infty$. The double signs $\mp$ correspond to an apparent decrease ("evolution") or increase ("anti-evolution") in cosmic time of $N(L,t)$ at *given* L, as indicated by eq. 4.1 in the text.

An important ingredient is also constituted by the shape of the local luminosity function $N(L,t_o)$. Eq. (2.2) is sensitive to its range and to its slope γ, which intervene both explicitly and in its moments $\langle \ell^2 \rangle$ and $\langle \ell^{3/2} \rangle$. The condition for a strong flattening of the counts soon outside the local neighborhood is $B > 0$ and substantial. Positive and large values of $H_o^{-1}[1/\tau_S + (\gamma-1)/\tau_{\dot L}]$, and large values of $\langle \ell^2 \rangle/\langle \ell^{3/2} \rangle$ corresponding to flat $N(L,t_o)$, enhance the flat behaviour.

APPENDIX B: THE SIMPLE HCS

We recall here the fundamentals of the hierachical clustering scenario (HCS, see Peebles 1974, 1980, White 1982) that proceeds bottom up from galaxies clustering into to groups, these into poor and eventually into rich clusters.

a. *Typical quantities*

Large scale structures form by gravitational instability from small density enhancements. When these emerge from recombination in the matter-dominated era are assumed to have random phases and a power spectrum of the form $|\delta_k|^2 \propto k^n$ where $k \sim 1/r$ in terms of the typical size, and $-3 < n \leq 1$ as we shall see. In a critical universe with density $\rho_u \propto t^{-2}$ the spectrum evolves in time simply as $\delta_k \propto t^{2/3}$.

In such conditions there is only one relevant scale, the mass of those fluctuations that are going non-linear at each epoch t, typically $M_c \propto \rho_c/k^3$. In terms of M_c, $\delta_c \propto M_c^{-a}$ and $M_c \propto t_c^{2/3a}$ follow, with $a \equiv (n+3)/6$.

The non-linear evolution of spherical perturbations starts with the detachment from the general Hubble flow, proceeds to a recollapse and eventually results into a virialized sphere with radius $\simeq 1/2$ the maximum radius and an internal, average density proportional to the density of the surrounding universe, $\rho_c \sim 170\,\rho_u$.

Summing up, for the average condensation gone non-linear at successive steps of the hierarchy – which in reality constitute a continuous development – the characteristic quantities ρ_c, M_c, the size r_c and T_c follow simple self-similarity scaling laws

$$\rho_c \propto \rho_u \propto (1+z)^3, \quad M_c \propto (1+z)^{-6/(n+3)}$$

$$r_c \propto (M_c/\rho_c)^{1/3} \propto (1+z)^{-(5+n)/(n+3)}, \quad T_c \propto \frac{GM}{r} \propto (1+z)^{(n-1)/(n+3)}. \quad (B.1)$$

Values of the effective index n in the range $-3 < n < 1$ preserve the sequential increase of the collapse times and that of the binding energies along the hierarchy. In the cold dark matter scenario the effective index n decreases from 1 to $\simeq -3$ with M decreasing: for top-hat shaped perturbations $n \simeq -1, -1.4, -1.7$ for $M = 10^{15}, 10^{14}, 10^{13} M_\odot$ respectively, and for bell-shaped ones slightly smaller values obtain (N. Vittorio, private communication).

b. The number change

The rates of object destruction and production $S(L,t)$ to be inserted in eq. 4.1 will quantify the reshuffling of smaller condensations into larger ones typical of the HCS. To describe the luminosity function even to the lowest order, one has to include the effects of the statistical distributions for the initial perturbation field and for the ensuing non-linear quantities around the typical values given by eq. B.1.

In the dynamical HC framework the multiplicity function of Press and Schechter (1974) is simple but representative of a Gaussian perturbation field. The corresponding explicit expression for the source function is

$$S(L,t) = \frac{dN(M,t)}{dt}\frac{dM}{dL} \propto N(M,t)\,[(\frac{M}{M_c})^{2a} - 1]\,\frac{\dot{M}_c}{M_c}\frac{dM}{dL} ,$$

$$N(M,t) \propto \frac{1}{M_c^2}(\frac{M}{M_c})^{-2+a}\,exp[-\frac{1}{2}(\frac{M}{M_c})^{2a}] . \tag{B.2}$$

This has negative (destruction) and positive (production) values separated by $M = M_c$, the typical mass virializing at t, with general time behaviour (Cavaliere, Danese and De Zotti 1978, White and Rees 1978)

$$\frac{M_c(t)}{M_c(t_o)} = [\frac{(1+A)}{(t_o/t)^{2/3} + A}]^{1/a}, \quad A = (1-\Omega_o)(\frac{2H_o t_o}{3\pi\Omega_o})^{2/3} . \tag{B.3}$$

In the positive range $M > M_c$ the time scale for formation of new structures is $\tau_S \equiv N/S \simeq M_c/\dot{M}_c$, evaluating the maximum value of S at $M \simeq 2M_c(t)$ for $n = -1.2$. The cut-off scale $M_c^a(t)$ in eq. B.2, corresponds to the variance δ_c of the assumed Gaussian distribution of the initial perturbation field (Press and Schechter 1974).

In deducing the luminosity dependence of the source S, use is made of the the relationship $M \propto L^{3/4}\rho^{-7/8}(t_i)$ given implicitly by eq. 1.1. Note that the small M behaviour of the multiplicity function $N(M) \propto M^{-3/2+n/6}$ corresponds to the behaviour $S(L) \propto L^{-(11-n)/8}$ which generates a similar behaviour for $N(L)$ at its faint end.

The Press and Schechter rendition of the HCS suffers of a number of shortcomings, notably: the mass reshuffled at each step into larger condensations is too literaly constant; biasing phenomena were not originally included (but see Silk and Schaeffer 1985); the index n is taken independent of M while in the cold dark matter scenario the fluctuation spectrum has a curvature (but see Occhionero and Scaramella, in preparation; Colafrancesco, Lucchin and Matarrese, in preparation). These points and the ensuing modifications (minor in the present context) will be discussed elsewhere.

REFERENCES

Abell, G.O. 1958, *Ap. J. Suppl.*, **3**, 211.
Abramopoulos, F., and Ku, W.H.M. 1983, *Ap. J.*, **271**, 446.
Arimoto, N., and Yoshii, Y. 1987 a, *Astr. Ap.*, **173**, 23.
Arimoto, N., and Yoshii, Y. 1987 b, *preprint*.
Bahcall, J.N., and Bahcall, N.A. 1975, *Ap. J. Lett.*, **199**, L89.
Baron, E., and White, S.D.M. 1987, *preprint*.
Biermann, P., Kronberg, P.P., Madore, B.F. 1982, *Ap. J. Lett.* **256**, L37.
Bleeker, J., et al. 1984, *Physica Scripta*, **T7**, 224.
Blumenthal, G.R., Faber, S.M., Primack, J.R., and Rees, M.J. 1984, *Nature*, **311**, 517.
Butcher, H., and Oemler, A. 1978, *Ap. J.*, **219**, 18.
Cavaliere, A., Santangelo, P., Tarquini, G., and Vittorio, N. 1986, *Ap. J.*, **305**, 651.
Cavaliere, A., Morrison, P., Wood, K. 1971, *Ap. J.*, **170**, 223.
Cavaliere, A., Danese, L., and De Zotti, G. 1978, *Ap. J.*, **221**, 399.
Cavaliere, A. 1980 in *X-ray Astronomy*, R. Giacconi and G. Setti eds., (Dordrecht, Reidel).
Cavaliere, A., Gursky, H., Tucker, W.H. 1971, *Nature*, **231**, 437.
Cowie, L.L., and McKee, C.F. 1977, *Ap. J.*, **211**, 135.
Cowie, L.L., and Songaila, A. 1977, *Ap. J.*, **211**, 501.
Dekel, A. 1987, in Proc. IAU Symposium 124 *Observational Cosmology*, in press.
Dekel, A., and Silk, J. 1986, *Ap. J.*, **303**, 39.
Fabian, A.C., and Rees, M.J. 1978, *M.N.R.A.S.*, **185**, 109.
Fall, S.M., and Efstathiou, G. 1980, *M.N.R.A.S.* **193**, 189.
Forman, W., and Jones, C. 1978, *Ap. J.*, **224**, 1.
Forman, W., and Jones, C. 1982, *Ann. Rev. Astr. Ap.*, **20**, 547.
Giacconi, R., et al. 1974, *Ap. J. Suppl.*, **27**, 37.
Giacconi, R., et al. 1980, *The Advanced X-ray Astrophysics Facility - Science Working Group Report*, NASA Report No. **TM-78285**.
Gioia, I.M., Maccacaro, T., Schild, R.E., Stocke, J.T., Liebert, J.W., Danziger, I.J., Kunth, D., and Lub, J. 1984, *Ap. J.*, **283**, 495.
Gioia, I.M., Maccacaro, T. and Wolter, A. 1987, in Proc. IAU Symposium 124 *Observational Cosmology*, in press.
Gioia, I.M., Maccacaro, T., Morris, S.L., Schild, R.E., Stocke, J.T., and Wolter, A. 1987, *preprint*.
Gisler, G. 1979, *Ap. J.*, **228**, 385.
Gott, J.R., and Turner, E.L. 1977, *Ap. J.*, **216**, 357.
Gunn, J.E., and Gott, J.R. 1972, *Ap. J.* **176**, 1.
Gursky, H., Solinger, A., Kellog, E., Murray, S., Tananbaum, H., Giacconi, R., and Cavaliere, A. 1972, *Ap. J. Lett.*, **173**, L99.
Henry, J.P., Soltan, A., Briel, U., Gunn, J.A. 1982, *Ap. J.*, **262**, 1.
Holt, S., and McCray, R., 1982, *Ann. Rev. Astr. Ap.*, **20**, 323.
Ikeuchi, S., and Ostriker, J.P. 1986, *Ap. J.* **301**, 522.
Johnson, M.W., Cruddace, R.G., Ulmer, M.P., Kowalsky, M.P., and Wood, K.S. 1983, *Ap. J.*, **266**, 425.
Jones, C., and Forman, W. 1984, *Ap. J.* **276**, 38.
Kaiser, N. 1986, *M.N.R.A.S.* **222**, 323.

Kowalsky, M.P., Ulmer, M.P., Cruddace, R.G., and Wood, K.S. 1984, *Ap. J. Suppl.*, **56**, 403.
Kriss, G.A., Cioffi, D.F., and Canizares, C.R. 1983, *Ap. J.* **272**, 439.
Livio, M., Regev, O., and Shaviv, G. 1980, *Ap. J. Lett.* **240**, L83.
Maccacaro, T., et al. 1982, *Ap. J.*, **253**, 504.
Markert, T.P., et al. 1979, *Ap. J. Suppl.* **39**, 573.
Matteucci, F., and Tornambè, A. 1987, *preprint*.
McHardy, I.M., Lawrence, A., Pye, J.P., and Pounds, K.A. 1981, *M.N.R.A.S.* **197**, 893.
Mushotzky, R.F. 1984, *Physica Scripta*, **T7**, 157.
Norman, C., and Silk, J. 1979, *Ap. J. Lett.*, **233**, L1.
Nulsen, P.E. 1982, *M.N.R.A.S.* **198**, 1007.
Peebles, P.J. 1974, *Ap. J. Lett.* **189**, L51.
Peebles, P.J. 1980, *The Large Scale Structure of the Universe* (Princeton Univ. Press).
Perrenod, S.C. 1980, *Ap. J.*, **236**, 373.
Piccinotti, G., Mushotzky, R.F., Boldt, E.A., Holt, S.S., Marshall, F.E., Serlemitsos, P.J., and Shafer, R.A. 1982, *Ap. J.*, **253**, 485.
Press, W.H., and Schechter, P. 1974, *Ap. J.*, **187**, 425.
Rees, M.J. 1986, *M.N..R.A.S.*, **222**. 27p.
Sarazin, C.L. 1979, *Ap. Lett.* **20**, 93.
Sarazin, C.L. 1986, *Rev. Mod. Phys.*, **58**, 1.
Scalo, J.M., and Struck-Marcell, C. 1986, *Ap. J.*, **301**, 77.
Schmidt, M., and Green, R.F. 1986, *Ap. J.*, **305**, 68.
Schwartz, D.A., et al. 1980, *Ap. J. Lett.* **238**, L53.
Segal, I.E., and Nicoll, J.F. 1986, *Ap. J.*, **300**, 224.
Silk, J. 1985, *Ap. J.* **297**, 1.
Silk, J., and Schaeffer, R. 1985, *Ap. J.* **292**, 319.
Solinger, A.B., and Tucker, W.H. 1972, *Ap. J. Lett.*, **175**, L107.
Soltan, A., and Henry, J.P. 1983, *Ap. J.*, **271**, 442.
Takeda, H., Nulsen, P., and Fabian, A. 1984, *M.N.R.A.S.* **208**, 461.
Trümper, J. 1984, *Physica Scripta*, **T7**, 209.
Ulmer, M.P., et al. 1981, *Ap. J.* **243**, 681.
Ulmer, M.P., Cruddace, R.G., Fenimore, E.E., Fritz, G.G., and Snyder, W.A. 1987, *preprint*.
Wampler, E.J. 1987, *Astr. Ap.*, **178**, 1.
Weinberg, S. 1972, *Gravitation and Cosmology* (New York, Wiley).
White, S.D.M. 1982, in *Morphology and Dynamics of Galaxies* ed. by Martinet, L. and Mayor, M. (Geneva Observatory, Geneva), p.289.
White, S.D.M., and Rees, M.J. 1987, *M.N.R.A.S.*, **183**, 341.
Yahil, A., and Ostriker, J.P. 1973, *Ap. J.* **185**, 787.
Zel'dovich, Ya.B., and Novikov, I.D., 1975, *Structure and Evolution of the Universe*, (Nauka, Moscow).

EXOSAT OBSERVATIONS OF THE VIRGO CLUSTER

A. C. Edge and G. C. Stewart,
X-Ray Astronomy Group,
University of Leicester,
Leicester LE1 7RH, England.

A. Smith,
Space Science Department of ESA, ESTEC,
Postbus 299, 2200 AG Noordwijk,
The Netherlands.

ABSTRACT

Results are presented for observations of the Virgo cluster made using the EXOSAT Medium Energy Experiment. Limits are obtained on the temperature profile of the gas within 100 arcmin. These results provide an improved estimate for the mass surrounding M87.

1. INTRODUCTION

The Virgo cluster was one of the first extragalactic X-ray sources to be identified. Early UHURU (Kellogg et al., 1975) and OSO-8 (Mushotzky et al., 1978) observations indicated a two component spectrum comprising a thermal component with T of 2 to 3 keV and a "hard tail". Later ARIEL-V results suggested that these two components were not coincident (Davison, 1978; Lawerence, 1978). HEAO-1 observations (Lea et al., 1981) also showed marginal evidence of variation in intensity of the hard component over a period of six months. Very little information on the spatial variation of temperature or density could be obtained from these large field of view proportional counters. The IPC experiment on the EINSTEIN observatory was used to measure the surface brightness in soft X-rays with a resolution of a few arcmin (Fabricant, Lecar and Gorenstein, 1980, hereafter FLG; Fabricant and Gorenstein, 1983, hereafter FG). FG were able to determine the surface brightness out to 100 arcmin and the temperature out to 15 arcmin. From these results they were able to estimate the mass around M87 assuming hydrostatic equilibrium.

The EXOSAT ME experiment was an Argon/Xenon filled proportional counter with a relatively small field of view of 45 arcmin FWHM. Thus for very extended objects, such as clusters, it could be used to give coarse spatial resolution if several pointing positions were used. This method was used in EXOSAT observations of Virgo (reported here), Coma (Hughes, 1987) and Perseus (Branduardi-Raymont et al., 1985).

2. OBSERVATIONS

The observations were performed in four stages using the ME experiment. Two single pointings centred on M87 were made on day 199 of 1983 and day 360 of 1984, and two stepped scans across the cluster were made on days 194 to 196 of 1983 and days 147 to 150 of 1984. The pointing positions for the two scans are shown in Figure 1.

3. ANALYSIS

The data were analysed using standard techniques taking the background from pointings well away from M87. Data obtained during periods of solar contamination were discarded. The resulting spectra were fitted with a thermal bremsstrahlung model with an iron line at 6.67 keV (this energy was determined from the best S/N spectrum) and a fixed HI column of 2.5×10^{20} cm^2 (Heiles, 1975). Results for the flux are shown in Figure 2 and are compared with the expected flux from the IPC surface brightness. The results for temperature are shown in Figure 3 plotted against the count weighted radial distance. This count weighted distance was calculated from the IPC surface brightness integrated over the ME detector field of view for each pointing position.

A least squares fit for a line through the temperature data points gives the best fit logarithmic temperature gradient of -0.005 ± 0.035 with a temperature a 1 arcmin of 2.46 ∓ 0.20 keV. The temperature values determined are consistent with those derived from subsets of this data previously analysed by Smith and Stewart (1985) and Edge, Stewart and Smith (1986) and also with the temperatures derived from previous wide FOV. experiments. The temperature measurements from the IPC are quoted in FLG as 2.5 ± 0.3 at 10 arcmin, but after recalibration of the IPC the temperature is quoted in FG as 4.0 ± 1.0. This latter value is inconsistent with the EXOSAT results.

The results for the iron line equivalent width are shown in Figure 4a and show no strong variation with radius. The results within 40 arcmin. gave an average value of 1.0 ± 0.1 keV. The inner 7 pointings the spectra were also fitted with Raymond and Smith spectra (1977) and the results are shown in figure 4b. The derived iron abundance from these spectra was 0.48 ± 0.12 relative to solar overall and was constant out to 40 arcmin.

The poor sensitivity of EXOSAT above 15 keV prevented any investigation of the "hard tail" reported in previous observations.

4. GRAVITATIONAL MASS ESTIMATE

For gas in hydrostatic equilibrium the gravitating mass, within a radius R, required to bind the gas is given by:

$$M_{(<R)} = -\frac{k}{G\mu m_p}T(R)\left(\frac{d\log\rho}{d\log r}\bigg|_{r=R} + \frac{d\log T}{d\log r}\bigg|_{r=R}\right) \quad (1)$$

For the gas around M87 we assume that:

$$\rho(r) = \rho_0\left(1+\left(\frac{r}{a}\right)^2\right)^{-\frac{\Delta_1}{2}} \quad \text{and} \quad T(r) = T_1 r^{\Delta_2}$$

where ρ_0 is the gas density at $r = 0$ and T_1 is the temperature at 1 arcmin. These relationships simpify equation (1) to:

$$M_{(<R)} = \frac{k}{G\mu M_p}T_1 R^{\Delta_2+1}(\Delta_1 - \Delta_2) \quad \text{for} \quad R \gg a \quad (2)$$

where Δ_1 and Δ_2 are the logarithmic gradients of density and temperature. These can be calculated from the surface brightness and temperature profile respectively.

Taking $\Delta_1 = 1.3 \pm 0.15$ from FG and $\Delta_2 = -0.005 \pm 0.035$ and $T_1 = 2.46 \mp 0.20$ keV from this study, upper and lower mass estimates can be calculated from equation (2) for radii between 10 and 100 arcmin. The limits for the mass calculated using the above limits are shown in Figure 5 with those from FG, optical data from Sargent et al. (1978) and Mould et al. (1987) and a mass profile from Stewart et al. (1984).

The mass limits are consistent with those in FG although their estimates are poorly constrained by the lack of good temperature measurements. The models of FLG in which $M_{(<r)} \propto r$ and Stewart et al., $M_{(<r)} \propto r - \arctan r$ are both consistent with the measured temperature profile. However the FLG model is inconsistent with the FPCS line measurements and the optically determined mass at small radii.

The results presented here can preclude two component King models such as that of Binney and Cowie (1981), who predict a steep temperature gradient beyond 10 arcmin.

The above constraints on the gravitating mass can be used to derive the variation of the mass to light ratio and the distribution of 'Dark Matter' within the cluster. The mass of $5 \pm 1 \times 10^{13} M_\odot$ within 100 arcmin (or 440 kpc) of the centre of M87 shows that M/L_B ratio from ≈ 10 within M87 to ≈ 500 at 440 kpc taking the visible luminosity from de Vaucouleurs and Nieto (1978).

5. CONCLUSIONS

From these EXOSAT observations it is clear that the gas surrounding M87 is isothermal between 10 and 100 arcmin and there is no strong gradient in iron abundance. Taking limits on any temperature gradient and results from EINSTEIN for the density profile we calculate a gravitating mass of $5 \pm 1 \times 10^{13} M_\odot$ with 440 kpc

ACKNOWLEDGEMENTS

AS and GCS would like to thank the EXOSAT observatory staff for help with the observations.

REFERENCES

Binney, J. and Cowie, L. L. 1981, *Ap J.*, **247**, 464.
Branduardi-Raymont, G., Kellet, B., Fabian, A. C., McGlynn, T., Manzo, G.
    and Peacock, A. 1985, *A. Adv. Space Res.*, Vol **5**, No. **3**, P. 129.
Canizares, C. R., Clark, G. W., Jernigan, J. G., and Market, T. H. 1982,
    Ap. J. (Letters), **262**, L33.
Davison, P. J. N. 1978, *M.N.R.A.S.*, **183**, 39P.
de Vaucouleurs, G. and Nieto, J.-L. 1978, *Ap. J.*, **220**, 449.
Edge, A. C., Stewart, G. C. and Smith, A. 1986, in *NRAO Greenbank Workshop #16, Radio Continuum Processes in Clusters of Galaxies*,
    ed. C. P. O'Dea and J. M. Uson, P. 105.
Fabricant, D. and Gorenstein, P. 1983, *Ap. J.*, **267**, 535.
Fabricant, D., Lecar, M. and Gorenstein, P. 1980, *Ap. J.*, **241**, 552.
Heiles, C. 1975, *A. & A. Suppl.*, **20**, 37.
Hughes, J. 1987, Preprint.
Kellogg, E., Baldwin, J. R. and Koch, D. 1975, *Ap. J.*, **191**, 299.
Lawerence, A. 1978, *M.N.R.A.S.*, **185**, 423.
Lea, S. M., Mushotzky, R. F. and Holt, S. S. 1982, *Ap. J.*, **262**, 24.
Lea, S. M., Reichart, G., Mushotzky, R. F., Baity, W. A., Gruber, D. A.,
    Rothschild, R. E. and Primini, F. A. 1981, *Ap. J.*, **246**, 369.
Mould, J. R., Oke, J. B. and Nemec, J. M. 1987, *A. J.*, **92**, 53.
Sargent, W. L. W., Young, P. J., Boksenberg, A., Shortridge, K., Lynds, C. R.,
    and Hartwick, F. D. A. 1978, *Ap. J.*, **221**, 731.

Schreier, E. J., Gorenstein, P. and Feigelson, E. D. 1982, *Ap. J.*, **261**, 661.
Smith, A. and Stewart, G. C. 1985, *Sp. Sci. Rev.*, **40**, 661.
Stewart, G. C., Canizares, C. R., Fabian, A. C. and Nulsen, P. E. J. 1984, *Ap. J.*, **278**, 536.

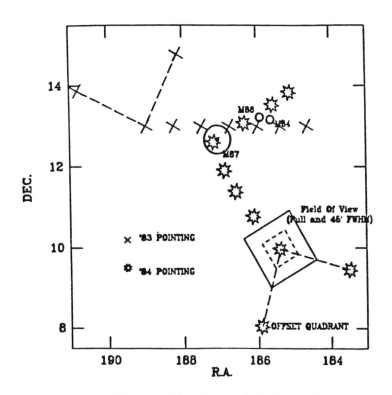

Figure 1. Schematic diagram of pointing positions.

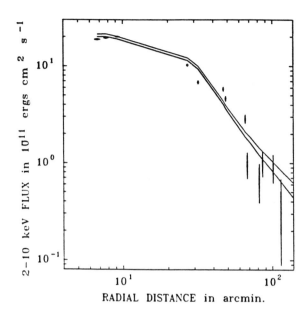

Figure 2. Plot of 2-10 kev flux vs. radial distance from M87. The points are the measured ME fluxes and the solid lines are the expected flux for the ME calculated from the IPC surface brightness quoted in FG including their errors in the normalisation.

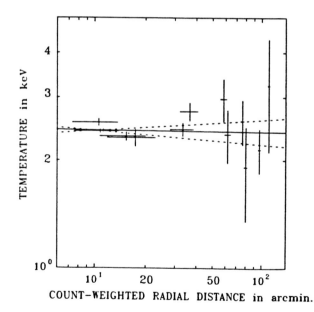

Figure 3. Plot of temperature vs. count weighted radial distance. The errors are 90% confidence. The solid line shows the best fit gradient of -0.005 and the dashed lines are the 90% confidence limits.

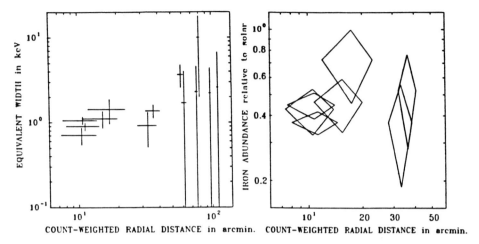

Figure 4.a and b Plots of iron line equivalent width (left) and iron abundance (right) vs. count weighted radial distance. The errors in both are 90% confidence.

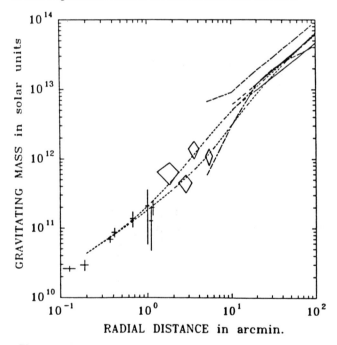

Figure 5. Plot of calculated gravitating mass vs. radial distance. The crosses and diamonds are optical measurements from Sargent et al., (1978) and Mould et al., (1987) respectively. The wide dashed lines are the limits from FG and the solid lines are the EXOSAT limits. The fine dashed lines are two-component mass models from Stewart et al., (1984) which agree with the FPCS line ratios. The two models are for core radii of 10 kpc (upper) and 40 kpc (lower).

8. FUTURE X-RAY MISSIONS

First Results from Ginga

Y. Tanaka
Institute of Space and Astronautical Science
4-6-1 Komaba, Meguro-ku, Tokyo 153
Japan

ABSTRACT. Ginga (Astro-C), the X-ray astronomy observatory of Japan, was launched on February 5, 1987. The main features and capabilities of Ginga are described. Some of the results of the observations of galactic and extragalactic sources obtained so far are presented.

1. X-RAY ASTRONOMY OBSERVATORY "GINGA"

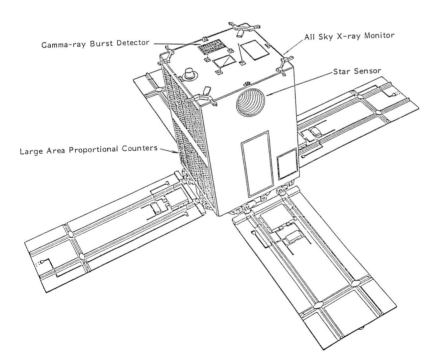

Fig. 1. X-ray astronomy observatory "Ginga"

The Japan's third X-ray astronomy satellite Astro-C was successfully launched on February 5, 1987, and was renamed "Ginga", meaning galaxy (Fig. 1). Ginga is a three-axis stabilized, orbiting X-ray astronnmy observatory. Ginga carries three experiments on board. Capabilities of these experiments are summarized in Table I. The main X-ray instrument of Ginga is a large-area proportional counter array (LAC) with a 4000 cm^2 total effective area which was prepared by the collaborative Japan-U.K. team consisting of the groups listed in Table II. The other two experiments are a gamma-ray burst detector (GBD) which is a collaborative experiment with the Los Alamos National Laboratory group, U.S.A., and an all sky monitor (ASM). Detailed description of Ginga (Astro-C) and the instruments was published elsewhere (Makino 1987).

TABLE I **Experiments on board Ginga**

---

* **Large-area proportional counters (LAC)**

Effective area:	Total	4000 cm^2 (eight counters)
Field of view:	Individual	0.8° x 1.7° (FWHM)
	Total	1.08° x 2.0° (FWHM)
Energy range:	1.5 - 30 keV (>10 % efficiency)	
Energy resolution:	18 % at 5.9 keV (FWHM)	
Background rate:	Total	40 counts/s in 2 - 10 keV
Time resolution:	0.98 ms - 2 s, depending on the mode	

* **All sky monitor (ASM)**

Effective area:	Total 400 cm^2 (six proportional counters)
Field of view:	1° x 45° (FWHM)
	six different slant angles
Energy range:	1.5 - 30 keV
Sensitivity:	< 40 mCrab in one scan at 360°/20 min.
Position error:	< 1° x 1°

* **Gamma-ray burst detector (GBD)**

Detectors:	A proportional counter (63 cm^2)
	A NaI(Tl) scintillation counter (65 cm^2)
Field of view:	Hemisphere (no collimation)
Energy range:	1.5 - 480 keV
Time resolution:	31.25 ms
	(shortest 0.24 ms in time-to-spill mode)

Ginga possesses a total effective area of 4000 cm^2, one of the largest ever flown, with good background rejection capability. This enables us to obtain energy spectra of much fainter sources over a wide range of 2 - 30 kev and to measure time variabilities with a much higher sensitivity than previously achievable.

TABLE II Ginga LAC Team

Japanese Groups	U.K. Groups
* Institute of Space and Astronautical Science (ISAS) * University of Tokyo * Institute of Physical and Chemical Research * Rikkyo University * Nagoya University * Osaka University * Osaka City University	* University of Leicester * Rutherford/Appleton Laboratory

Since the launch, Ginga has been operating normally. During the first seven months, we conducted various observations mainly for the purpose of performance verification and calibration of the experiments. Among other sources, the Crab Nebula was used to calibrate the collimator response function and to verify detection efficiency as a function of energy of the LAC. A weak silver K-line (22.1 keV) generated in the silver coating of the collimator serves for an accurate calibration in flight. Besides, a considerable fraction of time was spent for the detailed examination of the background of the counters. The Crab Nebula is often utilized as a convenient X-ray flux standard. The measured count rate of the Crab Nebula with LAC is 10,500 counts/sec in the range 1 - 10 keV, hence 10 counts/sec in this energy range approximately corresponds to a flux of 1 mCrab.

The All Sky Monitor experimant surveys a large part of the sky typically once every day by rotating the spacecraft through 360°. In Fig. 2, an example of the All Sky Monitor record is shown.

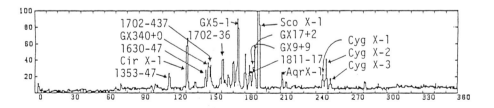

Fig. 2. Data of one of six counters of All Sky Monitor obtained from a single scan along the galactic plane.

Since October, Ginga has been in routine operation for observations of the targets from selected proposals. The Institute of Space and Astronautical Science (ISAS) invited U.S. scientists through NASA and European scientists through ESA to participate in Ginga observations in collaboration with Japanese scientists.

2. RESULTS

Among the results so far obtained, some of those which are relevant to the topic of "Hot Plasmas in Astrophysics" will be presented below. These results are, however, still considered to be preliminary.

2.1. Galactic sources

Various compact X-ray binaries involving neutron stars or possibly black holes for some cases, supernova remnants and stellar sources have been observed. We shall present here some results on supernova remnants and stellar sources.

Several faint supernova remnants have been observed in the course of a systematic search for asociated pulsars. New pulsar detection has so far been negative. However, wide-band energy spectra of these supernova remnants were obtained and the iron line intensities were determined. For example, Fig. 3 shows the spectrum of Kepler's supernova remnant. The energy of the iron emission line is determined to be 6.5 keV, excluding 6.7 keV with 90 % confidence. This indicates a non ionization equilibrium. The equivalent width of the line is found to be approximately 2.6 keV. These facts imply a significant overabundance of iron compared with the cosmic abundance.

Stellar sources include cataclysmic variables, RS CVn stars, Wolf-Rayet stars, star-forming regions, etc.. Figure 4 shows the spectrum of a Wolf-Rayet star HD 193793. Presence of the iron emission line at 6.7 keV is evident. This spectrum is not explicable in terms of a single-temperature thermal bremsstrahlung spectrum, but rather a power-law spectrum with a photon-number index of about 2.2 gives a better fit. This might suggest the presence of multiple components of different temperatures, or a hard tail due to the effect of up-Comptonization in a neighboring thin hotter plasma.

An RS CVn system UX Ari exhibited a flare activity during our observation. Of particular interest is the change in the intensity of the iron emission line. Figure 5 shows the energy spectra at two intensity levels which are different by a factor of 2. The spectrum for higher intensity level (Fig. 5a) shows kT of 6.1 keV with the iron-line equivalent width of about 0.6 keV, while the one at lower intensity (Fig. 5b) shows kT of 4.4 keV without a significant iron line present. The apparent disappearance of the iron line is not expected from this much drop of temperature. Some subtle effect may be present.

---

* The energy spectra shown in this paper are all pulse-height spectra as observed. No correction has been made for the energy-dependent detection efficiency, absorption in the entrance window and for energy resolution of the detector.

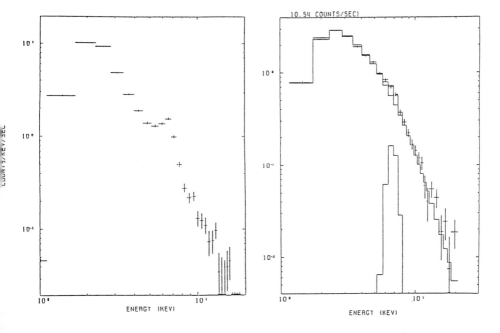

Fig. 3. Spectrum of Kepler's SNR   Fig. 4. Spectrum of HD193793

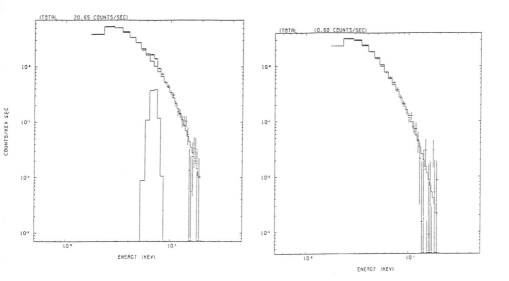

Fig. 5. Spectra of UX Ari at (a) high- and (b) low-intensity levels.

Several star-forming regions have been observed. The Orion Trapezium region (Agrawal et al. 1986) and the Rho Ophiuchi molecular cloud (Koyama et al. 1987) were observed from Tenma. Observed spectra of these regions are shown in Fig. 6. Another molecular cloud L 1457 was observed from Ginga and its spectrum is shown in Fig. 7. All of them commonly exhibit thermal spectra with kT of several keV and characteristic iron line at about 6.7 keV. The origin of such a high temperature plasma in star-forming regions is still unknown. X-ray spectra of single early-type stars are generally of much lower temperature.

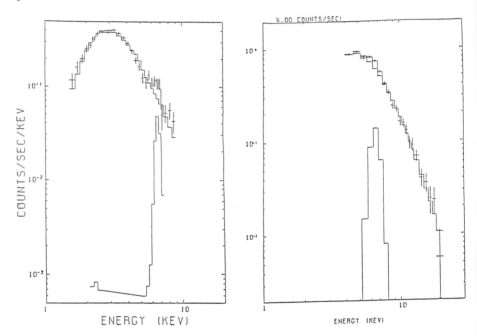

Fig. 6. Spectra of Rho Ophiuchi molecular cloud

Fig. 7. Spectrum of L 1457

The intensity distribution of unresolved X-ray emission extended along the galactic ridge was studied in detail from HEAO-1 observations by Worrall et al. (1982). Tenma observations at several source-free regions along the galactic ridge clearly revealed that this emission is of thin thermal origin with varying kT from position to position in the range 3 - 8 keV and characterized by an intense iron emission line (Koyama et al. 1986 a).

The origin of this emission is not known as yet. This emission cannot come from diffuse plasma of galactic scale, because such a high-temperature plasma is not gravitaionally bound by the galactic plane. A hypothesis that this emission comes from a number of yet unidentified supernova remnants (Koyama et al. 1986b) requires an order of magnitude higher supernova rate than currently accepted. Another possibility is to attribute to emissions from cataclysmic variables and RS CVn type stars both of which are numerous in our galaxy. However, the average intensity and spectrum of these sources are not yet well known. Future Ginga observations will hopefully enables us to obtain this information and to clarify the origin of the emission along the galactic ridge.

2.2. Extragalactic sources

Major emphasis has been put on the observations of extragalactic sources, which include normal galaxies, cluster of galaxies, and active galactic nuclei.

For "normal" galaxies, we observed M31, M49 and M82 so far. The M31 spectrum is explained qualitatively as being mostly a superposition of emission from a number of low-mass type binary sources. The Einstein observation resolved the individual sources in M31 (van Speybroeck et al. 1979). The elliptical galaxy M49 shows a power-law tail with little evidence of the iron line, a 90% upper limit of the equivalent width being 0.2 keV. The apparent lack of the iron line raises an interesting question regarding the origin of X-rays from M49. M82, a star-burst galaxy, shows a thin thermal spectrum of kT = 7.2 ± 0.2 keV with a significant iron emission line of a 0.2 keV equivalent width. Spectra of M49 and M82 are shown in Fig. 8.

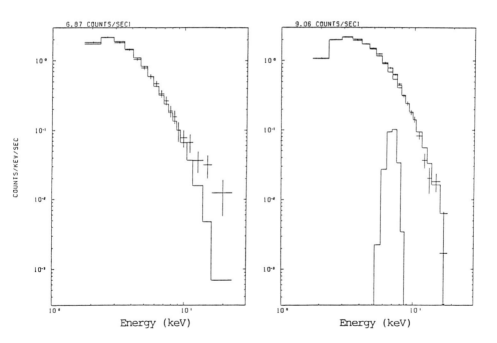

Fig. 8. Observed spectra of M49 (left) and M82 (right). Histograms are the best-fit thermal bremsstrahlung spectra.

Wide-band spectra of several clusters of galaxies have been obtained and their abundances of iron have been determined. Besides, the faint intracluster emission within Virgo Cluster was detected and the measured spectrum is expressed by a thermal bremsstrahlung spectrum with kT = 3.1 keV.

A number of AGN's, including Seyfert galaxies, BL Lac objects and QSO's, have been observed. In particular, of nine QSO's observed so far, eight QSO's were positively detected and their energy spectra over the range 2 - 30 keV were obtained. As a matter of fact, this is the first time that wide-band spectra of a meaningful number of QSO's have become available, which allows us to begin a systematic study of their spectral characteristics. More QSO spectra will hopefully be obtained with Ginga in future observations. Though still preliminary, the characteristics of the AGN spectra so far obtained from Ginga are listed in Table III. Some interesting results will be presented below.

TABLE III Parameters of the observed AGN's

Source	z	I_{obs}(c/s)	power index	$logN_H$	Line(keV)	Eq.W.(keV)
NGC 4051	0.0023	18	1.0±0.1	21.7		
		7.5	0.7±0.1	21.7		
NGC 4151	0.0033	33	0.56±0.05	23.0	6.4±0.1	0.5±0.1
NGC 4593	0.0087	20	0.7±0.1	21.7	6.5±0.2	0.3±0.1
Mrk 348	0.014	7.3	0.7±0.2	23.1		
Mrk 421	0.03	21-56	1.8±0.2	22.1		
3C 371	0.050	2.2	0.7±0.2	<21		
PKS2155-30	0.118	57	1.71±0.05	20.8		
1E1352+18	0.152	2.1	0.2±0.2	<21		
3C 273	0.158	35	0.50±0.03	21.4	5.5±0.3	0.1±0.05
1E1821+64	0.297	10	0.9±0.1	21.1	5.1±0.2	0.3±0.1
3C 279	0.536	7.0	0.6±0.1	<21		
4C 29.45	0.729	1.8	0.7±0.3	<21		
1407+265	0.944	1.7	0.9±0.3	<21		
CTA 102	1.037	3.6	0.1±0.2	20.5		
3C 446	1.404	4.9	0.4±0.2	20.8		

* The errors quoted are 90 % confidence limits.

The nearest Seyfert II galaxy NGC1068 exhibits a remarkable spectrum as shown in Fig. 9. An iron emission line at 6.4 keV stands out above a weak continuum with an equivalent width as large as 2 keV. This result is consistent with the picture that the central source is totally hidden by the surrounding material (possibly the inner accretion disk) and that only scattered emission and fluorescent lines can be observed.

A fairly intense iron emission line is present in the spectra of two Seyfert I galaxies, NGC4151 and NGC4593, that we observed. The line energy is consistent to be 6.4 keV expected for the fluorescent line from relatively cool matter. The 6.4 keV iron line was also detected previously from Cen A in the observation from Tenma (Wang et al. 1986). Remarkably, we detected significant emission lines from two bright QSO's, 3C273 and 1E1821+64. The spectrum of 1E1821+64 is shown in Fig. 10. These lines are most probably the redshifted iron line, since the observed amount of redshift is in good agreement for both cases with the optically determined z-values. These line energies are consistent with 6.4 keV, which suggests the same origin of the emission as in the case of Seyfert I galaxies, NGC4151 and NGC4593. However, 6.7 keV cannot be excluded at present. Apart from the origin of the iron line, it is very important to note that the iron line, if commonly present in all QSO's, permits determination of z from X-ray observation alone in the future missions with higher line-detection capabilities. On the other hand, no evidence for the iron line was obtained from two bright BL Lac objects, Mrk 421 and PKS2155-30.

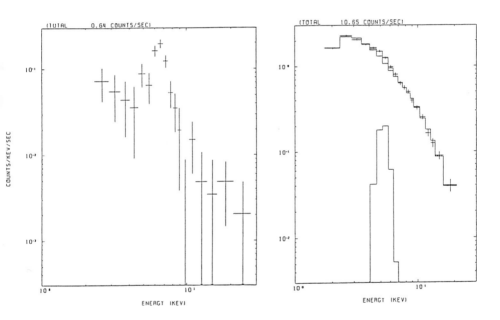

Fig. 9. Spectrum of NGC 1068   Fig. 10. Spectrum of 1E1821+64

As mentioned earlier, we were able to obtain the spectra of eight QSO's. As noted in Table III, the indices of power-law spectrum for QSO's determined essentially in the range 2 - 10 keV are widely distributed between 0.1 and 0.9. This is quite distinct from the case of Seyfert galaxies for which the power indices are concentrated within a narrow range around 0.7 (Mushotzky 1984). In contrast, BL Lac objects Mrk 421 and PKS2155-30 exhibit much steeper spectra than Seyfert galaxies and QSO's. Some of the observed spectra of BL Lac objects and QSO's are shown in Fig. 11.

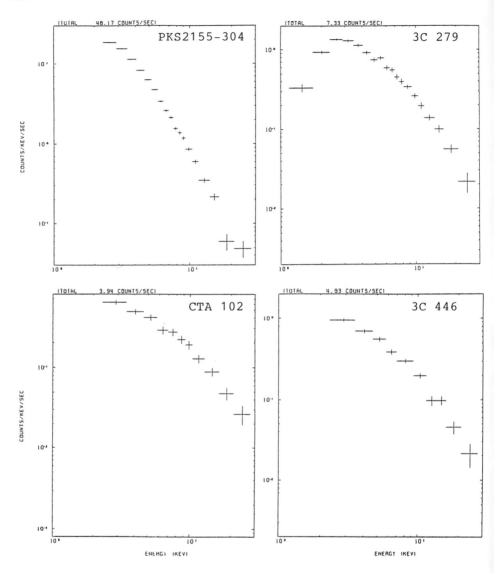

Fig. 11. Spectra of PKS2155-30 (BL Lac), 3C279, CTA102 (QSO), and 3C446 (QSO)

It is to be cautioned, however, that the measured fluxes of several QSO's are less than 3 counts/s (roughly equivalent to 0.3 mCrab). For the field of view of Ginga, 0.3 mCrab is close to the confusion limit based on the estimation from the Einstein log N-log S relation. Therefore, these faint sources may be subject to source confusion. A separate study of the fluctuation of the X-ray background, with respect to the intensity and spectrum, is underway.

REFERENCES

Agrawal,P.C. et al. (1986), Publ. Astron. Soc. Japan, **38**, 723.
Koyama,K. et al. (1986a), Publ. Astron. Soc. Japan, **38**, 121.
Koyama,K. et al. (1986b), Publ. Astron. Soc. Japan, **38**, 503.
Koyama,K. (1987), Publ. Astron. Soc. Japan, **39**, 245.
Makino,F. (1987), Astron. Letters and Communications, **25**, 223.
Mushotzky,R.F. (1984), NASA Tech. Memorandum 86071.
Van Speybroek,L. (1979), Astrophys. J. Letters, **234**, L45.
Wang, B. (1986), Publ. Astron. Soc. Japan, **38**, 685.
Worrall et al. (1982), Astrophys. J., **255**, 111.

THE ROSAT MISSION

J. Trümper
Max-Planck-Institut für Physik und Astrophysik
Institut für Extraterrestrische Physik
8046 Garching, W.-Germany

ABSTRACT. The scientific payload of the ROSAT spacecraft consists of a large X-ray telescope (6 - 100 Å) and a smaller XUV-telescope (60 - 300 Å) which are looking parallel. A primary objective of the mission is to perform the first all-sky survey with an imaging X-ray telescope leading to an improvement in sensitivity by several orders of magnitude compared with previous surveys. The spectral band covered in the survey mode by both telescopes ranges from 6 to 190 Å and can be divided into 6 "colours". A large number of new X-ray sources ~ 10^5) is expected to be discovered and located with an accuracy of 1 arcmin or better, depending on source strength. The sources discovered will represent almost all types of astronomical objects, ranging from nearby normal stars to distant quasars.

After completion of the sky survey which will take half a year, the instruments will be used for detailed investigations of selected sources with respect to spatial structures, spectra and time variability. In this pointing mode, which will be open for guest observers, ROSAT is expected to provide substantial improvements over the imaging instruments of the Einstein observatory.

The talk will give an overview of the scientific and technical aspects of ROSAT which is scheduled to be launched in early 1990 with a Delta II rocket.

1. INTRODUCTION

During the last 20 years X-ray astronomy has become one of the major disciplines in the exploration of our universe. A large variety of phenomena have been discovered in this area which comprise almost all kinds of astronomical objects - from the nearby stars to the most distant quasars at the edge of the known universe. In many cases, X-ray emission is the primary and dominating emission process (X-ray binaries with neutron stars or black holes, cooling neutron stars, hot gas in clusters of galaxies etc.), in other objects, the observation of X-rays gives very important complementary information (e.g. stars, degenerate stars, active galaxies and quasars). In X-rays we see the "hot universe", explosive phenomena, large concentrations of relativistic electrons in magnetic or intense radiation fields. It is clear today that these high energy phenomena play a very important role in the evolution of matter in our universe.

2. ROSAT X-RAY TELESCOPE

ROSAT is a free flying satellite to be launched by a Delta II rocket in early 1990. Its anticipated lifetime is 3 years. The scientific payload consists of two instruments, a large X-ray telescope (6 - 100 Å) and a smaller XUV-telescope (60 - 300 Å) which are oriented in parallel. In this talk I'll concentrate on the X-ray aspects. The XUV part of ROSAT will be covered by Barstow and Pounds in these proceedings.

The ROSAT X-ray telescope consists of a fourfold nested mirror system with 83 cm aperture and three focal instruments. Two of them will be position-sensitive proportional counters (PSPC, 6 - 100 Å, 0.1 - 2 keV) having a field of view of $2°$. In the pointing mode the angular resolution will be ~ 30". The positions of point sources may be determined with higher accuracy (to ~ 10 arcsec). The spectral resolution of the PSPC's will be 45 % FWHM at 1 keV. Both PSPC's are being developed and built by MPE Garching. The third focal instrument will be a high resolution imager (HRI, ~ 3") which is an improved version of the corresponding EINSTEIN instrument, provided by SAO/NASA. The X-ray mirror system will have a half power width of ~ 3". It is manufactured by Carl Zeiss Company in Oberkochen near Stuttgart. The calibration of mirrors and the complete X-ray telescope is done in the 130 m X-ray test facility of MPE Garching.

3. THE ALL SKY X-RAY SURVEY

A primary objective of the mission is to perform the first all-sky survey with an imaging X-ray telescope which will lead to an improvement in sensitivity by several orders of magnitude compared with previous "counter" surveys.

The sky survey is performed by continuously scanning great circles perpendicular to the earth-sun-line. This will result in full sky coverage in half a year. The scan of one great circle takes one orbit and thereby avoids earth occultations. At the ecliptic equator a particular source is scanned ~ 30 times during two consecutive days with an integral exposure of ~ 600 sec. The corresponding survey sensitivity (5 σ) is 2×10^{-13} erg/cm²s (0.1-2 keV). The ecliptic pole is crossed during every orbit which results in ~ 60.000 s of total observation time and a sensitivity of 1.5×10^{-14} erg/cm²s. We note that this latter figure is close to that of the Einstein deep surveys.

The sky survey will be done with the PSPC only because of its sensitivity and its relatively large field of view ($2°$ circular). The spectral resolution of the instrument allows to distinguish four "color" bands. Positions will be accurate to 1 arcmin (90 % confidence radius) for point sources near the sensitivity limit S_{min}. For stronger sources the position determinations will be more accurate due to the increased photon statistics. It is expected that for sources at ~ 10 S_{min} a limiting resolution of 0.5 to 0.1 arcmin (90 % confidence radius) will be reached depending on the final instrument and spacecraft characteristics.

The number of extragalactic sources to be discovered by the survey can be estimated to $\sim 10^5$. Of particular interest is the fact that large unbiased samples of various classes of sources will be collected: clusters of galaxies, quasars, Seyfert and BL galaxies, radio galaxies, normal galaxies. At the same time a large number ($\geq 10^4$) of low luminosity galactic X-ray sources such as normal stars and cataclysmic variables will be detected.

4. THE POINTED OBSERVATIONS

The second main objective of the ROSAT mission is to perform detailed investigations of selected sources. These will be carried out in the pointing mode and will improve our understanding of spatial structures, spectra and time variability of X-ray sources. In this mode the ROSAT X-ray telescope is expected to provide a substantial improvement over the imaging instruments of the EINSTEIN-observatory, particularly in terms of sensitivity (factor $\sim$ 5 to 10), spectral resolution (PSPC versus IPC factor $\sim$ 3) and angular resolution (factor $\sim$ 3 both for the PSPC and the HRI). In addition, the XUV telescope will allow the observations to be extended to long wavelengths. In the pointing mode there will be four "X-ray colours" and three "XUV colours".

5. THE ROSAT SATELLITE GROUND OPERATIONS

ROSAT is a three-axis stabilised satellite with CCD-type star trackers (two for the X-ray telescope, one for the XUV telescope) and attitude control by momentum wheels which are desaturated by magnetic torquing. Two tape recorders are used for on-board data storage. The satellite which has a mass of more than 2,5 tons will be launched by a Delta II into an orbit of $\sim$ 560 km height and 57° inclination. Mission control will be performed by the German Space Operations Center (GSOC) at Weilheim near Munich from where the data will be sent to the ROSAT science operation center at MPE Garching which is equipped with a dedicated VAX 8600 facility and a Microvax cluster.

The technical and scientific quicklook activities will be placed at GSOC and MPE as well. Date obtained during the pointed programme will be routed directly to the ROSAT data centers in the US and UK.

6. SCIENTIFIC DATA RIGHTS

The analysis of the ROSAT all sky surveys will be the responsibility of MPE and of the UK WFC consortium for the X-ray data and for the XUV data, respectively.

The pointed programme will be devoted entirely to a guest observer programme. The first call for proposals will be issued one year before launch, viz. in spring 1989, in the US, UK and Germany.

7. SPECTROSCOPIC FOLLOW-UP MISSION - ROSAT 2 ("SPEKTROSAT")

The basic philosophy of this project is to build a copy of ROSAT with some modifications which greatly enhance its spectroscopic capabilities. The principal objective of this followup mission which we call SPEKTROSAT is to perform emission line diagnostics on hot plasmas and to detect absorption features.

Spectroscopic resolution will be achieved by means of an objective transmission grating to be placed between the mirror system and the focal instruments. This will yield a resolution of 0.1-0.3 Å viz. $\lambda/\Delta\lambda \sim 100$. Using high efficiency transmission gratings which have already been developed at MPE the throughput of such a SPEKTROSAT will be larger by a factor of 20 to 200, depending on photon energy, compared with the Einstein observatory. A large number of interesting investigations of stellar and extragalactic sources can be made with such a powerful spectrometer. The total number of sources which could be observed in the spectroscopic mode would be ~ 1000.

8. CONCLUSIONS

With the termination of the Einstein and EXOSAT observations, the ROSAT mission will provide the only X-ray telescope in space until the mid 1990's. It will be evident from the preceding discussion, that because of its extended waveband coverage, its sensitivity, and its spectral resolution, ROSAT will provide the scope for a wide range of novel investigations on known astrophysical objects as well as an exciting capability for new astrophysical discoveries. This is true for the pointed observations, but also for the all-sky survey which will have a great impact on the work in other regions of the electromagnetic spectrum. Last, but not least, the survey will help to guide the observations of missions like SPEKTROSAT and SAX and of the large permanent X-ray observatories to be launched in the late 1990's, AXAF and its European complement XMM.

9. REMARK

A more detailed description of the project including references is given by the present author in Physica Scripta, T7, 1984.

THE WIDE FIELD CAMERA FOR ROSAT: OBSERVING STARS

M.A.Barstow and K.A.Pounds
X-ray Astronomy Group, Physics Department
University of Leicester
University Road
Leicester, LE1 7RH, UK

ABSTRACT. An imaging XUV telescope, the Wide Field Camera (WFC), will be flown on board the West German satellite ROSAT. The primary scientific objective of ROSAT is to perform an all-sky survey, over a 6 month period, in the X-ray (6-80Å) and XUV (60-200Å) wavebands. During the survey the WFC will, by the use of two filters, perform broad band photometry. In the subsequent pointed phase of the mission, more filters will be available to extend the spectral coverage to longer (up to $\approx$ 700Å) wavelengths. The WFC is most sensitive to material with temperatures from $\approx 3 \times 10^4 - 10^6$K and consequently is well suited to observing the photospheres of hot degenerate stars and the transition regions and coronae of cool stars. Current predictions suggest that $\approx$ 3500 such sources may be detected in the survey.

1. INTRODUCTION

An imaging XUV telescope, the Wide Field Camera (WFC), is being built by a UK consortium of X-ray astronomy and space research groups and will be flown aboard the West German satellite ROSAT. The main ROSAT instrument (Trümper, 1988) is a Wolter I X-ray Telescope (XRT), designed to perform the first imaging all-sky survey in the energy range 2-0.15keV (6-80Å). The WFC complements the X-ray telescope by extending the energy range of the survey to 0.21-0.062keV (60-200Å), into the XUV band. This region of the electromagnetic spectrum is largely unexplored, and the WFC will perform the first all-sky survey in this band, locating point sources to $\approx 1'$ and mapping extended regions of emission.

The first six months of operation of ROSAT will be dedicated to the sky-survey, after which a minimum of 1 year of pointed observations are planned, with possible extensions for up to 2 years, depending on the lifetime of the satellite. During the survey, the WFC will image the sky in two wavebands, by use of selected filters. Two more filters will be available for pointed observations, extending the spectral coverage to $\approx$ 700Å and allowing selected sources to be studied in more detail.

The WFC is most sensitive to material with temperatures from $\approx 3 \times 10^4 - 10^6$K. Objects of particular astrophysical interest include hot degenerate stars, coronae of cool stars, dwarf novae, faint extended objects such as supernova remnants and clusters of galaxies, and the diffuse XUV background.

In this paper we summarise current expectations of the WFC performance, discuss the limitations imposed on source visibility by the interstellar medium (ISM) and outline the impact that the WFC will have on the study of white dwarfs and stellar coronae, which will probably comprise $\approx 80\%$ of the total sources detected in the survey. Using specific examples, the kind of information that the WFC will provide is illustrated.

2. DESCRIPTION OF THE WIDE FIELD CAMERA

The WFC and its individual components have been discussed in much detail in other publications (e.g. Barstow et al, 1985a & 1985b; Willingale et al, 1987). Consequently, we give here only a brief description of the hardware, concentrating on the scientific performance.

Figure 1 is a schematic of the WFC showing the main optical components - grazing incidence mirrors, filters and microchannel plate (MCP) detectors. A nested set of 3 Wolter-Schwarzschild Type I mirrors, fabricated from aluminium and coated with gold for maximum reflectance provide a geometrical collecting area of 475cm^2 with a common focal length of 525mm. The grazing incidence angles chosen (typically $\approx 7.5°$) allow the collecting area to be optimised whilst retaining a wide (2.5° radius) circular field of view and a low energy reflectivity cut-off at 0.21keV (60Å). An on-axis resolution of $\approx 2.3'$HEW is expected but the response degrades to $\approx 4.4'$HEW at 2.5° off-axis due to inherent optical aberrations. Hence, the average resolution for the survey will be $\approx 3.5'$HEW.

Figure 1. Schematic diagram of the WFC showing its principal optical components.

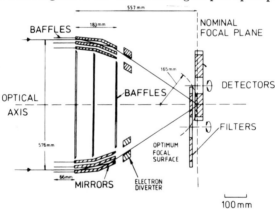

In order to take full advantage of the telescope resolution the pair of MCPs in the detector are both curved, like a watchglass, to match the optimum focal surface, as is the resistive anode readout system. A CsI photocathode is deposited directly onto the front face of the front MCP to enhance the EUV quantum efficiency. The detector resolution is substantially (a factor > 2) better than that of the mirror nest and consequently does not contribute significantly to the net response of the WFC. A focal plane turntable can be used to select one of two identical detector assemblies in flight.

The filter wheel assembly contains eight filters, any one of which can be selected to define the wavelength passbands and suppress geocoronal background radiation which would otherwise saturate the detector count rate. Another function of the filters is to prevent the detection of UV radiation from hot O/B0 stars which would otherwise be imaged indistinguishably from EUV sources. There are four each of two types of filter, large diamter (5° field of view) for the survey (S) and small diameter (2.5° field of view) for pointed (P) observations. S filters will also be used in the pointed phase of the mission. A UV calibration system, mounted on the mirror support, permits in-flight monitoring of detector gain drifts and thermally induced misalignment of the telescope axis. One position in the filter wheel is occupied by a UV interference filter for use with this system.

The WFC is sensitive to particle background, particularly soft electrons. A high percentage of incident electrons ($\approx 97\%$) are deflected away from the detector aperture by a magnetic diverter. However, the remaining particle flux can still be significant. Two particle detectors, a geiger tube and a channel electron multiplier (CEM), monitor this background and the former can provide a signal to switch off the MCP detector during passage through high background regions (ie. South Atlantic Anomaly, Auroral Zones). The CEM may be used, in conjunction with an 'opaque' filter (ie. one that excludes all radiation but the particle flux) mounted in front of the MCP, to establish the electron contribution to the total observed background. This is important if we wish to study the diffuse XUV background, and therefore the local interstellar medium, since the electron flux is not well determined and also exhibits large temporal variations. Studies of the electron background expected in the WFC have not yet established the feasibility or necessity of this technique but an opaque filter would occupy one of the P filter slots.

Table I summarises the main performance characteristics of the science filters and Figure 2 shows the effective area of the WFC for each EUV filter design.

Figure 2. The effective area of the WFC for each filter: [a] C+B+Lexan, [b] Be+Lexan, [c] Al+Lexan and [d] Sn+Al.

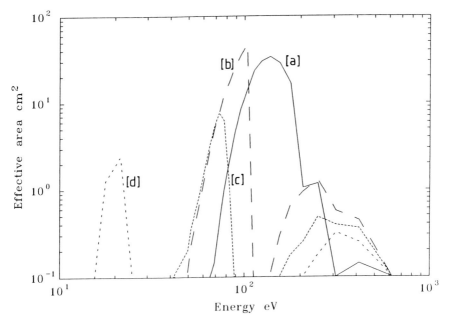

TABLE I. ROSAT WFC Filters, Wavebands and Sensitivity

Filter Type	Survey (S) or Pointed (P)	FOV Diam. (deg.)	'Mean' Wavelength (Å)	Bandpass (Å) (at 10% of peak efficiency)	Point-source Sensitivity[b] (μJy)	(HZ43^{-1})
C/Lexan/B ($\times$2)	S + P	5	100	60–140	1.0	3000
Be/Lexan ($\times$2)	S + P	5	140	112–200	1.4	7000
Al/Lexan	P	2.5	180	150–220	7.3	1400
Sn/Al [b]	P	2.5	600	530–720	220	185

[a] Provisional. [b] For 5σ significance, exposure time of 2000 s (a typical value for each filter for the survey and for pointed observations) and 'typical' background. The right hand column expresses the sensitivity in terms of the flux from the white dwarf HZ43, the brightest known EUV source.

3. INTERSTELLAR ABSORPTION

Attenuation of the incident flux from a source by the ISM is extremely important in the EUV. This results from the fact that most elements have electron binding energies in the range 10-100eV, corresponding to this range of wavelengths. Cruddace et al (1974) have calculated the effective interstellar absorption cross section (σ) per H atom in the range 1-2000Å from the weighted sum of all individual components. Absorption in the EUV is dominated by H and He and is a steep function of wavelength, increasing by a factor $\approx$ 500 from 100Å to the Lyman edge at 912Å. To some extent source visibility will be dependent upon the intrinsic spectrum. However, a good general indication of the limitations of the absorbing matter can be obtained by calculating the effective column density for unit photon mean free path (MFP) at the mean wavelength of each pass band and the MFP for a mean local density of 0.07 atoms cm^{-3} (see Paresce, 1984; Table II). In reality, as demonstrated by the data of Frisch and York (1983) and Paresce (1984), there are large variations in the line of sight column density with direction through the galaxy.

TABLE II. WFC Passbands: Absorption of XUV Radiation by the ISM

Wavelength (Å)	ISM Effective Cross-section (10^{-20}cm^2) [a]	ISM Effective Hydrogen Column Density (10^{19}cm^{-2}) for unit mean free path	Mean Free Path (pc) [b]
100	3.2	3.1	190
140	7.8	1.3	70
180	15.0	0.7	35
600	200	0.05	2.3

[a] From Cruddace et al (1974). [b] For ISM with constant volume number density of neutral hydrogen of 0.07 cm^{-3} (Paresce 1984).

Although when looking out of the galactic plane data is sparse and the column density/distance poorly determined, the integrated HI column density through the galaxy (Heiles and Jenkins, 1976) may be low enough ($< 10^{20}$cm^{-2}) to see through at high latitudes.

4. OBSERVING HOT WHITE DWARFS

Sion (1986) has recently reviewed our current understanding of the formation and evolution of white dwarfs (WDs). It is clear that there are many problems concerning the relationships between different groups of objects still to be understood. An unbiased survey of the higher temperature objects is fundamental in answering some of these questions. Most stars can be divided into two broad sub groups, those whose atmospheres are dominated by H and those where He is the main constituent. Some stars may contain significant fractions of CNO metals (eg. PG1159 objects) but their abundances are not well determined as yet, although some model atmospheres do exist for solar abundances (Hummer and Mihalas, 1970). Figure 3 compares the intrinsic spectra for a 60000K WD (log g $\approx$ 8), having an atmosphere ranging between the extremes of pure H (Wesemael et al, 1980) through a homogeneous He/H mixture (Petre et al, 1984; He/H= 10^{-4} in this case) to pure He (Wesemael, 1981), with the WFC bands. Clearly, the WFC survey is much more sensitive to H dominated stars of given temperature than for those with He, as would be expected given the relatively high He opacity at higher energies. The relative sensitivity of each filter to atmosphere and column density is illustrated in figure 4, where the minimum detectable temperature (T_{min}) is displayed as a function of the mean wavelength for each band pass. As might be expected, T_{min} for the three shortest wavelength filters is fairly insensitive to column but shows a marked increase for the transition from a pure H to pure He atmosphere. The reverse is true at the longer wavelength, although a high column density increases the size of the T_{min} gap between the atmospheres.

Figure 3. White dwarf model atmospheres for a 60000K star comprising [1] pure H, [2] He/H=10^{-4} and [3] pure He. The solid vertical lines indicate the mean wavelengths of each filter, designated as in figure 2.

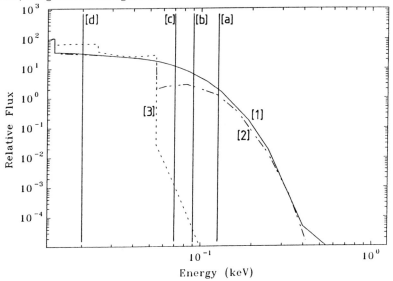

Hot white dwarf stars are sufficiently luminous in the EUV to be potentially observable out to distances approaching several thousand pc. On comparing this figure with the average MFPs of Table II, it is clear that in practice the effect of the ISM will largely determine the viewing horizon. Taking the approximate temperature limits of figure 4 and the mean horizon allows the total number of WDs that may be detected in each band to be estimated, provided the space density is known. Fleming, Liebert and Green (1986) present a summary of space densities for DA and DO/DB WDs subdivided by temperature. We expect, from this data, that the WFC will detect ≈ 2500 WDs during the sky survey. Most of these stars will be detected in C+B+Lexan filter with $\approx 5\%$ and $\approx 1\%$ also appearing in Be+Lexan and Al+Lexan respectively. The low average MFP (≈ 2.3pc) at lower energies implies that very few if any stars will be seen in the other filter. However, the extreme non-uniformity of the ISM is likely to make quite large volumes accessible to this filter along some viewing directions.

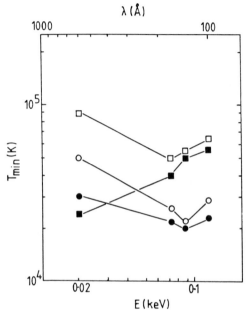

Figure 4. The minimum white dwarf temperature to which the WFC is sensitive as a function of photon energy and wavelength. The circles represent a pure H atmosphere and the squares a pure He atmosphere. Filled and open shapes are for column densities of 10^{18} and 10^{19} respectively.

A total of approximately 1500 WDs is known to exist, of which $\approx 30\%$ (about 450) have temperatures to which the WFC is sensitive. Comparing this figure with the number of WDs that the WFC is expected to see (≈ 2500) shows the statistical value of the survey. Indeed, given that space densities are generally determined from optical identifications this prediction may well be an underestimate.

5. OBSERVING COOL STARS

Pye and McHardy (1987), hereafter P&M, have discussed in some detail the capabilities of the WFC with respect to cool stars. However, discussion of cool star observations is relevant to these proceedings and some of their results are summarised here and compared with those for WDs.

The group of cool stars includes late type main sequence stars, dMe flare stars and active binary systems (eg. RS CVns). They vary widely in X-ray luminosity ranging from $\approx 10^{26}$erg s^{-1} up to $\approx 10^{31}$erg s^{-1} in the most active RS CVns (eg. Rosner et al, 1985; Walter et al, 1980). Such X-ray emission arises in hot thin plasma lying in the transition regions and coronae of these stars with temperatures lying in the range 10^5 to a few times

10^7K. Figure 5 shows the spectra of an optically thin thermal plasma, based on the data of Raymond and Smith (1977), for temperatures in this range. The WFC filter mean energies are indicated showing how plasma of different temperatures will be sampled.

Figure 5. Spectra of an optically thin thermal plasma at three temperatures: [1] 10^5K, [2] 10^6K and [3] 10^7K. Filter bands are marked as in figure 3.

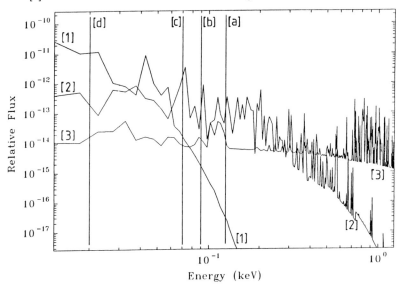

The quiescent EUV luminosity of main sequence stars can be estimated by scaling X-ray luminosities (see P&M) using data from EXOSAT grating spectra and a typical solar spectrum. It is expected that these stars will only be detected out to distances $\approx$ 25pc. The absorbing column is still quite low at 25pc ($\approx 5 \times 10^{18}$cm^{-2} on average) and the limiting factor is simply the low intrinsic luminosity of these objects. Active binaries, such as RS CVns are much more luminous and likely to be observed out to 100pc or more. From known space densities (see P&M), it is expected that the WFC will detect $\approx$ 1000 cool main sequence stars and $\approx$ 10 RS CVn systems. When divided into F, G, K and M spectral types, the stars detected in the survey should be 50% , 30% , 30% and 20% respectively of the total number of each class in the volume out to 25pc. Many cool stars are variable sources, as a result of flaring and rotational modulation of active regions. In binary systems eclipses can also lead to variability and are, in fact, a useful diagnostic tool for determining the spatial distribution of coronal plasma, as demonstrated by EXOSAT (eg. White et al, 1987). The typical orbital periods of RS CVn systems lie in the range 1 - 14 days. During the survey the viewing axis of ROSAT will scan in ecliptic latitude (θ) at a rate of $\approx 4°$min^{-1} and in longitude (ϕ) at $\approx 1°$day^{-1}. A source will come within the 5° WFC field of view for $\approx 5/\cos\theta$ days for durations up to $\approx$ 1min every orbit ($\approx$ 1.5hrs). As the minimum coverage of a target is 5 days ($\theta = 0$) it can be seen that the WFC coverage is well matched to monitor RS CVn emission throughout at least a substantial fraction of a binary period.

Cool stars are the major identified class of flaring X-ray source. The repeated observing of individual objects, as described above will lead to some flare detections. P&M have estimated the sensitivity of both the WFC and the ROSAT XRT to events detected in a single pass over a source to be $\approx 15\mu$Jy (at 100Å) and $\approx 0.5\mu$Jy (at 1 keV) respectively. Convolving these figures with transient event frequencies in previous instruments yields estimates of the number expected in the ROSAT sky survey. They estimate that the XRT will see ≈ 180 flares and the WFC ≈ 12, ranging in duration from a few seconds to a few hours.

6. THE WFC AS A DIAGNOSTIC TOOL

Hot white dwarfs (and related objects) and cool stars comprise $\approx 80\%$ of the sources that the WFC will see. We discuss here what the WFC can tell us about the objects detected. The survey will be performed in two 'colours' with 2 extra, longer wavelength bands, available during the pointed phase. However, it will probably be possible to plan follow-up observations for only $\approx 20\%$ of the survey sources. For most sources the WFC will provide a count rate measured in one survey filter and either a count rate or upper limit measured in the other. The parameters that we might wish to determine using the data are temperature (T), column density (N_H) and, in the case of WDs, the atmospheric composition to some degree.

The ratio of Be+Lexan and C+B+Lexan count rates (R) are shown, as a function of temperature, in figure 6, for the WD pure H and pure He atmospheres and optically thin thermal plasma. Data is only displayed for temperature components that can be detected in both filters. Except for a narrow temperature range ($\approx 5 \times 10^4 - 10^5$K) of pure He atmospheres, R has no unique value that can determine the object being observed (and these plots do not include other possible types, eg. AGNs). Hence, the survey must be well supported by a programme of identifications. This should be quite easy for cool stars as they are well observed and catalogued. However, a significant fraction ($> 50\%$) of WDs seen in the survey will have no known optical counterpart.

Given an a priori identification, the problem of interpretation of the data is similar to that of many EXOSAT observations. An example is the 1984 observation of the RS CVn HD155555 (Barstow, 1988; figure 7). Using the Raymond and Smith (1977) thin plasma model a curve of T/N_H can be generated where the predicted count rate ratio (Lexan:Al/parylene) agrees with that observed (2.6). The temperature range can only be further constrained (to $5 \times 10^5 - 1.3 \times 10^6$K) with other information, in this case N_H limits from Paresce (1984) and Frisch and York (1983). If some normalisation constant is available for the source model spectrum, such as the optical magnitude for a WD, a unique solution might be determined. This is illustrated in figure 8 where T/N_H curves are displayed for the hot He rich WD K1-16 (Barstow, 1988), corresponding to the observed Lexan and aluminium/parylene count rates (0.019 and 0.0019 cs^{-1} respectively) with a pure He model spectrum normalised to $m_V = 15.03$. The curves meet at the point corresponding to 127000K and 1.7×10^{20}cm^{-2}. Note however, that additional data (in this case optical spectra) are needed in order to choose a model atmosphere. Data available from other filters will improve the constraints on T and N_H etc. and also decrease the dependence on supplementary data from other wavebands.

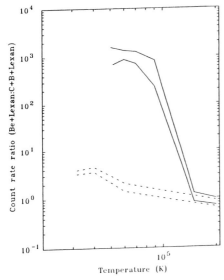
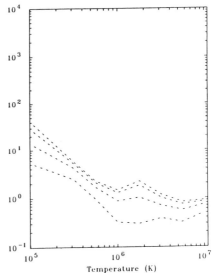

Figure 6. The ratio of Be+Lexan to C+B+lexan count rate for white dwarfs (left hand panel) and cool stars (right hand panel) as a function of temperature. The white dwarf atmospheres are pure H (dashed lines) and pure He (solid lines) with upper and lower curves representing column densities of 10^{18} and 10^{19} repectively in each case. The column densities for the Raymond and Smith cool star model are 10^{18}, 3×10^{18}, 10^{19} and 3×10^{19} from top to bottom.

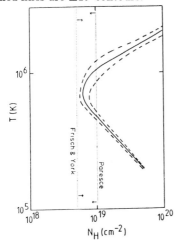

Figure 7. Plasma temperature (T) as a function of column density (N_H) for the observed mean filter count rate ratio (Lexan:Al/parylene, 2.6) of HD155555. Dashed lines are $\pm 1\sigma$ contours.

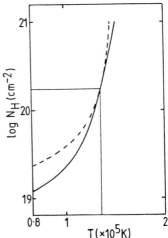

Figure 8. T/N_H curves of constant count rate for K1-16, corresponding to Lexan (solid line) and Al/parylene (dashed line) filters.

7. CONCLUSION

We have described the WFC outlining its spectral coverage, in both survey and pointed modes, and summarising its point source sensitivity. In our discussions of the likely results of the survey we have concentrated on the detection of white dwarfs and related objects and of cool stars, showing how the absorption of the ISM affects this. We expect that > 3500 sources will be detected in these two categories. However, we note that a significant fraction (> 1000) will probably be uncatalogued and that a major programme of optical observations will be needed to make the optimum use of the survey data.

8. ACKNOWLEDGEMENTS

The WFC project is supported by the UK Science and Engineering Research Council. The instrument is being developed by a consortium of five institutes: University of Leicester, Rutherford Appleton Laboratory, Mullard Space Science Laboratory, University of Birmingham and Imperial College of Science and Technology. Many personnel from these institutes have made invaluable contributions to this project. In particular we would like to thank Drs. J.P. Pye and I.M. McHardy for the use of their cool star data.

9. REFERENCES

Barstow,M.A., 1987, *Mon.Not.R.astr.Soc*, **228**, 251.
Barstow,M.A., 1988, these proceedings.
Barstow,M.A., Fraser,G.W., and Milward,S.R., 1985a, *Proc. SPIE*, **597**, 352.
Barstow,M.A., Willingale,R., Kent,B.J., and Wells,A., 1985b, *Optical Acta*, **32**, 197.
Cruddace,R.G., Paresce,F., Bowyer,S., and Lampton,M., 1974, *Ap.J.*, **187**, 497.
Fleming,T., Liebert,J. and Green,R.F., 1986, *Ap.J.*, **308**, 176.
Frisch,P.C., and York,D.G., 1983, *Ap.J.*, **271**, L59.
Heiles,C. and Jenkins,E.B., 1976, *Astr.Astrophys.*, **46**, 333.
Hummer,D.G., and Mihalas,D., 1970, *Mon.Not.R.astr.Soc.*, **147**, 339.
Paresce,F., 1984, *Astron.J.*, **89**, 1022.
Petre,R., Shipman,H.L. and Canizares,C.R., 1986, *Ap.J.*, **304**, 356.
Pye,J.P. and McHardy,I.M., 1987, Proceedings of the **Midnight Sun conference on cool stars**, in press.
Raymond,J.C. and Smith,B.W., 1977, it Ap.J.Suppl., **35**, 419.
Rosner,R., Golub,L. and Vaiana,G.S., 1985, *Ann.Rev.Astr.&Astrophys.*, **23**, 413.
Sion,E.M., 1986, *Publ.astr.Soc.Pac.*, **98**, 821.
Trümper,J., 1988, these proceedings.
Walter,F., Cash,W., Charles,P.A. and Bowyer,C.S., 1980, *Ap.J.*, **236**, 212.
Wesemael,F., 1981, *Ap.J.Suppl.*, **45**, 177.
Wesemael,F., Auer,L.H., Van Horn,H.M. and Savedoff,M.P., 1980, *Ap.J.Suppl.*, **43**, 159.
White,N.E., Shafer,R., Parmar,A.N. and Culhane,J.L., 1987, Proceedings of the 5th Cambridge workshop on **Cool Stars Stellar Sytems and the Sun**, in press.
Willingale,R., 1987, in **Grazing Incidence Optics for Astronomical and Laboratory Applications**, *proc. SPIE*, in press.

THE SAX X-RAY ASTRONOMY MISSION

G. C. Perola
Istituto Astronomico
Università di Roma "La Sapienza"

ABSTRACT. The SAX satellite, whose launch is foreseen for the second half of 1992, will be devoted to spectral and variability studies of celestial sources over a wide energy range, from 0.1 to 200 keV. This paper gives a synthetic description of the instruments and of the scientific objectives of the mission.

1. OUTLINE OF THE MISSION

SAX, acronym of "Satellite for Astronomy in the X-rays", proposed in 1981 in response to an AO and selected in 1982, is a project of the italian National Space Plan (PSN) of the National Research Council (CNR), in collaboration with the Netherlands Agency for Aerospace Programs (NIVR). At present (September 1987) the development program is near the end of Phase B, and the launch of the satellite is foreseen in the second half of the year 1992.

The SAX mission aims to expand our knowledge and understanding of celestial X-ray sources by carrying out spectral and variability studies with a set of instruments covering a wide energy range, from about 0.1 to about 200 keV. To attain its scientific objectives, the mission is conceived with the following capabilities:

- Imaging with moderate angular resolution (1 arcminute) and broad band spectroscopy ($E/\Delta E \sim 10$) in the range 1-10 keV, with the spectroscopic capability alone extended down to 0.1 keV.
- Continuum and line spectroscopy ($E/\Delta E \sim 5 \div 20$) over the range 3-200 keV.
- Time variability studies of bright source spectra both on short (msec) and large (days to months) time scales.
- Systematic studies of the long term variability of sources brighter than 1 mCrab equivalent flux through periodic surveys of preselected

regions of the sky.

To obtain these capabilities, the following set of instruments (to be described in Sect. 2) has been chosen for the payload:

- Concentrator/ Spectrometer - 4 units (C/S)
- Phoswich Detector System (PDS)
- High Pressure Gas Scintillation Proportional Counter (HPGSPC)
- Wide Field Camera - 2 units (WFC).

The satellite is a free flyer of the one ton class, stabilized on three axis and capable of pointing at any source in the sky in every single year of its operational lifetime - the main constraint on the portion of the sky accessible at any one time being posed by the requirement that the solar array should normally remain within 30° of the sun line. The pointing stability will be better than 2 arcminute, and the attitude reconstruction post facto better than 1 arcmin. The orbit will be circular and low altitude (550 km), with an inclination $\leq$ 12° with respect to the Equator and a period of approximately 94 minutes. This type of orbit was chosen to optimize the environment (particle background) particularly for the high energy, non imaging instruments. Consequently the Ground Station for uplink telecommand and downlink telemetry will be situated near the Equator (San Marco/Malindi, Kenya), while the Operation Control Centre, connected to the station by a communication satellite link, will be located in Italy. The minimum lifetime of the mission will be 2 years.

Originally designed for a launch with the Shuttle-IRIS (Italian Research Interim Stage) combination, due to the failure of the Shuttle program to assure a launch date in the early 90's, the SAX satellite is now being redisigned for an expendable vehicle. Apart from this forced change in the program, the detailed descriptions of the payload and the scientific objectives in Spada (1983) and Perola (1983), along with the updates in Scarsi (1986) are still valid.

The scientific objectives and the instrumental payload have been conceived and planned under the responsibility of a consortium of Institutes in Italy together with Institutes in Holland and with the partecipation of the Space Science Department of the European Space Agency. The composition of this consortium is listed below:

- Istituto per le Tecnologie e lo Studio delle Radiazioni Extraterrestri, ITESRE/CNR - Bologna
- Istituto di Astrofisica Spaziale, IAS/CNR - Frascati
- Istituto di Fisica Cosmica e Tecnologie Relative, IFCTR/CNR and Unità GIFCO - Milano

- Istituto di Fisica Cosmica e di Applicazioni dell'Informatica, IFCAI/CNR and Unità GIFCO - Palermo
- Istituto dell'Osservatorio Astronomico, Università di Roma "La Sapienza"
- Space Research Laboratory - Utrecht
- Space Science Department, SSD, of ESA - Noordwijk

In the following sections I give a brief account of the instrument parameters and performances (Sect. 2), of the scientific objectives and observational strategy (Sect. 3), and close with a quick assessement of the role of SAX in the context of past and future mission.

2. THE INSTRUMENTS

The main parameters of the instruments are summarized in Table 1. The three instruments with a narrow field of view (NFI) will point in the same direction to provide a simultaneous coverage of the celestial targets over the energy band 0.1 - 200 keV. The two WFC units will point in nearly opposite directions and at 90° to the axis of the NFI.

The C/S is indeed an assembly of four identical grazing incidence telescopes, each made of 30 nested double cone mirrors. The angular resolution achievable (~1 arcmin) with the conical approximation to the Wolter 1 geometry matches rather well the spatial resolution of the position sensitive GSPC chosen as the focal plane instrument. The mirrors are thin (0.2 - 0.4 mm) shells of gold coated nickel; the tecnique developed for their production is electroformation on superpolished (or polished and lacquered) mandrels, and is described, along with the successful tests on prototype mirrors, in Citterio et al (1987). The telescopes were designed to maximize the effective area around 7 keV, compatibly with the focal length allowed by the satellite structure: Fig. 1 emphasizes this aspect by comparing the total effective area of the four SAX units with that of the single telescope on the Einstein Observatory and on AXAF. While three of the focal plane GSPC will have a Be window opaque to photons with $E < 1$ keV, the fourth GSPC will have a plastic window, which allows photons to be detected down to $E \simeq 0.1$ keV (the development of the low energy GSPC is responsibility of SSD). The C/S will perform spatially resolved spectroscopy on extended sources, although this applies only above 1 keV because of the poor spatial resolution of the detector at lower energies.

The HPGSPC is a new development, where the high pressure of the Xenon extends to more than 100 keV the excellent energy resolution of the GSPC (better than in proportional counters by about a factor of 2). Both the HPGSPC and the PDS, which together cover the range 5 - 200 keV,

Table I. The SAX scientific instruments

Concentrator/spectrometer - 4 units

Optics (per unit):
Focal length	185 cm
Number/length of double cone mirrors	30/30 cm
On axis resolution (HPR)	1 arcmin
Field of view (FWHM)	30 arcmin

Focal plane detectors:
GSPC with 25μ Be window (3 units)
Energy range	1-10 keV
Effective area per unit at 8 keV	40 cm^2
Energy resolution $\Delta E/E$	$0.08/\sqrt{E/6\text{ keV}}$
Position resolution	$0.5/\sqrt{E/(6\text{ keV})}$ mm = $0.9/\sqrt{E/(6\text{ keV})}$ arcmin

GSPC with 1μ polyprop. window (1 unit)
Energy range	0.1-10 keV
Effective area at 0.55 keV	40 cm^2
8 keV	40 cm^2

High Pressure GSPC

Exposed geometrical area	450 cm^2
Field of view	1° (collimated)
Energy range	3-120 keV
Energy resolution $\Delta E/E$	$0.25/\sqrt{E(\text{keV})}$; 0.03 at 60 keV

Phoswich Detector System

Exposed geometrical area	800 cm^2
Field of view	1.5° (collimated)
Energy range	15-200 keV
Energy resolution $\Delta E/E$	$1.40/\sqrt{E(\text{keV})}$

Wide Field Camera -2 units

Effective area per unit	250 cm^2 (through mask)
Field of view per unit	20°x20°
Angular resolution	5 arcmin
Energy range: centre of field	2-30 keV
full field	2-10 keV
timing (no image)	2-35 keV
Energy resolution	20% at 6 keV

are equipped with a rocking collimator in order to monitor directly and continuously the particle induced background and minimize the effect of its orbital modulation on the ultimate sensitivity.

The WFC's are position sensitive proportional counters watching the sky through a coded mask, and therefore combine a large field of view with an angular resolution in the order of a few arcmin. This instrument will be supplied by SRON/Utrecht, which is responsible together with the University of Birmingham, for a similar device (TTM) now operating on the Qvant module attached to the russian space station MIR.

Figg. 2 through 10 offer a synthetic illustration of some of the performances. Fig. 2 shows the broad band sensitivity of the NFI in a typical exposure time of 10^4 sec, while Fig. 3a,b shows the potentiality in the study of spectral variations over the 0.1 to 200 keV range. Figg. 4 and 5 illustrate the ultimate sensitivity respectively at low (C/S) and high (PDS) energies: in particular the confusion limit of the C/S is well below the limit of the ROSAT (E < 2 keV) All Sky Survey, hence the observational program could fully exploit this survey in the selection of representative samples of faint objects for detailed spectral studies up to 10 keV. Fig. 6 gives the sensitivity for the detection of the 6.7 keV iron line, and Fig. 7 represents a simulation of the expected counts from the central region (2' radius) of M87, illustrating the quality of both spectral resolution and count statistics than can be achieved. Fig. 8 is a simulation of the Her X-1 cyclotron line in absorption as measured by the HPGSPC. Finally Figg. 9 and 10 illustrate detection limits and ability to measure intensitivy variations of the WFC.

3. SCIENTIFIC PROGRAM: STRATEGY AND OBJECTIVES

SAX will be operated like an observatory in space, capable of 2000-3000 pointings in its minimum lifetime of 2 years. While the NFI will be prime most of the time, the WFC will be periodically used to scan the galactic plane to monitor the temporal behaviour of sources above 1 mCrab and to detect transient phenomena; thanks to the large field of view, the WFC will also be used to provide long term monitoring of bright selected sources while the NFI will carry out their sequence of pointed observations. The observational program will be held flexible enough to accomodate targets of opportunity, in particular those arising from the WFC surveys. The observations will be organized on the basis of a "core program", mainly devoted to systematic studies, and of a "guest observer program".

Given the instrumental capabilities and the ample spectral coverage, SAX can be profitably used in practically every one of the

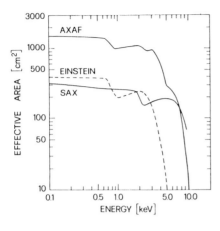

Fig. 1. SAX Concentrator total effective collecting area, compared with those of the Einstein Observatory and AXAF

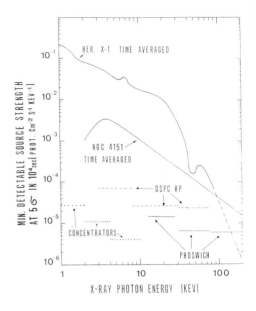

Fig. 2. SAX NFI broad band sensitivity for an exposure time of 10^4 s

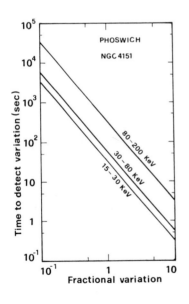
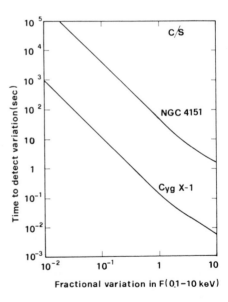

Fig. 3 a,b. Examples of ability of the C/S and the PDS to detect variations

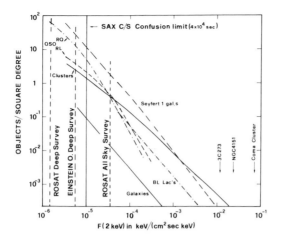

Fig. 4. Sensitivity-confusion limit of the C/S compared with the Einstein Deep Survey and the ROSAT All Sky Survey. The predicted source counts are from Setti and Woltjer (1982)

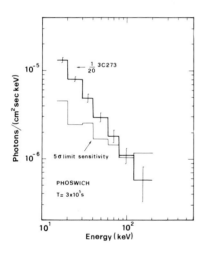

Fig. 5. High energy spectrum of the equivalent of 1/20 of 3C 273 obtainable with the PDS in 3×10^5 sec

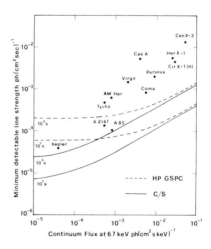

Fig. 6. Sensitivity of the C/S and the HPGSPC to the 6.7 keV iron line

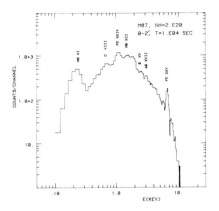

Fig. 7. Expected C/S counts in 10^4 sec from the centre of M87 (2' radius)

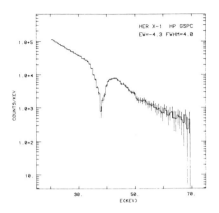

Fig. 8. Expected HPGSPC counts in 10^5 sec from Her X-1, showing the cyclotron line in absorption

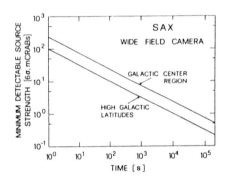

Fig. 9. WFC sensitivity

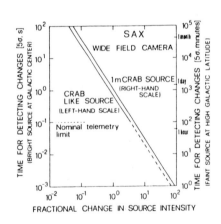

Fig. 10. WFC ability to detect variations

numerous astronomical areas where the previous missions have shown the
importance of X-ray measurements. What follows is merely a list of these
areas, with just a mention of some of the prominent aspects on which
SAX can give significant contributions:

- Compact galactic sources, containing a degenerate or collapsed stellar object: continuum shape and variability; temporal behaviour of spectral features (such as Fe fluorescence line, cyclotron lines, absorption phenomena) as a function of rotation and orbital phase; study of transient phenomena.
- Stars: spectra of coronal emission with sensitivity and spectral resolution comparable to those achieved with the SSS device on the Einstein Observatory, but with extension to 10 keV.
- Supernova remnants: spatially resolved spectra of galactic very extended (>> 1') remnants; spectra of the Magellanic Clouds remnants, a class of objects particularly suited for a systematic study because placed at the same distance from us.
- Active galactic nuclei: spectral dynamics of variable continuum; spectral shape up to 200 keV for objects down to 1/20 of 3C273, spectra of very distant (z = 3.2 for the equivalent of 3C273) objects up to 10 keV; soft X-ray excess and photoelectric absorption; Fe fluorescence line.
- Clusters of galaxies: spatially resolved spectra of the numerous nearby ($z \lesssim 0.1$) clusters; study of temperature gradients.
- Normal galaxies: spectra of the extended X-ray emission discovered with the Einstein Observatory.

Seen in the framework of past and future X-ray missions, SAX stands out for its wide spectral coverage, the balance in its performances between the low-medium (0.1 - 10 keV) and the medium-high (10 - 200 keV) energy bands, and it is fair to say that it will offer ample opportunities to make further steps in several of the directions opened and/or enlightened by the missions up to the Ginga satellite. Concerning future missions, ROSAT will fly (1989) before SAX: with its imaging and spectral capabilities below 2 keV it will have different goals and, at the same time, with the All Sky Survey offer a much wider basis for the selection of targets in the SAX observational program. Together with the japanese Astro D mission (spectroscopy below 10 keV), the russian Spectrum X-gamma mission (a complex one with emphasis on spectroscopy in the 0.1 - 10 keV band), both of which are foreseen for the time frame 1992/93, and the american XTE (timing in the 2-200 keV band), whose launch date has yet to be decided, SAX is a mission with a wide range of applications, which will be capable of significant contributions in the multiform field of X-ray astronomy before the next generation of satellites, AXAF and XMM, will become available in the second half of the 90's.

REFERENCES

Citterio, O., Bonelli, G., Conti, G., Mattaini, E., Santambrogio, E., Sacco, B., Lanzara, E., Brauninger, H., Burkert, W., <u>Proceedings of SPIE Conference</u>, San Diego, Aug. 1987, in press

Perola, G.C. 1983, <u>Proc. of Workshop on "Non-thermal and Very High Temperature Phenomena in X-ray Astronomy"</u>, eds. G.C. Perola and M. Salvati, Rome, 19-20 Dec. 1983, p. 175

Scarsi, L. 1986, <u>Proc. of Symposium on "Variability of Galactic and Extragalactic X-ray Sources"</u>, ed. A. Treves, Como, 20-22 Oct. 1986, p. 233

Setti, G., Woltjer, L. 1982, <u>Astrophysical Cosmology</u>, eds. H.A. Brück, G.V. Coyne and M.S. Longair, Pontif. Acad. Scient. Scripta Varia N. 48, p. 315

Spada, G.F. 1983, <u>Proc. of Workshop on "Non-thermal and Very High Temperature Phenomena in X-ray Astronomy"</u>, eds. G.C. Perola, and M. Salvati, Rome, 19-20 Dec. 1983, p. 217

THE ADVANCED X-RAY ASTROPHYSICS FACILITY

Harvey Tananbaum
Harvard-Smithsonian Center for Astrophysics
Cambridge, Massachusetts

1. Introduction

The Advanced X-ray Astrophysics Facility (AXAF) will be a long-lived observatory intended to carry out X-ray studies from space. Much of the design for AXAF has been predicated on the highly successful HEAO-2/ Einstein X-ray observatory that NASA operated from 1978 until 1981. AXAF has been under study for several years under the direction of Marshall Space Flight Center assisted by our group at Smithsonian Astrophysical Observatory. AXAF has recently completed the preliminary design/definition stage, Phase B. Two teams of industrial contractors have studied the overall observatory to understand the requirements, to develop design concepts, and to formulate an overall technical basis for building AXAF. Moreover, four focal plane scientific instruments and two sets of gratings have been selected and studied for nearly three years. AXAF is a prime candidate for a formal new start to begin construction in FY89, with a planned launch date in 1995.

Fig. 1 - Artist's Sketch - AXAF

An artist's sketch of AXAF is shown in Figure 1. The X-ray telescope is indicated at one end with an array of scientific instruments at the other end, connected by a long metering structure or optical bench. Since only one focal plane instrument at a time can use the telescope, the mirror will be moved to direct X- rays to the selected instrument. The overall length is 14 meters, the diameter is about 4 meters, and the weight is approximately 12,000 kg. The figure shows the solar arrays used to generate $\sim$ 2500 watts of electrical power and the antennae for communicating with the ground through the Tracking and Data Relay Satellites. There is also a spacecraft module which contains standard support hardware.

AXAF will probe the nature of cosmic violence on all scales from solar system objects to stars to superclusters of galaxies. It will address fundamental questions, such as the details of the dynamo process believed to power stellar coronae; the specifics of stellar evolution and supernova explosions; the properties of supermassive black holes believed to provide the power at the centers of active galaxies and quasars; the contribution of hot gas to the mass of the universe; the presence of dark matter in galaxies and clusters of galaxies; and the age and size of the universe.

The capability of AXAF to carry out such broad objectives and to contribute to many other fields was key in obtaining the recommendation of the Astronomy Survey Committee (Field Committee) for AXAF as the number one priority, major new program for all of ground-based and space astronomy in the U.S. for the next decade. In the next section, we present an overview of AXAF as an observatory. Then several examples of science projects are discussed to illustrate the scientific importance and power of this facility.

2. Overview of AXAF Observatory

Perhaps the most technically challenging side to AXAF is the high resolution X-ray mirror. The AXAF design consists of 6 nested sets of grazing incidence X-ray mirrors with the largest having a 1.2 m diameter (twice that of HEAO-2). Figure 2 shows the effective mirror area in cm^2 as a function of energy from 0.1 to 10 keV for AXAF and HEAO-2. At low energies AXAF has about 1700 cm^2 or about 4 times the area of HEAO-2, and about 1.5 times that of ROSAT (the FRG/UK/US mission scheduled for launch in 1990). The AXAF area still exceeds 1000 cm^2 at 3 keV. The 10m focal length of AXAF (vs. 3 m for HEAO-2) leads to greatly increased area at higher energies, extending AXAF beyond 8 keV, whereas the HEAO-2 high energy cutoff was around 4 keV. This opens opportunities for carrying out plasma diagnostics on iron emission and absorption features in the 6-7 keV range and also extends the energy band available for making broadband, continuum spectral measurements.

AXAF will be more sensitive than HEAO-2/Einstein by a factor of 50 to 100 in source detection (and up to 1000 for high resolution spectroscopy). The increased collecting area is one of three major factors which lead to this increase in sensitivity. A second factor is the improved detector performance, particularly with higher quantum efficiency and lower background. The third factor is higher angular resolution due to improvements in mirror figure and surface smoothness.

Figure 3 shows the predicted AXAF performance and actual HEAO-2 angular resolution at 2.5 keV, by plotting the fraction encircled energy versus radius. For AXAF 60% of the signal will be concentrated in a 1/2" radius, a factor of almost 20 times better than HEAO-2 and $\sim$5 times better than ROSAT. This concentration of signal means that for point X-ray sources the detection-cell size and its associated background can be greatly reduced, thereby increasing sensitivity.

To develop the required technology and to demonstrate that the desired performance can be achieved, we have built (at Perkin-Elmer) an X-ray test mirror (scaled to about 2/3 the dimensions of the second most inner AXAF pair). A first X-ray test of this mirror showed it to have the best figure ever achieved, but also uncovered a mid-frequency (mm scale) ripple which is being corrected in a second polishing cycle. In this cycle, we have also corrected a low frequency software problem and expect to improve the surface finish to between 5 and 10Å, thereby further reducing scatter at higher energies. X-ray test data should be available in mid-1988 to confirm this improved performance (and demonstrate achievement of the levels required for AXAF).

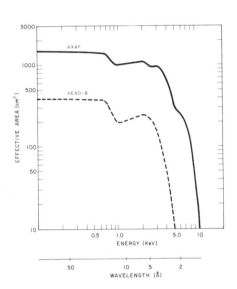
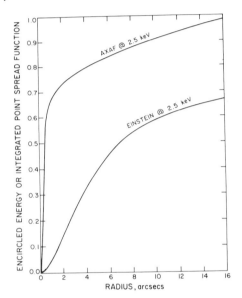

Fig. 2 - Area vs Energy   Fig. 3 - Encircled Energy vs Radius

There are four "candidate" focal plane instruments (the initial launch complement may only contain 3 of these), along with two sets of objective gratings. Brief summaries of these instruments are provided below; more detailed descriptions are available in *Astrophysical Letters and Communications (1987, Vol. 26, Nos. 1-2)*

Figure 4 shows the baseline layout for the AXAF CCD Imaging Spectrometer or ACIS instrument (PI: G. Garmire, Penn State). Six chips are configured to cover most of an 11.6' by 13' region in the center of the telescope field of view. (Chips numbered 3 and 4 are placed over part of the readout section(s) of the other 4 chips). The readout uses the frame store technique in which 1/2 the chip is totally blocked from incoming X-rays. For readouts, a rapid transfer occurs from the observing half to the blocked half and then a slower, lower noise readout occurs while data accumulation resumes in the active section. The CCD of choice is a virtual phase device of 850 x 750 (375 useable for detection) pixels of 22μ dimension, corresponding to a pixel 0."45 on a side. The current baseline is a deep depleted derivative of the TI4849 chip, and this is a primary area of ongoing development work. The energy resolution already achieved is $\sim$ 120 eV at 1.5 keV, rising to $\sim$ 150 eV at 6 keV. Currently a quantuum efficiency of 30% has been obtained at 1.5 keV, and the deep depletion ($\sim 40\mu$) is intended to raise this to ~ 85 %. The deep depletion also increases the ACIS sensitivity by reducing the background due to cosmic rays and by reducing charge spreading effects. The planned CCD array for reading out the gratings is also shown in the figure.

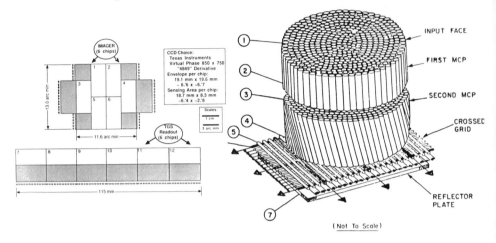

Fig. 4 ACIS - Layout   Fig. 5 - HRC Schematic

The High Resolution Camera (or HRC) (PI: S. Murray, SAO) is a derivative of the Einstein and ROSAT HRI's. The main elements of the detector - the two sets of channel plates and the crossed grid readout are shown in Figure 5. This 100 mm x 100 mm detector has a 32' x 32' field of view with a spatial resolution of 25μ or 0.5" FWHM. Its quantum efficiency will be substantially improved over Einstein through use of CsI (or KBr) coatings, with efficiencies ranging from 20% to 50% in the 0.1-3 keV band and from 10% to 20% in the 3-8 keV band. The AXAF HRC will also have modest energy resolution ($\Delta E/E \sim 1$ at 1 keV). The large FOV makes the HRC well suited for reading out low-energy transmission gratings, and in fact two detector units may be used on the diagonal to get the maximum linear coverage needed. The HRC also has significantly lower background than Einstein due to improvements in the channel plates and due to use of plastic anticoincidence to reduce the charged particle induced background.

Figure 6 illustrates the principles of operation of the X-ray Spectrometer (XRS) or calorimeter (PI: S. Holt, GSFC). X-rays are stopped in a thin layer of high Z material (currently mercuric cadmium telluride). The electrons produced interact in the doped silicon substrate raising its temperature by an amount proportional to the energy of the incident X-ray photon. Thermister implants are used to sense this temperature rise. An array of detectors provides the XRS with a FOV of 1' x 1'. The XRS covers the energy range from 0.1 to 10 keV, with an efficiency of over 90% for energies above 0.5 keV. The technique promises to achieve energy resolution $\prec 10$ eV FWHM. Developments to date have substantially reduced both the energy conversion noise and the readout noise, with a best resolution, so far, of about 17 eV FWHM. There are also engineering challenges involved in cooling the XRS to temperatures approaching 0.1°K, for extended times (years) in orbit, but the scientific promise of the XRS makes these worth pursuing. The XRS also is accompanied by a silicon detector derived from the Einstein SSS.

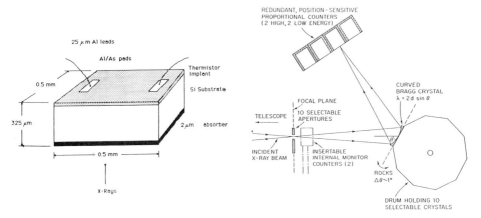

Fig. 6 - XRS Schematic   Fig. 7 BCS Schematic

The fourth focal plane instrument in the Bragg Crystal Spectrometer, or BCS (PI: C. Canizares, MIT), which also has substantial heritage from Einstein. The operation of the BCS is indicated schematically in Figure 7. X-rays in the converging beam from the telescope pass through a selectable aperture and the diverging beam impinges on the selected curved crystal diffractor. The crystal parameter (lattice spacing d) and the selected angle θ determine the wavelength λ for X-rays that will be coherently reflected and refocused onto the position sensitive proportional counter. Resolutions ($E/\Delta E$) up to 2000 can be achieved at some energies.

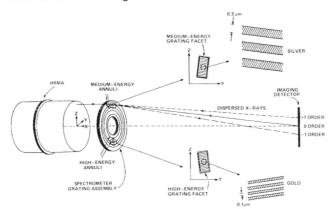

Fig. 8 - Grating Schematic

Figure 8 schematically shows the operation of the transmission gratings. The High Energy Transmission Grating, or HETG, and M(edium) ETG (PI: C. Canizares, MIT) form one set, and the L(ow) ETG (PI: A. Brinkman, Utrecht) a second set; either set can be inserted behind the AXAF telescope. The gratings are optimized for different energies by proper choice of grating material and thickness. The HETG emphasizes $E \succ 4$ keV, covers the inner three mirrors, and uses 1.0μ thick gold (with a period of 0.2μ or 5000 lp/mm). The METG emphasizes $E \prec 4$ keV, covers the outer three mirrors, and uses 0.5μ

thick silver (with a period of 0.6μ, or 1700 lp/mm). The LETG uses 0.6μ thick gold bars (1024 lp/mm) and is designed for optimized performances at still lower energies. The gratings can be read out by either ACIS or HRC, with advantages to specific pairings. Resolutions (E/ΔE) can be expected to range from 100 to 1000 as a function of energy.

3. AXAF SCIENCE

Deep exposures with AXAF, aimed at resolving as much of the all-sky background as possible, require large collecting area and high angular resolution for the detection and identification of very faint sources. Cosmological studies on the formation and evolution of quasars, galaxies, and clusters involve young and therefore distant objects. This usually translates into small numbers of photons reaching us, driving the AXAF mirror diameter (1.2m) and angular resolution (0.5 arc sec). Long focal length (10m) is needed for the mirror to reach beyond 7 keV for study of iron lines and to measure continuum temperatures and spectral indices.

3.1 Deep Surveys

Figure 9a shows a piece of the deepest Einstein exposure lasting 300,000 sec; the picture covers 1/100 sq. deg. and shows two faint X-ray sources, which have been identified as distant galaxies or groups of galaxies. We know about the mystery of the all-sky X-ray background - is it comprised totally of discrete sources, or is there significant diffuse emission? Sources such as these - to the limit of Einstein - add up to $\sim$ 25 or 30% of the X-ray background. Figure 9b shows an AXAF simulation for the same piece of sky, and same exposure time. A conservative model was used with integrated source numbers N($\succ$S) increasing as $S^{-1.15}$. With AXAF we may expect to see a few hundred sources as shown in the simulation. The important issue is what AXAF will actually see. X-ray surveys are very efficient at selecting distant quasars and clusters. QSO's are thought to have formed at redshifts from 3 to 10. Einstein could just reach brightest quasars at redshift 3.5. AXAF will be able to see QSO's, if they exist, out to redshifts of order 5 or 6 and may be able to determine when QSO's first appear and how early QSO's behave. The situation is similar for clusters, which are thought to have formed at redshifts from 1 to 3. Einstein couldn't see this far away or back in time; but AXAF will. AXAF also has the potential for discovering new classes or subclasses of objects - such as X-ray bright protogalaxies which have been predicted by some to explain the X-ray background.

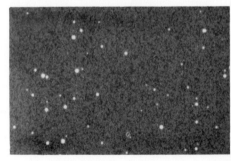

Fig. 9a - Einstein Deep Survey   Fig. 9b - AXAF Simulation

3.2 Mass Determinations in Galaxies and Clusters of Galaxies

The use of X-ray observations to measure mass is one of the most important capabilities of AXAF. To illustrate how X-ray observations are used to determine mass in galaxies and clusters, we take the Einstein data obtained for M87, the giant elliptical galaxy near the center of the Virgo cluster of galaxies. (See Fabricant, Lecar, and Gorenstein 1980, and Fabricant and Gorenstein 1983 for a detailed discussion of the M87 X-ray observations and analysis.) The Einstein image of M87 shows smooth, extended, approximately azimuthally symmetric X-ray emission. This emission extends beyond 60' in radius covering at least 1/2 Mpc in diameter. The key is that the X-ray emitting gas traces the underlying gravitational potential arising from the total mass present.

We average azimuthally to simplify the calculations to be a function of radial coordinate only. (Note: this is not necessary in order to apply the method, in general.) The gas is assumed to be in hydrostatic equilibrium, with the outward directed force of thermal gas pressure gradient at radius r balanced by the inward directed gravitational force due to mass interior to r. This assumption is supported by the radial profile of the gas which is significantly flatter than expected either for gas freely falling in or flowing out of the galaxy. It is also supported by the fact that the gas cooling time is much longer than the free-fall time and by the fact that there is no observed temperature increase towards the center which would be expected if the gas were settling or expanding adiabatically. (See Fabian, 1988, for a further discussion of these points.) This situation is described by:

$$\frac{dP_{gas}}{dr} = -\frac{GM(\prec r)\rho_{gas}}{r^2} \tag{1}$$

where P_{gas} is the gas pressure, ρ_{gas} is the gas density, G is the gravitational constant, and $M(\prec r)$ is the total mass interior to radius r. We can also relate gas pressure, density, and temperature through the ideal gas law:

$$P_{gas} = \frac{\rho_{gas}\, k\, T_{gas}}{\mu m_H} \tag{2}$$

where μ is the mean molecular weight of the gas (0.6 here), m_H is the mass of the hydrogen atom, and k is Boltzmann's constant. Differentiating the second equation and substituting into the first and simplifying, we express $M(\prec r)$ in terms of the gas temperature and the logarithmic gradients of gas density and gas temperature:

$$M(\prec r) = -\frac{kT_{gas}}{G\mu m_H}\left[\frac{d\log\rho_{gas}}{d\log r} + \frac{d\log T_{gas}}{d\log r}\right]r \tag{3}$$

The required quantities on the right-hand side of equation (3) can be determined from the X-ray observations.

Quantitative results from the Einstein observations of M87 are presented in Figure 10, showing the surface brightness (summed in azimuth) in erg cm^{-2} s^{-1} arcmin^{-2} versus angular radius in arc min. The data are shown along with representative error bars. The conversion from observed counts to 0.2 - 4 keV flux makes use of the average temperature measured and is not very sensitive to the value used. An analytic expression, introduced by Cavaliere and Fusco-Femiano (1976) to describe a hydrostatic, isothermal gas in a spherical gravitational potential, can be conveniently fit to the observations as shown by the solid curve and represented by:

$$S(r) \propto (1 + (r/a)^2)^{-n} \tag{4}$$

A χ^2 fit to the data determines a = 1'.62 ± 0.'28 and n = 0.81 ± 0.01 (90% error limits). Note that an analytical fit is not really necessary since a numerical analysis also can be carried out. The surface brightness profile S(r) is then deprojected to determine the density profile $\rho_{gas}(r)$ which, when folded through the telescope and detector would produce the observed S(r). For these data an isothermal gas profile is sufficiently accurate, but in a more general case the temperature versus radius can affect the deprojection and an iterative process may be needed. The resultant gas density profile is given by:

$$\rho_{gas}(r) \propto (1 + (r/a)^2)^{-(n+1/2)/2} \tag{5}$$

and the logarithmic density gradient is computed from this expression as

$$\frac{d \log \rho_{gas}}{d \log r} = \frac{-(n+1/2)(r/a)^2}{1 + (r/a)^2} \tag{6}$$

These expressions are evaluated for M87 using the values of n and a given above.

Figure 11 presents the measured temperature data versus radius observed for M87 with two Einstein instruments. The dashed diamonds give the results for the non-imaging solid state spectrometer (from Lea, Mushotzky, and Holt 1982), and the solid lines give the temperature and 90% confidence limits obtained with the imaging proportional counter (Fabricant and Gorenstein, 1983). In overlapping regions, the agreement is quite good. Data in the inner few arc minutes show evidence for cooling. While this is interesting scientifically, it is not very relevant here since most of the mass is found at larger radii. The available temperature data extend out to almost 30' radius and suggest temperatures of ~ 3-4 keV fit the data. In any case, these data are used to set limits on temperature gradients,

$$-0.4 \preceq \frac{d \log T_{gas}}{d \log r} \preceq 0.3 \tag{7}$$

which are used in equation (3) along with the density gradient to determine M(≺ r).

The result for the gravitational mass (in solar units) interior to radius r is plotted versus r (in arc min) in Figure 12. The optical measurements of Sargent et al (1978) extend to ~ 1' and find a few x $10^{11} M_O$ in the central part of the galaxy, but the velocity-dispersion data provide little information at larger radii. The X-ray data constrain the total mass to lie between the two solid lines shown in Figure 12 which incorporate the uncertainties in the measured gradients and the temperature normalization. At r = 50', or about 220 Kpc, the mass interior to r is (3 - 5) x $10^{13} M_O$, or more than 100 times that seen in the visible. The X-ray emitting gas, itself, accounts for ~ 3 - 5% of the mass.

Most of the mass is not radiating in the X-ray or the optical bands. underlying potential. This analysis could be carried out using Einstein data for M87 because it is a bright and relatively cool X-ray source. However, many clusters have higher gas temperatures (above the high energy cutoff of Einstein) and require the added bandwidth coverage of AXAF. With its increased area and bandwidth, AXAF can greatly improve on measurements of surface brightness and temperatures, which will extend these studies to much fainter and more distant objects. For M87, a factor of 10 improvement in precision will be possible - mass will be measured to an accuracy of ~ 10% which is spectacular by astronomical standards.

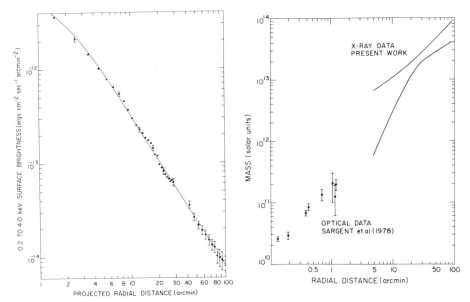

Fig. 10 - M87 Surface Brightness vs Radius

Fig. 12 - M87 Mass vs Radius

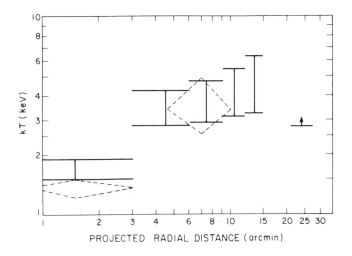

Fig. 11 - M87 Temperature vs Radius

This technique can be applied to galactic halos and to groups and clusters of galaxies, to determine radial mass distributions as well as total amounts of matter present. This will be useful for determining the distribution of dark matter on different size scales, which can provide critical information for choosing between hot, low mass candidate particles and cold, high mass candidates for the composition of the dark matter.

3.3 Cosmological Measurements

AXAF can be used to make cosmological measurements via the Sunyaev-Zel'dovich effect, combining X-ray and microwave observations of clusters of galaxies to obtain cosmology independent distance determinations. This is schematically indicated in Figure 13.

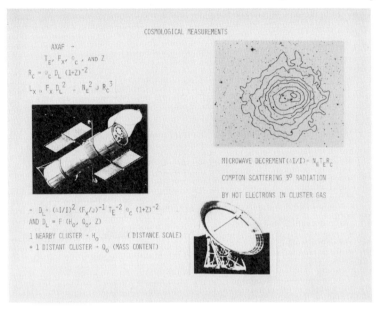

Fig. 13 - Sunyaev-Zel'dovich Effect

X-ray observations provide measurements of cluster temperature T_E, X-ray flux F_x, and angular radius θ_c. Redshifts can be determined from optical or X-ray data. The linear radius, R_c can be expressed in terms of the angular radius times the distance. Here D_L is the luminosity distance. Note the $(1+z)^{-2}$ factor relating the angular diameter distance to the luminosity distance. Clusters emit by thermal bremsstrahlung, and therefore luminosity (proportional to flux times distance squared) can be determined as proportional to electron density, N_E, squared times volume times j, where j is the X-ray emission per unit emission measure for the source temperature.

Ground-based microwave observations provide data on the effect of Compton scattering of the 3° background radiation by the hot electrons in the cluster gas, with the change in the microwave intensity $\Delta I/I$ proportional to the product of electron density times cluster temperature times radius. These observations then can be combined to determine the distance to the cluster. The key element in the use of this method to determine absolute distances is the different functional dependences of the measured quantities on N_E and R_c.

Once we have cosmology independent distance measures, the results can be compared to values obtained for a particular cosmological model such as a Friedmann Universe characterized by Hubble constant H_o and deceleration parameter q_o (and cosmological constant $= 0$). The data on one cluster can give H_o and, in principle, for one more distant cluster q_o as well. This provides key information on the size, age, and ultimate fate of the Universe.

What factors are likely to limit the precision of these measurements? Van Speybroeck (1984) has carried out a thorough analysis to assess this question via simulations based on Einstein observations and currently available microwave data. A few of the results are summarized in Table 1. Two clusters, A2218 and 0016+16, at redshifts of 0.174 and 0.541 respectively, have planned AXAF exposures of 3×10^4 and 10^5 seconds. The table lists $1\,\sigma$ measurement errors (statistical only) for the X-ray flux normalization Fx/j, temperature T_E, and angular radius θ_c. The fractional contributions to the distance measurement uncertainty ($1\,\sigma$) from the AXAF X-ray data only are then given.

SUNYAEV-ZEL'DOVICH EFFECT

Cluster	A2218	Cl 0016+16
Z	0.174	0.541
AXAF Exp Time (s)	3.10^4	10^5
Fractional Errors (1σ)		
[AXAF CCD]		
Fx/j	.014	.020
T_E	.022	.030
θ_c	.012	.022
[Current μwave]*		
ΔI	.072	.083
Fractional Distance Error (1σ)		
AXAF X-ray only (CCD)	.04	.06
AXAF X-ray + current μwave	.15	.18
AXAF X-ray + improved μwave**	.08	.10

*Moffet and Birkinshaw (1987): 1.5cm; 106" FWHM beam; fits μwave decrement including off-center measurements.

**Assumes a factor of 2 improvement over current cm measurements.

Table 1

With respect to the microwave decrement measurements, there has been significant progress in the past few years, and the results obtained for these two clusters by Moffet and Birkinshaw (1987), at 1.5 cm, with a 106" FWHM beam give fractional errors (statistical) at 7 and 8% of the observed decrements. Combining these in-hand measurements with projected AXAF X-ray data, leads to predicted precisions of 15% and 18% in the determination of distances to these 2 clusters. A factor of 2 improvement in the precision of the microwave results would reduce the distance uncertainties to 8% and 10% respectively (and high precision measurements of the decrement at 1 mm could do even better). Note that the required precision of the X-ray data will require AXAF calibrations of order 2% which is also a challenging requirement.

At z=1, the difference in distance between the two most interesting cases of $q_o = 0.5$ and $q_o = 0.0$ is about 20%, so either we need data for some clusters at $z \geq 1$ (along with the better microwave data) or we need to use a number of clusters to reduce the overall statistical uncertainty to obtain useful estimates of q_o. Estimates of q_o allow us to carry out calculations of the age of Universe ($1/H_o$ vs. $(2/3)(1/H_o)$ for $q_o = 0.0$ vs $q_o = 0.5$) and to predict the ultimate fate of the universe (open for $q_o \succ 0.5$).

4. Conclusion

These are just a few examples of the exciting science which will be done with AXAF. The technology required for this program is well in-hand, and we are optimistic that approval to proceed will soon be forthcoming.

REFERENCES

1. Cavaliere, A. and Fusco-Femiano, R. 1976, *Astr. Ap.* **49**, 137.
2. Fabian, A. 1988, *this volume*.
3. Fabricant, D. and Gorenstein, P. 1983, *Astrophys. J.* **267**, 535.
4. Fabricant, D., Lecar, M., and Gorenstein, P. 1980 *Astrophys. J.* **241**, 552.
5. Lea, S., Mushotzky, R., and Holt, S. 1982 *Astrophys. J.* **262**, 24.
6. Moffet, A. and Birkinshaw, M. 1987, *in preparation*.
7. Sargent, W.L.W., Young, P.J., Boksenberg, A., Shortridge, K., Lynds, C.R., and Hartwick, F. D.A., 1978, *Astrophys. J.* **221**, 731.
8. VanSpeybroeck, L. 1984, SAO Proposal P1394-2-84 to NASA.

THE XMM-MISSION

J.A.M. Bleeker
SRON - Laboratory for Space Research Utrecht
Beneluxlaan 21, 3527 HS Utrecht, The Netherlands

and

A. Peacock
ESA - Space Science Department ESTEC
P.O. Box 299, 2200 AG Noordwijk, The Netherlands

ABSTRACT. The XMM-mission is one of the four cornerstone projects in the ESA long term programme for space science. The satellite observatory comprises a high throughput facility for X-ray spectroscopy with the aid of dispersive and non-dispersive imaging spectrometers. The areas of major impact for astrophysical research are briefly reviewed, subsequently the diagnostic power of such a facility for the study of several sites of hot thin astrophysical plasma is illustrated with a few computer simulations, based on the configuration of a model payload.

1. INTRODUCTION

The X-ray Multi-Mirror (XMM) mission is one of the cornerstone projects identified in the ESA long term programme for space science: Horizon 2000[1]. The objective of XMM is largely focussed on X-ray spectroscopy, with sufficient sensitivity for acquiring high quality X-ray spectra of large samples from practically all classes of astronomical populations. The requirements for such a mission on throughput, bandwidth, spatial and spectral resolution were discussed by the scientific community at an X-ray Spectroscopy Workshop in 1985 at Lyngby[2]. Since then, various adjustments have taken place in the course of a detailed definition study. XMM has now been scheduled by ESA as the second cornerstone mission to be executed, leading to a launch in 1998 with a planned operational lifetime of ten years. The XMM-observatory is envisaged as a space-borne facility in astronomy accessible by a wide user community engaged in astrophysical research work.

In the following paragraphs we first summarize the major impact of the mission, then briefly describe the present configuration of the model payload and finally highlight a few research areas dealing with the astrophysics of hot thin plasmas, viz. the subject of this ASI.

2. AREAS OF MAJOR IMPACT

The value of X-ray spectroscopy as a powerful diagnostic in the study of many astrophysical phenomena has been outlined in the main lecture series during this course. Potentially, spectral measurements in the X-ray domain provide the basis for a quantitative assessment of the physical state and evolution of the hottest objects in the universe and for the study of the dynamics of many of the most energetic phenomena like flares, bursts, pulsations, beaming and shocks.

Although we have known since the first X-ray imaging mission (the Einstein Observatory), that X-ray emission is a common feature of virtually all classes of astronomical objects, detailed spectral information is as yet only available for a handful of the brightest sources. Nevertheless the little X-ray spectral data that has been acquired has proven to be of major importance for the understanding of the physical processes operating in these objects.

The high throughput imaging, the large bandwidth and the spectral resolving power of the XMM-observatory have been tailored and optimized to provide the next major leap forward in high energy astrophysics. Following after the ROSAT-all sky survey XMM will have the capability to provide high quality X-ray spectra of millions of objects ranging from the nearest stars to the most distant quasars.

Although XMM will undoubtedly contribute to almost every conceivable topic in high energy astrophysics, some outstanding contributions are now highlighted:
- The study of the large and medium scale structure of (hot) matter in the universe, i.e. its formation and evolution in density, temperature and composition. Owing to its imaging and spectroscopic capability XMM can measure the temperature and density profiles of X-ray emitting hot gas in clusters of galaxies out to very large radii, thereby tracing the total mass distribution out to a distance where most of the dark matter resides. Its sensitivity is sufficiently great to measure spectra and redshift of distant clusters and to explore when heavy elements were injected into the intracluster medium. This is a crucial issue for the study of the formation of galaxies and clusters. In addition, the presence of cooling flows in clusters can be measured out to high redshift ($z > 1$), which will indicate whether many galaxies could have been formed in this manner. Similarly the distribution of dark matter, the presence of cooling flows and the abundance of high Z-elements in halos of massive galaxies can be studied.

- The establishment of the evolution of and the physical structure in the inner regions of Active Galactic Nuclei. Hundreds of thousands of wide band spectra of distant AGNs down to the confusion limit of $2 \cdot 10^{-15}$ ergs cm^{-2} sec^{-1} will become available.
    This large sample of AGN spectra will be totally serendipitous. With such an unbiased survey questions can be addressed like: is the canonical index of 0.7 for the continuum slope independent of

luminosity and redshift (and therefore an intrinsic feature of how the central engine and the core region are constituted), what are the characteristic variability time scales and their relation to AGN luminosities and central mass, what is the origin of the low-energy X-ray excess observed in some AGN's, what is the nature of the absorbing medium observed in certain Seyfert galaxies and BL-LAC objects? In addition, coupling these data to the wide band spectra for distant clusters and galaxies will unequivocally determine the discrete source contribution to the hard X-ray diffuse background. Most likely, therefore, XMM will be capable of determining the origin of the diffuse X-ray background, thereby solving one of the major outstanding problems in observational cosmology. Apart from the new insights in the evolution of quasars and galaxies this could result in the identification of a new major contributor to the mass of the universe.

The understanding of the physical characteristics of degenerate stars, black holes, neutron stars or white dwarfs, and their environment in close binary systems. Time resolved X-ray spectroscopy is a unique diagnostic tool for probing the properties and morphology of accretion flows under extreme physical conditions with respect to gravity, radiation pressure and magnetic field. This applies in particular to the boundary layers near the inner accretion disk (column, torus) and the surface or magnetosphere of the degenerate star. In the particular case of the high luminosity X-ray binaries, which comprise either a neutron star or a black hole as the degenerate component, XMM offers the unprecedented capability of time resolved spectral studies of large homogeneous samples of these objects in galaxies even outside of the Local Group. Apart from a dramatic increase of the number of systems which can be analyzed, this introduces the opportunity of comparative studies with respect to luminosity function, distribution of orbital periods, fraction of population I and II sources and intercorrelation with specific galaxy features such as spiral arms, OB associations, distribution of globular clusters.

Determination of the dynamics and physical structure of various sites of hot, optically thin, plasma like stellar coronae, stellar supernova remnants and the interstellar medium.

Regarding supernova remnants, XMM provides, in addition to the capability for spatially resolved spectroscopy on the galactic remnants, adequate sensitivity for statistical spectral studies of SNR-samples in other members of the Local Group (e.g. as far as M33). Since these galaxies are at well defined distances and possess rather uniform absorption measures, accurate determination of spectral luminosities and emission measures (abundances) of supernova remnants are possible. Owing to its high throughput, XMM is ideally suited for imaging and spectroscopy of low surface brightness sources, such as the hot component of the interstellar medium in our own and other galaxies, and should for example determine the nature of the X-ray emission observed from the Galactic Ridge.

Due to its high spectral resolution and capability for time resolved spectroscopy, coupled to the availability of a huge (and complete) sample of stars, outstanding questions with respect to stellar coronae which can be addressed by XMM, include the investigation of the heating mechanism and structure of coronae by analyzing the dependence of the X-ray properties on the radiation field, mass flow, convection, rotation and age. This includes the measurement of the variability of coronal X-ray emission over a wide range of time scales (minutes to days) which is related to the principal heating mechanism and structural properties.

3. THE MODEL PAYLOAD

The XMM model payload comprises three co-aligned high-throughput imaging telescopes equipped with dispersive and non-dispersive spectrometers. The present configuration of the model payload is the result of an extensive trade-off study made by the XMM-Telescope Working Group[3] and the XMM-Instrument Working Group[4]. Figure 1 shows a schematic of a typical telescope configuration. The Wolter-I-type grazing incidence optics has been optimized for maximum throughput in the 2-8 keV energy range while retaining an adequate angular resolution of equal or better than 30 arcseconds half-energy-width (HEW) at 7 keV. Each optical module comprises 58 densily packed confocal mirror shells with an outermost diameter of 70 cm and a focal length of 750 cm. The mirror shells are composed of carbon-fibre-reinforced expoxy (CFRE) with a high quality X-ray reflecting surface (Iridium) replicated onto it. Figure 2 displays the throughput of an assembly containing three modules as a function of X-ray energy. Comparison with other, passed and planned, missions is also shown.

At the exit of the optical module (see Figure 1) a stack of flat blazed reflection gratings, replicated from a single master, intercepts half of the converging X-ray beam. This grating-stack provides the wavelength dispersion required to meet a spectral resolution adequate to resolve the triplets of He-like oxygen and several heavier He-like ions. The grating grooves are oriented normal to the plane containing both incident and diffracted radiation (in-plane mounting), the convergent-beam grating aberrations are eliminated by a smooth variation in the groove spacing across the grating plane. The spectrum is read-out by means of a dedicated CCD-strip detector, a wavelength range of 4-50 Å can be covered with the necessary resolution by employing the minus first, second and third grating orders (see Figure inset).

The remaining half of the X-ray beam converges onto an imaging detector in the telescope focal plane. As shown in Figure 3, this detector comprises a butted array of six CCD's with negligible dead space in between, covering a field of view of 30 arcminutes. The optics-CCD combination provides the basic low-resolution spectroscopy mode of the payload, which is used simultaneously with the medium resolution spectroscopy mode provided by the reflection grating spectrometer. Figure 3 shows the spectral resolution of the CCD-imager as a function of X-ray energy, line complexes of interest are superimposed for visualization. An objective Bragg-crystal assembly in front of one telescope is presently being studied to accommodate for high resolution imaging spectroscopy in the iron line complex (6-8 keV).

A free-standing Optical Monitor, bore-sighted with the X-ray telescope viewing direction, is incorporated to allow for simultaneous optical and X-ray coverage of various sources, in particular for the study of spectral variability on a wide variety of timescales. The optical monitor covers a field of view of 8 arcminutes with a resolution of 1 arcsecond and can detect a star with a magnitude $m_B = 24.5$ in a thousand seconds.

The XMM-payload will be launched by an Ariane 4 and put in a highly excentric orbit for reasons of high operational efficiency (> 70%), long uninterrupted observing periods and real time observatory operations.

4. SCIENTIFIC PERFORMANCE

In this paragraph we shall highlight a few examples of the results to be expected from potential observation programmes to be carried out with the XMM-observatory.

4.1. The diffuse X-ray background.

The problem of the "diffuse" background (DXRB) is the oldest observational puzzle of X-ray astronomy. Because of its relative uniformity it is clear that the DXRB originates at high redshift. There are two general models of this emission (i) that it is made-up of weak unresolved "point-like" sources or (ii) that the emission is actually diffuse on relatively large angular scales ($\sim$1 arcmin). A point source interpretation implies a substantial evolution both in the luminosity and spectrum of active galactic nuclei on a cosmological scale or the existence of a new population of objects at high redshifts. A diffuse source model implies that a substantial fraction of the total baryons in the universe have been heated to a high temperature at a relatively early epoch, this important contributor to the mass density of the universe can only be observed in the X-ray band.

Figure 4 illustrates the capability of XMM to resolve the DXRB, as lower flux thresholds are reached (a useful reference point is the Einstein Deep Survey Flux limit, DSF). It is clear that if the log N-log S curve flattens by 1/2 at a flux just below the Deep Survey level (as indicated by recent work) then one makes up only 2/3 of the diffuse flux at levels of 0.01 DSF! In this case it would be difficult, if not impossible, to totally resolve the DXRB with any "forseeable" mission. The XMM spatial resolution of 0.2 square arcminutes implies that source confusion sets in at $\sim$120 sources per XMM field (0°.5). This is rather nicely compatible with the XMM sensitivity at exposures of

$3 \cdot 10^4$ seconds (see Figure 4), a typical exposure necessary to obtain high quality spectra. Thus for direct imaging of the DXRB a "deep survey" with XMM is not needed. XMM will in fact obtain high quality spectra (e.g. measure a power-law index to an accuracy of ±0.1) of five thousand serendipitous sources during its lifetime (10^4 sky fields) as well as obtain broad-band spectra of over a million of such sources. Now if the DXRB is not composed of point sources at the flux level XMM is capable of resolving, the spectrum of the residual component can be well determined, since the spectra of AGN's, clusters and galaxies obtained serendipitously during every observation will form a template of the spectra of weak objects to be subtracted from the known spectrum of the DXRB.

In addition to the direct resolution of the point-like objects XMM will be able to determine if the DXRB is composed of small, but extended objects, such as primeval galaxies or clusters. The existence of such objects will be revealed either by direct imaging, if they are larger than 30 arcseconds (1 Mpc at a redshift of 2), or by a fluctuation analysis if they are smaller (in a manner similar to that of determining the deep radio counts). An object whose surface brightness is twice that of the DXRB and whose size is 1 square arcminute will be detected at 3 sigma in $6 \cdot 10^4$ seconds. At a redshift of 1 this would correspond to a galaxy of luminosity of $3 \cdot 10^{42}$ ergs sec^{-1}, quite comparable to that of the nearby elliptical galaxies studied by Einstein, but now having a size of 1/2 Mpc - truly a protogalaxy.

Fluctuations in the DXRB depend in detail on the composition of the sky. If the DXRB is composed of point sources and the log N-log S curve has a 3/2 slope then one expects 50% fluctuations on bins of 10 square arcminutes. XMM will be able to see such fluctuations at the 10 sigma level in exposures of $6 \cdot 10^4$ seconds and obtain 180 independent measurements of the fluctuations in each field. The simple "conventional" wisdom is that the fluctuations allow one to measure log N - log S about a factor of 10 below the source confusion limit. Thus XMM will be able to constrain source counts at levels two orders of magnitude below the Einstein Deep Survey limit. At this flux level, either the X-ray background will be totally resolved, or the source counts will be seen to roll over indicating that the point source contribution to the background is bounded and less than 100%. The resolution of this problem will give new insights into the evolution of quasars and galaxies and perhaps the measurement of a new major contributor to the mass density of the universe.

4.2. Hot gas in superclusters, clusters and galaxies.

The measurement of the temperature and the density profile of the X-ray emitting hot gas residing in clusters of galaxies (the intracluster medium ICM) will allow the determination of the total mass distribution in the cluster. Such observations alone would yield the gravitational potential and thus determine the mass distribution of dark matter in the cluster. In addition they may be used in combination with optical studies to examine the dynamical state of clusters through careful measurement of both the galaxy and gas distribution. XMM, with its imaging spectroscopic capability, will be able to make such measurements for a large sample of clusters, covering a wide range of cluster properties out to a cluster redshift of ~ 0.2.

Figure 5 shows the sensitivity of XMM for a cluster with $L_x = 4.10^{44}$ ergs sec^{-1} with a core radius of 300 kpc at $z=0.1$. The X-ray flux is that contained in 1 arcminute pixels, the temperature measurement can be extended to several core radii (see position 2 in Figure 5) by lumping the spectral data in 1 arcminute annuli in the low-brightness outer regions. This capability is important in studying the evolution of clusters. Evolution is evident not just in cluster galaxy distributions but also in the distribution of the ICM. Some clusters show a relaxed equilibrium state, where the ICM has a temperature of $T \sim 10^8 K$ and has a sharply peaked distribution, while others contain cooler, more irregularly distributed gas. In addition, there are variations in the measured iron abundance, which may be correlated with cluster properties. No clear picture has emerged of the sequence of events in which galaxies form and then relax dynamically and in which primordial and galactic gas react to the cluster potential. Testing of evolutionary models requires a systematic study of cluster luminosity, mass distribution, temperature and iron abundance in a large representative sample of clusters extending to a redshift limit of at least 1. Figure 6 illustrates the ability of XMM to observe the Helium-like iron line in clusters over a range of redshifts. In addition to such a study, XMM should be able also, in a sample of the nearer, brighter clusters, to measure the iron abundance as a function of radius. This is essential for studies of galaxy gas ejection and stripping, as well as in setting limits to the proportion of primordial gas in the ICM.

XMM will be able to observe cooling flows in clusters out to high redshifts ($z \sim 1$), and thus determine the role such giant galaxy formation plays in cluster evolution. The XMM data will constrain the temperature and density profiles in nearby cooling flows and thus directly measure the mass accretion rates. Measurements of the variations of accretion rate with radius in the nearby clusters should identify the regions in the galaxy where star formation is occurring.

XMM measurements of the density and temperature distribution in clusters will allow unambiguous interpretation of the Zel'dovich-Sunyaev effect (in which the hot intracluster gas distorts the microwave background by inverse Compton scattering). This will allow determination of the Hubble constant by combining X-ray and microwave measurements.

Because of its large collecting area, its high sensitivity to low surface brightness features, and its wide field of view, XMM may be able to determine if superclusters of galaxies contain X-ray emitting gas. This gas represents the most likely candidate for a major baryonic contributor to the mass of superclusters and might be one of the sources of the diffuse X-ray background. It might also represent material "leftover" after galaxy formation has ceased. The present upper limits are considerably higher than the expected surface brightness of the gas, if it exists. If superclusters are filled with such gas it is likely to be primordial and be representative of the material created in the Big Bang. Our only opportunity for detecting such "unprocessed" matter may lie in long-exposure observations of selected superclusters.

With an angular resolution of 30 arcseconds XMM will be able to determine the density and temperature distribution of the gas in elliptical galaxies out to a distance of 100 Mpc, i.e. the distance of the Coma cluster and thus derive the binding mass distribution of these objects. If one assumes a space density of 0.002 per Mpc^3 for elliptical galaxies, we find that XMM will be able to study the mass distribution in over 1000 elliptical galaxies. At a distance of 50-100 Mpc the "average" elliptical, with $L_x \sim 1.5 \ 10^{41}$ ergs s^{-1} will have its integral spectrum determined in less than 10^4 seconds and a spectrum of high quality derived in 10^5 seconds (see Figure 7). Moreover XMM can determine the spatially resolved temperature structure of the cooling flows in the 100 bright ellipticals within 20 Mpc. Because of the complex temperature structure in the central regions, where the gas is cooling rapidly, medium resolution X-ray spectroscopy is necessary to deconvolve the temperature structure. Here the reflecting grating spectrometer on XMM, because it is able to do a good job on sources of angular size up to 1-2 arcminutes, will be of tremendous use in measuring the integral spectra of the central regions of the cooling flows. With its broad band spectral coverage XMM should be able to distinguish, via their spectral signatures, galaxies whose X-ray emission is dominated by their X-ray binary population and those whose emission is dominated by hot gas.

4.3. Supernova remnants.

Supernovae play a dominant role in the energising and replenishment of matter (plasma and energetic particles) of the interstellar medium. X-ray observations of supernova remnants provide major inputs for the study of:
. the morphology of and the power ratio between non-thermal (electron acceleration and magnetic field amplification) and thermal (shock heated gas) radiation components.
. the heating process by collisionless shocks propagating through inhomogeneous media (e.g. temperature and ionization equilibrium, cloud evaporation).
. the degree of enrichment of the replenished material (young remnants).
. the physical properties and thermodynamics of the global interstellar medium.

Substantial progress has already been made with the high resolution X-ray images and spatially integrated low to moderate resolution X-ray spectroscopy, which became available with the EINSTEIN and EXOSAT telescopes. Moreover the Einstein Focal Plane Crystal Spectrometer has shown a first glimpse of the potential impact of medium and high resolution X-ray spectroscopy for SNR-studies with the data obtained for e.g. Puppis A, Cas A and N132 D. The spectral data available indicate that ionization equilibrium may not be present in many of the

galactic X-ray SNR's; however spatially resolved spectroscopy is
required to unequivocally assess equilibrium effects and elemental
abundances, since complications arise due to (1) reverse shock heating
of ejecta (young SNR's), which introduces a priori spatially distinct
multi-temperature components, and (2) the imhomogeneity of the circumstellar and interstellar medium through which the blast wave propagates, thereby introducing multi-temperature structures.

In order to obtain a realistic picture of the ionization and
temperature conditions in a supernova remnant, one should obtain
spatially resolved spectral data which cover the full range of
ionization conditions. An example of the wealth of information that can
be derived is the EXOSAT X-ray observation of the old galactic remnant
Puppis A. The low energy (< 2 keV) 4 colour images indicate that major
differences in the properties of the hot plasma exist from one part of
the remnant to the next. Future observations of remnants at different
stages of evolution will however need a wide band coverage to accommodate the appropriate temperature regimes, ranging from $2.10^6 - 5.10^7$.
This requires a pass band of 0.3-15 keV to allow adequate measurement
of continuum temperatures, and a relative spatial resolving power $\theta/\Delta\theta$
of 30 (θ angular extent of the remnant, $\Delta\theta$ the angular resolution).

The CCD imagers on XMM are well matched to these requirements and
can provide adequately resolved spectroscopy of many galactic supernova
remnants.

With respect to statistical spectral study of SNR's in external
galaxies, XMM provides the sensitivity to obtain medium resolution
spectra of tens of SNR's in the Large and Small Magellanic Clouds and
in other members of the Local Group (M31, M33). A simulation of the
spectral data to be expected from an SNR with $L_x = 5.10^{36}$ ergs sec^{-1} in
M31 is shown in Figure 8 Since SNR's in other galaxies are at a well
defined distance and since the line of sight absorption is rather
uniform, a unique opportunity exists to establish accurately spectral
luminosities and emission measures. Figure 9 shows a simulation of a 1
and 2 arcminute diameter SNR observed by the XMM reflection gratings
for an X-ray SNR located in the LMC with $L_x = 3.10^{36}$ ergs sec^{-1}. XMM is
unique in having the capability of a spectral resolving power R > 100
for extended objects of this size below 2.5 keV, by virtue of the high
dispersion reflection grating spectrometer. The wealth of spectral
information will allow unambiguous identification of SNR's among the
total X-ray source population (down to 5.10^{34} ergs sec^{-1}) in these
external galaxies, and allow for a detailed comparison of their
properties with those of the galactic X-ray SNR's, including the inter-
play with the surrounding media. XMM will be particularly well suited
to study the large diameter remnants in some nearby local group
galaxies.

4.4. Stellar Coronae

Among the many questions that stellar coronal physics needs to address,
three are particularly important and as yet largely unanswered by
previous space missions:

- What is the heating mechanism of stellar coronae, and how does it
  depend on parameters such as radiation field, mass flows, convection,
  rotation and age?
- What is the structure (both in temperature and density) of stellar
  coronae? Why and how is the plasma organised - as it seems - in
  spatially distinct features, presumably confined by magnetic fields?
- What is the time variability of stellar coronal sources on a variety
  of different timescales (from minutes to days and years)? How can the
  observed temporal variations be used to infer the mechanism of
  coronal heating and the structuring of stellar coronae?

XMM, with its combination of high throughput, wide bandwidth, medium
spectral resolution and long continuous look capability, can address
all of the above questions.

For the medium resolution instruments of XMM it will be possible to
fully resolve the temperature structure of stellar coronae, and to
determine the amount of coronal plasma present at each temperature.
Figure 10 shows a simulated spectrum of Capella as seen by the
reflection grating spectrometers on two XMM telescopes, in an
observation time of 5.10^4 sec. Individual lines of ions formed at
different temperatures are clearly resolved, including the He-like
triplets of abundant ions such as O VII and Ne IX. The inset shows an
expanded view of the region of the He-like triplet of O VII. The
resonance, intercombination and forbidden lines are clearly resolved.
This is important since the ratio of the latter two lines is a
diagnostic of coronal density for sufficiently high densities
(10^{11} cm^{-3}). There are indications from EINSTEIN and EXOSAT data that
such high densities might indeed occur in the coronae of many active
stars; these stars appear in fact to be covered by high temperature,
high density active regions whose physical conditions approach those of
the flaring Sun.

XMM will have the sensitivity to observe and obtain good quality
spectra with the CCD array and the reflection gratings, of sources
which are over three orders of magnitude weaker than a coronal source
like Capella (c.f. the inset of Figure 10), which also shows the
simulated spectrum obtained in only 1500 seconds by the CCD cameras,
indicating that broad band time resolved spectra can be obtained on
such sources). There are thousands of such sources (either weaker or at
greater distances than Capella) that could be observed with XMM. The
X-ray line emission from such sources would still be clearly detected
with an intensity sufficiently high to be used as diagnostics of the
temperature structure. It is important to note that even the XMM CCD
spectra will represent a major improvement with respect to existing
data, and will complement in an essential way the reflection grating
spectrometer data. The H-like and He-like lines of the most abundant
elements are prominent, and can be used for differential emission
measure analysis. The most prominent lines are still well resolved in a
source an order of magnitude weaker than Capella, by using a comparable
exposure time, or in fainter sources using correspondingly longer
exposure times. Hundreds of such spectra of nearby stars, covering a
wide range of different physical conditions, will also be obtained
serendipitously, in addition to the thousands of other stellar sources

that will just be detected. Moreover, time resolved spectra t ∿ 100 sec
with the CCD and t ∿ 1000 sec with the reflection grating) of flares on
nearby stars such as Algol can be obtained. This will permit the study of
the temperature evolution during transient events (see Figure 11).
Observations of temporal variations are also crucial to resolve the
spatial structure of eclipsing binaries or to infer the imhomogeneous
distributions of surface activity on single, rapidly rotating stars. An
essential requirement for this technique to be successful is continuous
monitoring for periods at least one orbital cycle and preferably longer.
The highly eccentric orbit of XMM will make this type of observation
possible.

The authors wish to acknowledge the other members of the XMM Science
Advisory Group who contributed to the XMM Science Report[5], on which this
paper is largely based.

6. REFERENCES

(1) Space Science Horizon 2000, December 1984, ESA SP-1070.
(2) Proceedings of a Workshop on a Cosmic X-ray Spectroscopy Mission,
    September 1985, ESA SP-239.
(3) The High Throughput X-ray Spectroscopy Mission, Report of the
    Telescope Working Group.
    B. Aschenbach, O. Citterio, J.M. Ellwood, P. Jensen, P. de Korte, A.
    Peacock and R. Willinggale, February 1987, ESA SP-1084.
(4) The High Throughput X-ray Spectroscopy Mission, Report of the
    Instrument Working Group.
    U. Briel, A.C. Brinkman, J.M. Ellwood, G.W. Fraser, P.A.J. de Korte,
    J. Lemon, D.H. Lumb, G. Manzo, A. Peacock, E. Pfeffermann, R. Rocchia
    and N.J. Westergaard, November 1987, ESA SP-1092.
(5) The High Throughput X-ray Spectroscopy Mission, Science Report.
    J. Bergeron, J. Bleeker, J.M. Ellwood, A. Gabriel, R. Mushotzky,
    J. van Paradijs, A. Peacock, R. Pallavicini, K. Pounds, H. Schnopper,
    J. Trümper, March 1988, ESA SP-1097.

FIGURE CAPTIONS

Figure 1: Schematic of a typical XMM telescope configuration.

The resolving power of the blazed X-ray reflection grating as a function of X-ray wavelength for the minus first and second orders is shown as an inset.

Figure 2: Effective area of the XMM optics (three modules) as a function of X-ray energy. Other missions are also indicated for comparison.

Figure 3: Configuration of the XMM imaging detector comprising a butted area of CCD's. A field of view of 0.5 degrees is covered. The spectral resolution as a function of X-ray energy is also shown, several line complexes of interest, all of equal strength, folded with the CCD spectral response function, are indicated.

Figure 4: The number of serendipitious sources found in the XMM field (1/2 degree) (blue) and the fraction of the DXRB which they comprise (red), as a function of limiting sensitivity in ergs cm^{-2} sec^{-1} as well as in units of the Einstein Deep Survey Flux limit (1 DSF = 2.10^{-14} ergs cm^{-2} sec^{-1}). Note the Einstein deep survey was performed using an effective exposure time of 3.10^5 seconds, i.e. 1 week of actual observing time, given the orbit efficiency. Two cases are shown for $\gamma = 1.5$ and 1.0 where γ is the slope of the log N - log S curve. The flux limits for the XMM CCD camera on three telescopes for an exposure time of 10^4 to 10^5 seconds are indicated. The flux levels of the AXAF ACIS (CCD array) and HRS (microchannel plate) experiments for a similar exposure time are indicated.
Also shown is a high latitude field ($\gamma = 1.5$) for a typical XMM exposure time of 3.10^4 seconds (limiting flux 2.10^{-15} ergs cm^{-2} sec^{-1}, 0.1 DSF) with the full CCD payload. About 200 sources are contained in this field.

Figure 5: The flux and temperature distribution across a cluster with $L_x = 4.10^{44}$ ergs sec^{-1} and a core radius of 300 kpc at a redshift of 0.1 ($q_0=0.5$). The flux is that which is contained in a 1 x 1 arcminute pixel. The sensitivity limits for the XMM CCD camera on three telescopes is shown for exposure times ranging from 3.10^4 seconds (typical) to 10^5 seconds.

Figure 6: The number of counts observed by the XMM CCD's from the Helium-like iron line in the cluster described in Figure 6, but now as a function of redshift. The line equivalent width varied from 100 to 400 eV. The 5-sigma sensitivity to the line is also shown, illustrating that in the example such observations are possible out to a redshift of 1.4 → 2.2.

Figure 7: The CCD spectra of an elliptical galaxy at 50 Mpc with $L_x \sim 10^{41}$ ergs sec^{-1} and at a gas temperature of 1 keV. The observation time was 10^5 seconds. The inset shows the grating spectrum, binned at a resolution of 15 eV such that the satellite and resonance lines from the various ions indicated are summed. A theoretical thermal plasma having solar abundances in ionization equilibrium was used as the basis in all cases for the input spectrum.

Figure 8: The simulated spectrum of a point-like SNR in M31 with $L_x \sim 5 \cdot 10^{36}$ ergs sec^{-1} as observed by the CCD cameras and the reflection gratings (inset), for an exposure of 10^5 seconds. A two temperature plasma ($5 \cdot 10^6$ K, $5 \cdot 10^7$ K) which was in collisional ionization equilibrium was assumed. The bin size was 45 eV and 15 eV for the CCD cameras and reflection gratings.

Figure 9: The simulated spectrum of an SNR in the LMC for the case of the reflection gratings when the effect of source angular extent is also incorporated. Two cases of SNR's with angular diameters of 1 and 2 arcminutes are shown for an observation time of $3 \cdot 10^4$ seconds.
The plasma conditions were the same as in Figure 8.

Figure 10: The XMM reflection grating spectrum of Capella obtained in an exposure time of $5 \cdot 10^4$ seconds. The oxygen He-triplet is also shown (inset) along with the CCD spectrum. In the case of the CCD broadband spectrum the exposure time was 1500 seconds.

Figure 11: The CCD spectrum of a flare in Algol. The exposure time was 10^3 seconds and the temperature was assumed to rise during the flare from 2.7 to 5 keV.

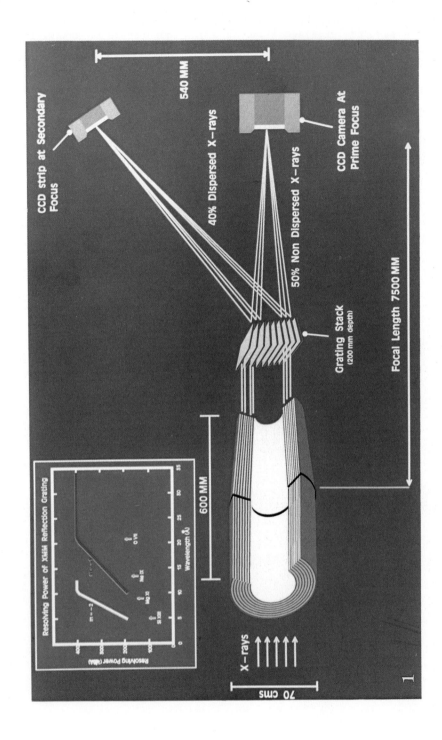

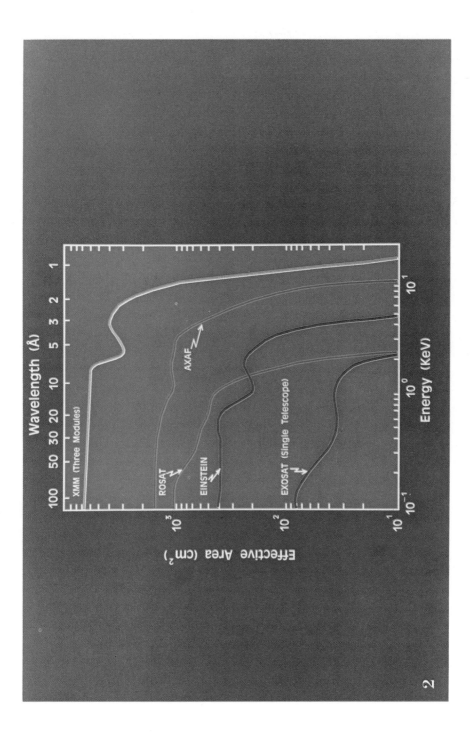

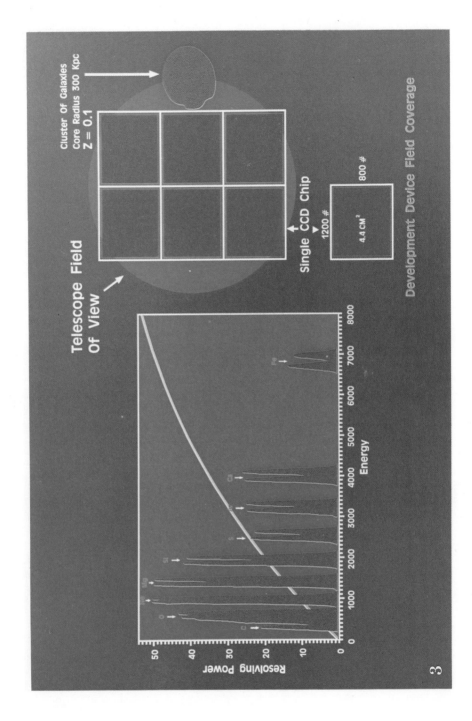

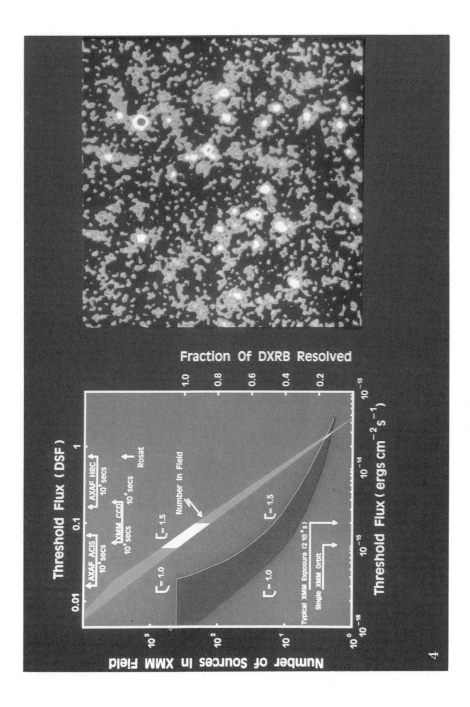

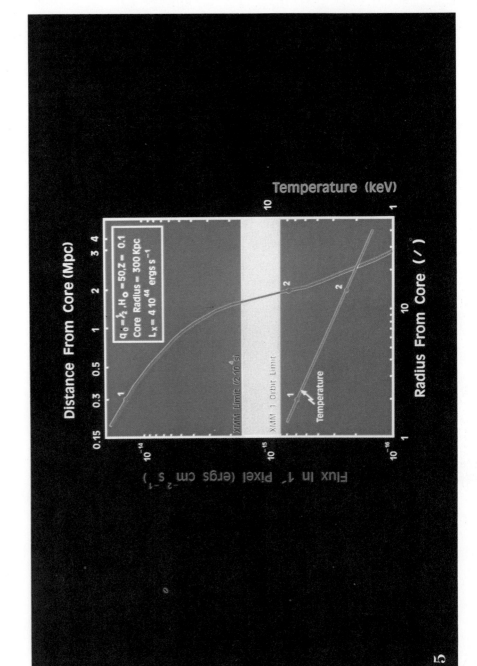

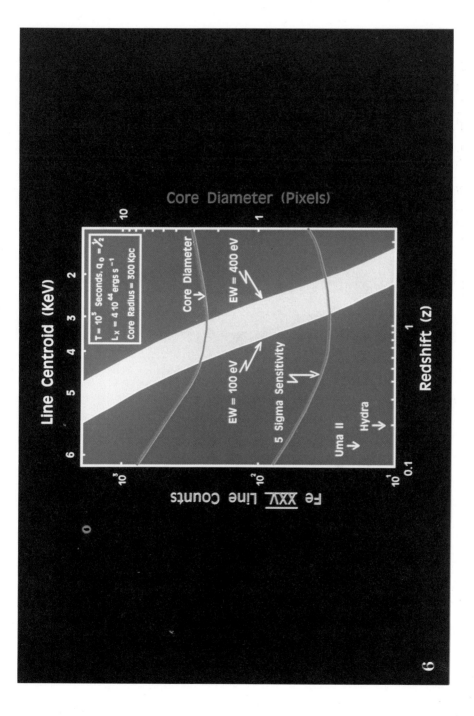

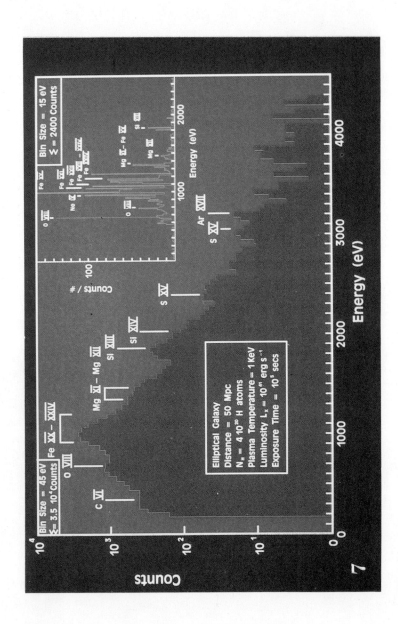

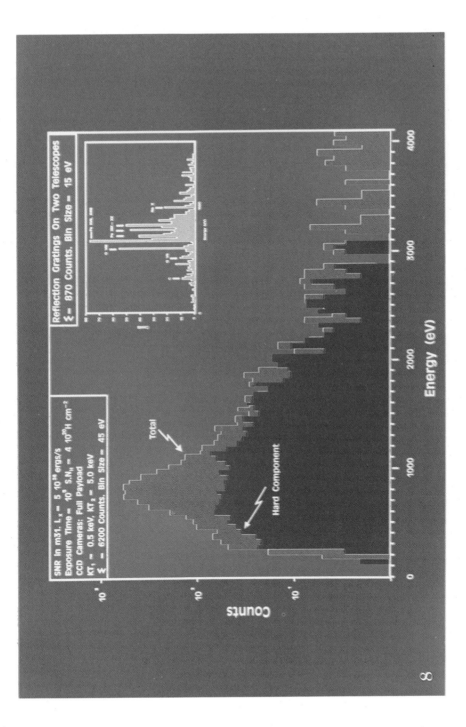

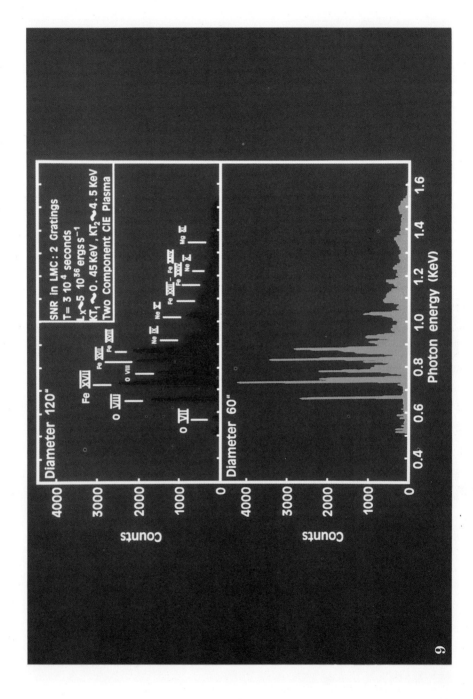

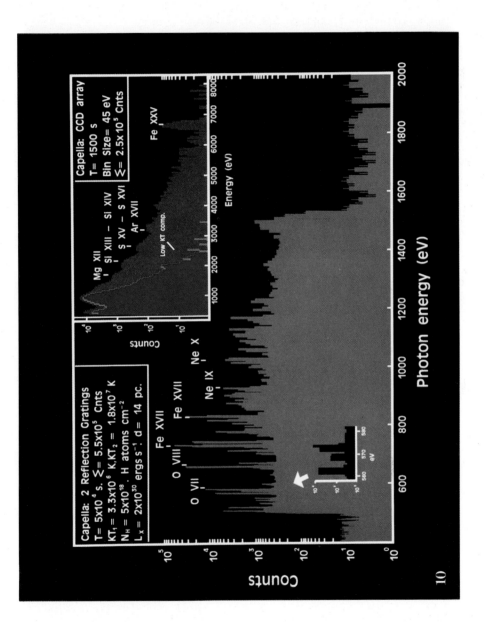

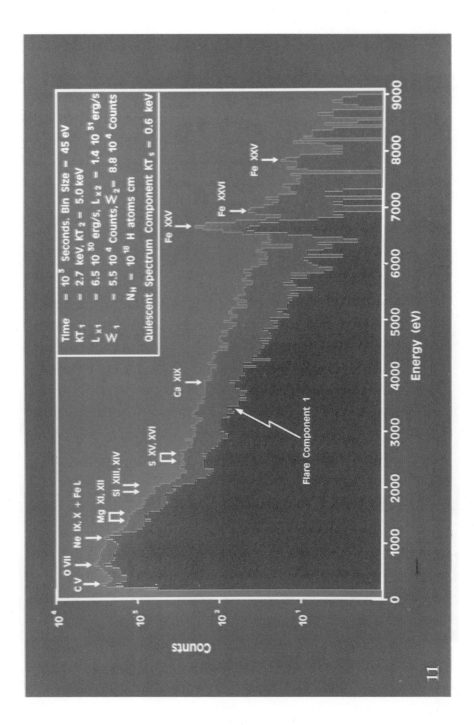

The USSR Space Astronomy Programme.

Alan Smith
Space Science Department of ESA
Postbus 229
Noordwijk
The Netherlands.

ABSTRACT. Presented here are the past, present and future programmes of USSR space astronomy. The presently operating missions ASTRON and ROENTGEN are reviewed. Future missions such as GAMMA 1, GRANAT, RELIC-2, SPEC X, RADIOASTRON etc are described. Emphasis is placed on observational astronomy.

1. Introduction

Surely the launch of Sputnik 1 on the 4th of October 1957 must be one of the milestones of human achievement. (1) provides a list of major unmmanned spaceflights up to 1984. Of the 53 flights listed, 29 are by the USSR.

With manned spaceflight it was again the USSR that took the lead. In this area the USSR has directed its efforts towards a manned space station with the series of SALYUT missions and, more recently, the new MIR space station.

Soviet space activities have always gone hand-in-hand with planetary and astronomical research, both by the direct exploration of the Moon, Venus and Mars, and through observational astronomy at almost all wavelengths.

In this work a brief history of Soviet space astronomy is given. The present status of two astronomical observatories in orbit is described (ASTRON and ROETGEN) and the planned missions for the future are reviewed.

2. The Past

2.1 Exploration of the Planets

No report of this kind would be complete without some mention of the Soviet achievements in planetary research.

On the 2nd march 1959 LUNA 1 became the first spacecraft to escape the gravitational pull of the Earth on its way to the moon. The first lunar impact came with LUNA 2 (12/9/59) and the far side of the moon was observed for the first time ever with LUNA 3 (4/10/59). 7 years later the first soft landing (LUNA 9 31/1/66) and first lunar orbits (LUNA 10 31/3/66) were made. A further 6 years later LUNA 16 made the first automatic return of lunar soil and LUNA 17 placed the Lunakod robot on the surface of the planet.

On the 1st of march 1966 VENERA 3 became the first spacecraft to reach another planet. This began a major investigation of Venus, a truly inhospitable world. Highlights were:- VENERA 7 (1970) - first transmissions from the surface; VENERA 13/14 Venusian soil analysis; VENERA 15/16 Radar imagers in orbit providing first high resolution images of planet. Even the VEGA missions to Halley's comet found time to release ballon probes into the Venusian atmosphere.

The study of Mars was rather more modest. MARS 2 and MARS 3 spacecraft reached Mars at the end of May 1971 during a planet-wide sand storm. Unlike MARINER 9, which was there at the same time, the MARS 2 and 3 missions were designed to land the surface probes at once and so the data from them were largely degraded by the sand storm which MARINER 9 could sit out in orbit. Nevertheless MARS 3 became the first spececraft to soft land on the planet while MARS 2 gave us the first direct data about the composition of the Martian atmosphere. After four failures in the launch campaign of 1973 the Soviet Union suspended their exploration of Mars in favour of Venus but will be returning in 1988 with their PHOBOS mission.

Most recently there has been the very successful VEGA 1 and 2 missions that rendezvoued with Halley's comet on 3-6/3/86 after a voyage via Venus. The mission to Halleys comet was co-ordiated with the other space programs involved and provided invaluable pathfinder information for the GIOTTO spacecraft that arrived a little later.

Table 1 SALYUT Astronomical Experiments

Mission	Launch	Experiments	
SALYUT 1	19/04/71	ORION-1 (UV spectrometer)	[a]
		Anna-3 (>100 MeV Gamma ray)	[b]
SOYUZ 13	18/12/73	ORION-2 (UV spectrometer, x-ray)	[a]*
SALYUT 3	25/06/74	None	
SALYUT 4	26/12/74	OST-1 (UV spectrometer)	+
		ITS-K (I/R detector) [c]	
		Filin X (X-ray 'telescope') [d]	**
		RT-4 (X-ray 'telescope') [c]	
		SSP-2 (Solar spectrometer) [e]	
SALYUT 5	22/06/76	ITS-5 (I/R detector) [c]	
SALYUT 6	29/09/77	BST-1M (sub mm/UV)	
		SRT-10 (radio)	++
SALYUT 7	19/04/82	EPHO-1 (Electrophotometer) [f]	
		XT-4M (X-ray 'telescope') [c]	***
		XS-02M (X-ray 'telescope') [d]	
		Piramig (photographic) [g,i], (3)	
		PCN (photographic chamber) [g]	
		Sirene (X-ray GSPC) [j,h]	
		Gamma (Gamma ray detector) [g]	

notes:-
* Wide angle version of ORION-1, UV spectra of stars down to 12 mag, 5" pointing. + Orbital Solar Telescope, 600 UV spectrographs of the Sun made. ** 1' angular resolution x-ray telescope, focal plane instrument failed early. ++ 10 metre radio antenna deployed in orbit, 200kg. Support leg failed and was cut away during an EVA. *** 3000 cm^2 x-ray detector (see ASTRON)

Institutes:-
a) Armenian SSR's Academy of Science. b) Moscow Institute of Enginering and Physics. c) USSR Academy of Science, P.N. Lebedev Institute of Physics. d) P.K. Shternberg State Astronomical Institute. e) Leningrad University, Laboratory of Physics, f) Checoslovakia Astronomical Institute. g) CNRS France, h) Space Science Department of ESA, ESTEC. i) I.K.I. Moscow. j) CESR -Toulouse.

2.2 Observational Astronomy

Compared with their achievements in the exploration of the nearby planets the area of purely observational astronomy has been less spectacular. However many Soviet orbitting spacecraft, space stations and interplanetary probes have included experiments devoted to observational astronomy.

In general the instruments have been relatively modest with short periods of observation.

Particularly successful were the gamma burst detectors flown on PROGNOZ 7, VENERA 11 and 12 (2) which were a collaboration between the Soviet Space Research Institute in Moscow (IKI) and CESR france. In fact a similarly designed instrument will constitute part of the GRANAT payload that will be launched in 1988 (see below). Also very successful was the microwave background experiment RELIC-1 on board the PROGNOZ 9 satellite which produced an 8mm radio brightness map of the sky with 5.8 deg. angular resolution and 0.5 mK brightness temperature resolution.

Table 1 provides a list of the astronomical instruments flown as part of the SALYUT space station programme. SOYUZ 13 is included even though the early failure of SALYUT 2 meant that the equipment was in space for only 7 days.

The SALYUT program provides some insights into the Soviet approach to space flight. A common theme is progressive re-use of proven designs rather than the expensive development of new concepts. For instance, the supply ship for the SALYUT stations (and now MIR) is the PROGRESS which is a modification of the SOYUZ manned spacecraft but without solar panels or heat shield, (since they are destroyed on re-entry). Although this approach means that equipement can be made on a production line basis, systems tends not to be ideally optimised to the task involved. The docking of the PROGRESS to the mother station is automatic and another common theme in Soviet developments.

3. The Present

The first example of a mission dedicated to observational high energy astrophysics came with the satellite ASTRON launched in March 1983 and still operating. Details of this mission are given in table 2.

The absence of on-board memory and the short periods of ground contact (3hr/day) limit the scientific return of the instrument. The proportional counter experiment is also

limited by a high background count rate which reduces its
sensitivity. As an example the ASTRON observations of Her
X-1 are presented in (6).

Table 2. ASTRON

Experiment Institute	Energy Range keV	Energy Resolution	F.o.V. deg.	Angular Resolution	Area cm^2
SKR-02M IKI	2 - 20	22% @ 6 keV	3 FWHM	none	1780
UFT LAS Crimea Ast. Obs.	1500 - 3500 A	0.4 A			80cm diam

Orbital characteristics   Operational Characteristics

Apogee   201,000 kms   Pointing accuracy: several
Eccentricity   0.916   Pointing or scanning modes
Inclination   51°   Observating period   3hs/day
Period   4.1 days   Direct transmission
Launch   23/3/83   Spacecraft - Venera type
Mass   3900 kg

Brief resume of experiments:-

SKR-02M - Collimated proportional counter with plastic
            anti-coincidence shield. See (4).

UFT   - UV telescope - Cassegrain design, wt 400 kg,
            Three slits (1, 10, 75") Spectrometer
            Toroidal concave grating + PMTs
            5m focal length, 400 kg, See (5)

After the SALYUT series came the MIR space station which,
although a modified SALYUT concept, is the corner stone for
the Soviet manned space programme. MIR will be built up
from various building blocks, one of which - KVANT -
contains a payload devoted to the study of x-ray astronomy.
This payload, called ROENTGEN, was launched on the 31 march
1987 and consists of instrumentation from the USSR, West
Germany, the Netherlands, Great britain and ESA. The
details of the mission are shown in table 3. A resume of
the observing programme up to the 23 August 1987 is given
in table 4.

Table 4. ROENTGEN

Experiment Institute	Energy Range keV	Energy Resolution	F.o.V. deg.	Angular Resolution	Area cm^2
TTM Utrecht Birmingham	2-30	25% @ 6 keV	7 * 7	1-2'	625
GSPC SSD/ESTEC	3-100	10% @ 6 keV	3 FWHM	none	315
HEXE MPI Tubingen	20-150	18% @ 60 keV	1.6 FWHM	none	4 x 120
PULSAR X1 IKI	20-800	12% @ 662 keV	3 FWHM	none	4 x 250
PULSAR V IKI	20-1500		2 pi	none	1 x 314

Brief resume of experiments:-

TTM - Position sensitive proportional counter + coded mask
    Also includes star tracker. Imaging. see (7)

GSPC - High pressure (3 atm) xenon filled gas scintillation
    proportional counter with escape gate. see (8)

HEXE - NaI - CsI phoswich including collimator. see (9)

PULSAR X1 - NaI-CsI phoswich.

PULSAR V - NaI-CsI gamma burst monitor.

Space Station Characteristics:-

Apogee   370 km   Inclination   51 degrees
Perigee   330 km   Mass   32 tons

KVANT launch date   31/03/87   Launcher PROTON SL-13
Start of Operations   02/06/87
Data storage - Tape recorder.

Since the whole of MIR has to be pointed for each target, observing time is restricted both due to attitude constraints and the requirements of other experiments on-board. This largely accounts for the small fraction of observing time obtained so far. In the future the situation is expected to improve as the operation of the instruments becomes more routine and the above fraction moves towards 10%.

Table 5. <u>ROENTGEN Observation Programme Summary</u>

2//6/87 - 23/8/87

Target	# Orbits	Time (minutes)	% mission
Blank	1	7	0.3
Cyg X-3	4	117	5.0
Cen A	10	202	8.6
Cyg X-1	15	307	13.0
Her X-1	25	426	18.1
SN 1987a	67	1297	55.1
Total	122	2356	100

Average exposure per orbit when operating :- 19.3 min.
Average exposure per day (2/6/87 - 23/8/87) :- 28.4 min.
Fraction of time actually observing :- 2.0 %

4. The Future

Before discussing future missions let us first mention both the Launch capabilities and the willingness for international collaboration.

As a rough average the USSR launch a satellite every three days! Even without the recent launcher difficulties of NASA this is still five times the rest of the world put together. The USSR have a large range of launchers, (designated SL in the west). The PROGRESS and SOYUZ spacecraft are normally launched with the SL-4 vehicle although the workhorse of the space programme is the SL-9 PROTON which can put 15 tons into low earth orbit. SALYUT and MIR components are launched on the larger SL-13.

More powerful still is the 'Medium Lift' SL-16 which has been recently developed. On the 15/5/87 the USSR had the first successful launch of their Energiya booster which is presently the world's largest launch vehicle, it has about 2/3 of the capability of the now defunct Saturn V and can put 100 tons into low earth orbit.

In summary the USSR launch capabilities are impressive and well matched both to the relatively heavy spacecraft and the wide range of orbits planned for the future.

As far as international collaboration is concerned much has been done already. (e.g CNES, France has a history of over 20 years of such joint ventures including the flight of a French cosmonaut). Many of the future missions mentioned here are largely international.

Let us now outline the future missions, concentrating on the area of observational astronomy although some mention of future plantary missions is made.

4.1 Short Term Plans

The details of GAMMA 1 (10) are given in table 6. This mission is devoted to gamma ray astronomy in the range 50 to 500 MeV. The main component of the payload is a 12 gap spark chamber in which electron/positron pairs are produced. These are then detected in scintillation detectors, a Cerenkov gas detector and a scintillation calorimeter. The instrument is designed to take over where SAS-2 and COS-B left off in this area of astronomy. Unfortunately the mission has been subject to some delays but is presently virtually ready to fly and has recieved some impetous from the supernova in the LMC.

A hard X-ray/Gamma ray mission with a similar timescale is GRANAT. See table 7

A mission devoted to the study of the microwave background is RELIC-2, (a follow-up of the successful RELIC-1). This will be sensitive in the range 2cm - 1mm and will have a 5 degree angular resolution. The spacecraft, which will be placed in orbit around the Earth/Sun Lagrangian point, is all USSR manufactured and will probably fly in 1990/91.

Table 6. __GAMMA - 1__

Experiment Institute	Energy Range	Energy Resolution	F.o.V. deg.	Angular Resolution	Area cm^2
Gamma 1 IKI*	50-500 MeV	70% @ 100 40% @ 500	10*10	20'	500
PULSAR X2 IKI	2-25		20*20		4 x 150

Launch 1988-89; Mass 1800 kg; Pointing accuracy - 3'

Brief resume of experiments:-

Gamma 1 - Spark chamber with Cerenkov gas detector and scintillation calorimeter readout. Includes a movable random mask and plastic anti-co.

PULSAR X2 - four offset proportional counters each $10 * 10°$ FWHM

* also Technical Institute of the Academy of Science, Leningrad, Moscow Physical Research Institute, CEA Saclay and CESR Touloues, France
Polish Academy of Science, Warsaw.

4.2 Longer Term Plans

The USSR have a number of missions envisaged for the first half of the 1990's.

A future mission devoted to the x-ray/gamma ray range is the so called SPECTRA X-gamma. The details of this mission are still being discussed but it will probably comprise of two spacecraft, the first being devoted to x-ray, the second to gamma ray astronomy.

The first, (SPECTRA X) may fly as early as 1992 and will be a USSR/European collaboration. The orbit will be highly eccentric with a northern appogee and the payload will comprise of about 2500 kg of scientific equipment. Potential instruments include an EUV telescope, two x-ray concentrators and a hard x-ray detector. Of the two x-ray concentrators one will be of modest angular resolution (2') but with a large area (>2000 cm^2), an objective Bragg crystal providing $\lambda/d\lambda$ of >1000 with solid state and gas imagers in the focal plane. The second x-ray concentrator

Table 7. GRANAT

Experiment Institute	Energy Range keV	Energy Resolution	F.o.V. deg.	Angular Resolution	Area cm^2
SIGMA CNES/CESR CEA SACLAY	15-2000	15% @ 60 keV	7*7 14*14*	10' FWHM	1000
ART-P IKI	4-100	23% @ 6 keV	1.8*1.8 FWHM	5' FWHM	4 x 600
ART-S IKI	4-150	20%,<10% @ 6,60 keV	2*2 FWHM	none	4 x 600
PHOEBUS CESR	100-100 MeV	10% @ 600 keV	All Sky		5 x 50
CONUS Inst. Phys. Tech. Leningrad	20-800	15% @ 600 keV	All Sky	1-2 deg.	7 x 300
Sunflower IKI	2-25	20% @ 6 keV	2.5*2.5	none	4 x 150
WATCH DSRI	6-180		All Sky		4 x 45

Orbit:-

Apogee	200,000 kms
Perigee	2,000 kms
Period	4.5 days
Inclination	51°

Operations:-

Typical operational period	3 days
Ground contact	3 hrs
Telemetry rate	64 K baud
Lifetime	1 +2 yrs
Attitude control	+/- 20'
Attitude reconstruct	+/- .5'

Breif resume of experiments:-

SIGMA	Coded Mask position sensitive NaI detector Includes two star trackers and 128 Mbit bubble memory. French. See (11)
ART-P	Position senstive MWPC with coded mask. Escape-gating above xenon k-edge.
ART-S	MWPC. 5 layers with rocking collimator.
PHOEBUS	French, BiGO, Gamma Burst monitor.
CONUS	Previously flown on VENERA and PROGNOZ series. NaI Gamma Burst Monitor. (see SIGNE - 2)
Sunflower	Proportional counters on rotating platform. Positioned to direction of gamma-ray burst in 2 seconds. Location of burst provided by CONUS.
WATCH	Danish, NaI, CsI Rotating Modulation Collimator. Introduced very late to the mission with only one year between instigation and the delivery of the flight hardware.

---

will be high resolution (<0.5') and >200 cm^2 with a CCD in the focal plane. Both instruments will be sensitive up to 10 keV in energy.

The second spacecraft, (SPECTRA gamma), is less well defined but will be sensitive in the hard x-ray soft gamma range. Although it may have a similar orbit to SPECTRA X it might become linked to the USSR space station or be a low earth free flier, (the most likely). It is expected to fly 2-4 years after SPECTRA X.

RADIOASTRON (6) is a set of three radio astronomy mission planned for the early 1990's. In the first mission two spacecraft are to be placed in highly eccentric northern orbits. VLBI between the spacecraft and ground based receivers is planned. Four wavelengths (1.35, 6, 18, 92cm) will be available and the antennae will be a 10 metre diameter deployable carbon fibre structures (developed from SRT-10 flown on SALYUT 6). The recievers are likely to be international collaborations involving the Netherlands, West Germany, Australia and Finland. The second mission will be a single sattelite devoted to mm wavelengths while the third will involve three spacecraft each having 30 metre diameter antennae. in this case the orbits will be geostationary, highly eccentric and solar providing a maximum baseline of 2 x 10^6 km.

Other missions envisaged include;- AELITA a sub-mm observatory sensitive in the range 300 microns to 3mm; SPECTRA A, a large UV telescope and a Schmidt telescope.

Table 8. <u>Future plans</u>

Mission	Year	Comments
GRANAT	1988	Hard X-ray/ Gamma Ray
GAMMA-1	1988/9	Gamma Ray
PHOBOS	1988/9	Mars mission including Phobos encounter
RELIC-2	1990/1	Microwave background
MARS	1992/4	Columbus - Orbitter + lander
	1996+8	Mars soil returns
	2002	Long term Mars rover
SPECTRA X	1992/3	X-ray Observatory
RADIOASTRON		Radio VLBI
cm	1991	stage 1 2 spacecraft
mm	1996	stage 2 1 spacecraft
K-K	2001	stage 3 3 spacecraft
SPEC Gamma	1994?	Gamma Ray Observatory
Luna Polar	1992/4	Luna Polar mission with down looking x-ray detectors for composition studies
VESTA	1994	Asteroid encounter via Mars
CORONA	1995	Solar Corona and Jupiter
	1996	Lunar far side soil return
	1999	Jupiter, Saturn+Titan lander
	2000	Unmanned Lunar laboratory
AELITA	?	Sub-millimetre
SPECTRA A	?	UV Telescope

4.3 Conclusions

An approximate timetable of future missions is shown in table 8. Dates are sometimes approximate. Clearly a very significant and ambitious programme is planned, which, if successful will bring the USSR into the front line in space borne astronomy.

References

1. Jane's Spacecraft Directory, Jane's publ. 26, 1984.
2. Diyachkov, A.V., et al, Adv. Sp. Sci., **3**, 211, 1983.
3. Levasseur-Regourd, A.C., et al, Adv. Sp. Sci., **5**, 27, 1985.
4. Golinskaya, I.M., et al, Adv. Sp. Sci., **3**, 539, 1984.
5. Hua, C.T. et al, Adv. Sp. Sci.,**5**, 201, 1985.
6. Andreyanov, V.V. et al, Sov. Astr., **30**, 504, 1986.
7. Brinkman, A.C., Dam, J., Mela, W.A., Skinner, G.K., Willmore, W.P., <u>Non-Thermal and Very High Temeperature Phenomina in X-ray Astronomy</u>, ed. Perola and Salvati, Rome, 263, 1983.
8. Smith, A., <u>Non-Thermal and Very High Temperature Phenomina in X-ray Astronomy</u>, ed. Perola and Salvati, Rome, 271, 1983.
9. Reppin, C. et al, Proc. 20th International Cosmic Ray Conf., Moscow, **OG 9**, 289, 1987.
10. Akimov et al, Sov. Astr. **30**, 508, 1986.
11. Paul, J. et al., Proc. 20th International Cosmic Ray Conf., Moscow, **OG 9**, 301, 1987.

Note:-
In order to keep this work as up to date as possible some information that was made available shortly after the oral presentation has been included.

Acknowledgments
I would like to thank Drs R. Sunyaev, N. Yamberenko and O. Babushkina of IKI for their useful comments.

FUTURE SPACE ASTRONOMY PROGRAMME OF JAPAN

Y. Tanaka
Institute of Space and Astronautical Science
4-6-1 Komaba, Meguro-ku, Tokyo 153
Japan

ABSTRACT. In the planning of the space research programme of Japan, continuation of activities in each sub-discipline has been considered of vital importance. In this paper, the future space astronomy programme of Japan is outlined, and the Solar-A mission (solar physics) and the Astro-D mission (X-ray astronomy) are described in some detail.

1. OUTLINE OF FUTURE PROGRAMME

The Institute of Space and Astronautical Science (ISAS) is an inter-university institute which is responsible for planning, coordination, and implementation of the space research programme of Japan. The Space Science Committee consisting of members representing the space scientists all over the country makes recommendation of future missions to ISAS. Recently, ISAS established its long-range planning through beginning of the 21st century, based on extensive discussions. Table I summarizes the past, current and future missions of ISAS. Once a mission is approved, ISAS assumes the management responsibility through launch, operation and data handling. ISAS has also been developing its own launch vehicles, and also conducts the satellite launches.

The ISAS missions are all small in scale compared to those of NASA, ESA and Soviet. Therefore, in order to remain competitive in science, we have considered it vital to maintain continuous activities in each subdiscipline by a proper allocation of mission opportunities. So far, we think we have achieved this goal by launching seventeen satellites in the last seventeen years. We hope to be able to maintain this pace also in the forthcoming years.

2. SPACE ASTRONOMY MISSIONS

2.1. Solar Physics: Solar-A mission

In 1981, ISAS launched "Hinotori" for the X-ray study of solar flares in the last solar maximum period. Hinotori carried a hard X-ray imager

Table I. ISAS Missions

Discipline	1977	1978	1979	1980	1981	1982	1983	1984	1985	1986	1987	1988
Earth		Kyokko Jikiken				Ohzora						
Planet										Sakigake (Halley) Suisei (")		
Sun						Hinotori						
Astronomy			Hakucho(X)					Tenma(X)			Ginga(X)	
Technology	Tansei III			Tansei IV								

Discipline	1989	1990	1991	1992	1993	1994	1995	1996	1997	1998	1999	2000
Earth	Exos-D			Geotail								
Planet							Planetary			Planetary		
Sun			Solar-A									
Astronomy					Astro-D(X) IR	VLBI				X-ray IR UV		
Technology		MUSES-A		SFU			Tech.			Tech.		

consisting of a set of rotating modulation collimators, broad-band X-ray and gamma-ray spectrometers, and high-resolution crystal spectrometers. This 180-kg satellite recorded seven hundred solar flares and yielded many results concerning flare dynamics, production of hot plasma, etc..

Solar-A is an orbiting solar observatory scheduled for launch in the summer of 1991 during the next solar maximum period. This mission is a natural follow-up of Hinotori, having much more enhanced capabilities. The scientific objectives of the Solar-A mission are to investigate solar flares in detail and also to study the quiet sun by means of X-ray imaging and X-ray spectroscopy.

Solar-A, weighing 420 kg, is three-axis stabilized with a stability of about 1 arcsec/s and 5 arcsec/min. It carries a soft X-ray imager and a hard X-ray imager both having full solar-disk coverage. Besides, wide-band spectrometers and a high-resolution spectrometer are on board. The experiments and their characteristics are summarized in Table II. The orbit of Solar-A will be approximately circular at an altitude between 550 and 650 km.

Solar-A will constitute one of the key elements for the multilateral solar physics study in the next solar maximum period, together with ground-based radio and optical observatories.

Table II. **Experiments on Solar-A**

* Hard X-ray imager
    - Instrument: Fourier synthesis telescope
    - Energy range: 10 - 100 keV
    - Angular resolution: 7 arcsec.
    - Effective area: 3 cm^2 x 64 elements
    - Detector: NaI(Tl) scintillator
    - Time resolution: 1 sec.

* Soft X-ray imager
    - Collaborative experiment with NASA (Lockheed Palo Alto Res. Lab.)
    - Instrument: Modified Wolter type I optics
    - Energy range: < 4 keV
    - Angular resolution: 2.5 arcsec.
    - Geometrical area: approximately 3 cm^2
    - Detector: CCD
    - Exposure time: 10 msec (flare) - 10 sec (quiet sun)

* Bragg crystal spectrometer
    - Collaborative experiment with MSSL. and Rutherford Appleton Lab.
    - Instrument: Bent crystal spectrometer
    - Energy range: S XV, Ca XIX, Fe XXV, Fe XXVI
    - Energy Resolution: 1/3000 - 1/8000
    - Detector: Position-sensitive prop. counters
    - Time resolution: 1 sec (partial data) - 4 sec (full)

* Wide-band spectrometer
    - Instrument:
        Soft X-ray / proportional conter
        Hard X-ray / NaI scintillator
        Gamma ray / BGO scintillator
    - Energy range: 2 keV - 50 MeV
    - Time resolution: Pulse height spectrum / 2 sec
        Broad-band count rate / 1/16 - 1 sec

2.2. X-ray Astronomy: Astro-D Mission

In the X-ray astronomy field, ISAS launched "Hakucho", the first X-ray astronomy satellite of Japan, in 1979. Hakucho weighed only 90 kg. "Tenma", weighing 220 kg, was launched four years later in 1983. "Ginga", the third of the series weighing 420 kg, was launched in February 1987, four years after Tenma. Some of the early results from Ginga are presented separately in this issue.

Each time, we made a substantial increase in the capabilities. We appreciate this stepwise growth with a constant pace of flight oportunities, by which our expertise have been accumulating and steady activities have been maintained.

Astro-D is the fourth X-ray astronomy satellite, following Ginga, to be launched in 1993. Astro-D is planned to be a high-capability X-ray observatory. It will be equipped with high-throughput X-ray optics covering a wide energy range exceeding 10 keV, and able to image the X-ray sky with a spatial resolution comparable to that of the IPC of the Einstein Observatory, but with much higher sensitivity and spectral resolution over a substantially wider energy range. The conceptual view of Astro-D is shown in Fig. 1.

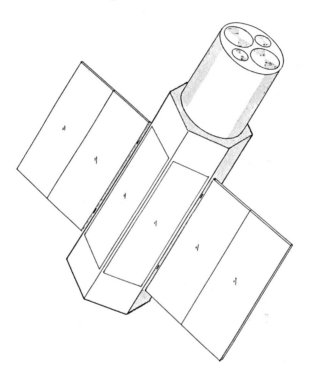

Fig. 1. Conceptual view of Astro-D

Astro-D is in the concept design phase at present. The main features of Astro-D as presently envisioned are given in Table III. The total weight is limited to be 430 kg for an approximately circular orbit of a 550 - 650 km altitude. The spacecraft will be three-axis stabilized with an accuracy of an arcminute.

The X-ray telescope will consist of four sets of multi-nested thin-foil mirrors such as developed at Goddard Space Flight Center (Serlemitsos 1981). As a matter of fact, possible collaboration with NASA in this area is under discussion. The design goal of the effective area is 1000 cm^2 below 2 keV and 500 cm^2 in the range 6 - 7 keV.

Figure 2 illustrates the simulated energy spectra of a supernova remnant (Tycho's SNR) measured with a conventional proportional counter (PC), a gas scintillation proportional counter (GSPC), and a solid state detector (SSD), respectively. A GSPC will resolve the emission line of

Table III. **Main Features of Astro-D**

1. High-throughput wide-band telescope

   Telescope: Multi-nested thin-foil mirrors
   Effective area: Approximately 1000 cm^2 in < 1 keV
                    Approximately 500 cm^2 in 6 -7 keV
   Focal length: Approximately 3 m
   Image size: Approx. 2 mm half-power diameter

2. Focal plane detectors under consideration

   Imaging gas scintillation proportional counters (IGSPC)
   & Solid State Detectors (SSD)

3. Imaging capability

   Field of view: 30 x 30 (arcmin)^2
   Point source resolution : < 1 arcmin.

4. Energy resolution

   < 8 %   (5.9 keV) with IGSPC
   2 - 3 % (5.9 keV) with SSD

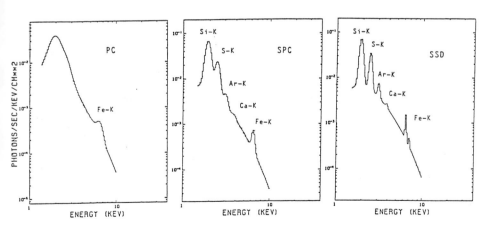

Fig. 2. Simulated spectra of Tycho's SNR measured with PC, GSPC and SSD, respectively

each element unambiguously with a detection efficiency of nearly 100 % over the energy range up to 10 keV. An SSD can resolve K-alpha and K-beta lines and also detect a Doppler shift for a velocity of the order of 1000 km/sec.

With the capabilities described above, a variety of scientific objectives will be achievable. Table IV contains the list of themes that are readily conceived. Aside from these current objectives, Astro-D will fill the gap of X-ray observations that are otherwise foreseen in early 1990's. The mere fact that Ginga happened to be just in time to catch SN1987A in the Large Magellanic Cloud demonstrates quite impressively the vital need of continued presence of at least one X-ray observatory in orbit, no matter how modest one it is.

Table IV. **Scientific Objectives of Astro-D**

---

* Active galactic nuclei
  - $\sim$10 times fainter AGN's than previously detectable
  - Spectra over a wide energy range
  - Detecton of characteristic lines
* Cluster of galaxies
  - Temperature and abundance structure up to z $\sim$ 1
  - Evolutionary effect
  - Zel'dovich-Sunyaev effect

* Cosmic X-ray background
  - Deeper than Einstein limit
  - New population or diffuse?

* Galactic sources
  - Spectroscopy: Emission and absorption lines, Absorption edges, Doppler shifts and broadening
  - Variability: Nature of QPO, flickering

* Supernova remnants
  - Spatially resolved spectra
  - Temperature and abundance structure
  - Search for neutron stars

* Emission along the galactic ridge
  - Spatial structure
  - Unidentified SNR's
  - New population sources?

REFERENCE

P.J.Serlemitsos, 1981, "Broad Band X-Ray Telescope (BBXRT)", X-Ray Astronomy in 1980's, NASA TM 83848, p441.